Lecture Notes in Computer Science 15765

Founding Editors

Gerhard Goos
Juris Hartmanis

The series Lecture Notes in Computer Science (LNCS), including its subseries Lecture Notes in Artificial Intelligence (LNAI) and Lecture Notes in Bioinformatics (LNBI), has established itself as a medium for the publication of new developments in computer science and information technology research, teaching, and education.

LNCS enjoys close cooperation with the computer science R & D community, the series counts many renowned academics among its volume editors and paper authors, and collaborates with prestigious societies. Its mission is to serve this international community by providing an invaluable service, mainly focused on the publication of conference and workshop proceedings and postproceedings. LNCS commenced publication in 1973.

Gidon Ernst · Matthias Güdemann ·
Alexander Knapp · Florian Nafz ·
Frank Ortmeier · Hella Ponsar ·
Gerhard Schellhorn · Alexander Schiendorfer
Editors

Go Where
the Bugs Are

Essays Dedicated to Wolfgang Reif
on the Occasion of His 65th Birthday

 Springer

Editors
Gidon Ernst 🆔
Ludwig-Maximilians-Universität München
Munich, Germany

Matthias Güdemann 🆔
Hochschule München
Munich, Germany

Alexander Knapp 🆔
University of Augsburg
Augsburg, Germany

Florian Nafz 🆔
Hochschule München
Munich, Germany

Frank Ortmeier 🆔
Otto-von-Guericke-Universität Magdeburg
Magdeburg, Sachsen-Anhalt, Germany

Hella Ponsar 🆔
Universität Augsburg
Augsburg, Germany

Gerhard Schellhorn 🆔
Universität Augsburg
Augsburg, Germany

Alexander Schiendorfer 🆔
Technische Hochschule Ingolstadt
Ingolstadt, Germany

ISSN 0302-9743   ISSN 1611-3349 (electronic)
Lecture Notes in Computer Science
ISBN 978-3-031-92195-7   ISBN 978-3-031-92196-4 (eBook)
https://doi.org/10.1007/978-3-031-92196-4

This Springer imprint is published by the registered company Springer Nature Switzerland AG
The registered company address is: Gewerbestrasse 11, 6330 Cham, Switzerland

If disposing of this product, please recycle the paper.

Preface

This Festschrift is dedicated to Wolfgang Reif on the occasion of his 65th birthday. Its title "Go Where the Bugs Are" reflects Wolfgang's intention and achievements in making a real difference to software and systems engineering: Develop theoretically grounded techniques, methods, and tools to avoid possible weaknesses in software and hardware systems or to detect the actual deficiencies of such systems, and then practically apply and improve them on concrete, hard, and critical instances of industrial size. In all his work in formal methods, safety and security, self-organizing systems, robotics and automation, as well as artificial intelligence methods, Wolfgang has successfully lived up to this intention and attitude.

This volume gathers fifteen contributions by Wolfgang's scientific friends, collaborators, and colleagues in all these research topics. Each paper was carefully reviewed by at least two reviewers. Additionally, the volume contains a laudatio recognizing Wolfgang's scientific life and results, and also a personal tribute by the scientific staff of his chair at the University of Augsburg. We thank all contributors and all reviewers for their time and effort in making this Festschrift a reality. We would also like to thank Easy-Chair for providing the reviewing infrastructure and to extend our gratitude to Springer for their willingness to publish this volume.

This Festschrift was presented to Wolfgang on June 6, 2025 at a scientific colloquium in his honor held at the University of Augsburg, Germany — right on his 65th birthday. The event was attended by numerous colleagues, friends, and former students of Wolfgang Reif. We thank the team of the Institute for Software and Systems Engineering for their support in organizing the symposium.

Congratulations, Wolfgang!

March 2025

Gidon Ernst
Matthias Güdemann
Alexander Knapp
Florian Nafz
Frank Ortmeier
Hella Ponsar
Gerhard Schellhorn
Alexander Schiendorfer

Organization

Editors

Gidon Ernst	Ludwig-Maximilians-Universität München, Germany
Matthias Güdemann	Hochschule München, Germany
Alexander Knapp	Universität Augsburg, Germany
Florian Nafz	Hochschule München, Germany
Frank Ortmeier	Otto-von-Guericke-Universität Magdeburg, Germany
Hella Ponsar	Universität Augsburg, Germany
Gerhard Schellhorn	Universität Augsburg, Germany
Alexander Schiendorfer	Technische Hochschule Ingolstadt, Germany

Additional Reviewers

Bernhard Beckert	KIT, Germany
Stefan Bodenmüller	Universität Augsburg, Germany
Brijesh Dongol	University of Surrey, UK
Benedikt Eberhardinger	CARIAD
Axel Habermaier	AMD
Dominik Haneberg	Universität Augsburg, Germany
Reiner Hähnle	TU Darmstadt, Germany
Oliver Kosak	Universität Augsburg, Germany
Carola Lenzen	Universität Augsburg, Germany
Jan-Philipp	Steghöfer XITASO GmbH IT & Software Solutions
Simon Stieber	Universität Augsburg, Germany
Sven Tomforde	Christian-Albrechts-Universität zu Kiel, Germany
Constantin Wanninger	Technische Hochschule Augsburg, Germany

Contents

A Tale of Formal Methods, Organic Computing, and Intelligent Robotics — The Scientific Life of Wolfgang Reif

Dominik Haneberg[1] , Alexander Knapp[1(✉)] , Oliver Kosak[1] ,
Florian Nafz[2] , Hella Ponsar[1] , Gerhard Schellhorn[1] ,
and Martin Wirsing[3]

[1] Universität Augsburg, Augsburg, Germany
{dominik.haneberg,alexander.knapp,oliver.kosak,hella.ponsar,
gerhard.schellhorn}@uni-a.de
[2] Munich University of Applied Sciences HM, München, Germany
florian.nafz@hm.edu
[3] Ludwig-Maximilians-Universität München, München, Germany
wirsing@lmu.de

Abstract. On the occasion of Wolfgang Reif's 65th birthday, we offer an outline of Wolfgang's remarkable scientific achievements. Beginning with his work on formal methods and, in particular, on the interactive theorem prover KIV, we look at his major results in the areas of security and safety, self-organizing systems, robotics, and AI methods. Additionally, we briefly discuss his impact on the research community and on academic study programmes. We conclude with our heartfelt thanks to him.

1 Introduction

Wolfgang Reif was born June 6, 1960 in Forbach im Schwarzwald. He studied computer science in Karlsruhe where he also obtained his PhD in 1991. In 1994 he became a professor at the University of Ulm, in 2000 he was appointed a full professor in Augsburg, where he first was director of the Institute for Computer Science and then founding dean of the Faculty for Applied Computer Science from 2004 to 2011. In 2008 he founded the Institute for Software & Systems Engineering inside the faculty and became its scientific director. Wolfgang has made significant contributions to the areas of formal methods, safety and security, software engineering for robotics, self-organizing systems, and artificial intelligence methods. Since 1986 he has published more than 280 scientific papers and led about 30 third-party funded research projects. From 2009 to 2017 he coordinated the research group OC-Trust investigating trustworthy organic computing and self-adaptive systems. He has also been a leading figure at the University of Augsburg, shaping the Faculty of Applied Computer Science and establishing several important study programmes in Augsburg. Up to 2025, he supervised

G. Ernst et al. (Eds.): Wolfgang Reif Festschrift, LNCS 15765, pp. 1–19, 2025.
https://doi.org/10.1007/978-3-031-92196-4_1

39 successful PhD students, five of them meanwhile having become professors themselves.

Structured by the prevalent research topics, we want to outline the main steps and achievements of Wolfgang's impressive scientific career together with the developments in his group and the accompanying projects; we include a brief account of his activities in the development of study programmes and his lasting impact on the science community. Two aspects of Wolfgang's research are particularly striking: Its breadth and constant innovation ranging from formal methods and security over safety and self-organization to robotics and artificial intelligence, and its leitmotif of practical applications of theoretical methods in competitive, realistic, large, and difficult case studies.

2   Formal Methods

Wolfgang began his scientific career at Wolfram Menzel's chair of Logic, Complexity, and Deductive Systems at the University of Karlsruhe. Driven by a strong interest in program verification, he wrote his diploma thesis on completeness and induction rules for Dynamic Logic [64] before starting the core development team of the Karlsruhe Interactive Verifier (KIV). This verification tool introduced a groundbreaking approach by using Dynamic Logic instead of Hoare logic and streamlining the formal verification process through heuristics and user-defined proof rules. In 1986 and 1987, the group, consisting of Maritta Heisel, Wolfgang, and Werner Stephan, published the first two papers on KIV at the top conferences CADE [24] and CSL [30]. In his 1991 PhD thesis [65], Wolfgang introduced modularization concepts into KIV, enabling the verification of larger programs. He subsequently published four outstanding papers, all in 1992: "Systematic Construction of Verified Software" (CADE [68]), "Verification of Large Software Systems" (FSTTCS [69]), "Correctness of Generic Modules" (LFCS [67]), and "Correctness of Full First-Order Specifications" (SEKE [66]).

Around the same time, Wolfram Menzel's chair was involved in a large collaborative project funded by the BMFT, bringing together 15 partners: the KORSO project, aimed at developing correct software through formal methods [10]. Among others, the research groups of Manfred Broy and Martin Wirsing also participated in this initiative. As Wolfram Menzel's representative, Wolfgang was responsible for coordinating the tool development in KORSO and presented the project (together with Heinrich Hußmann and Jacques Loeckx) at the 1993 annual conference of the GI [35]. Building on the results of his dissertation, he developed an elegant methodology for verifying large sequential software systems, well-aligned with KORSO's deductive program development approach. This methodology featured a modular structure, incremental error correction, and the reuse of typically over 90% of proof steps [70].

In subsequent work, first in Ulm and later in Augsburg, Wolfgang together with his group focused on the integration of interactive and automatic proof techniques [1] and applied these specification and verification techniques to sequential Java programs [88], both in collaboration with Peter H. Schmitt in Karlsruhe.

This modular approach has then been extended to the verification of concurrent systems, integrating temporal logic with thread-local rely-guarantee proof strategies [11,76], and applied to the verification of parallel, lock-free algorithms [95].

Over the years, the research has resulted in a number of outstanding contributions to the verification of complex systems using KIV [73], such as a fully verified Prolog compiler [72], the Mondex money card protocol, or a lock-free, concurrent stack with hazard pointers [96]. The Mondex protocol was a pilot project of the Grand Verification Challenge, initiated by Tony Hoare in 2003. Organized by Jim Woodcock, several research groups attacked the challenge and Wolfgang's group was the first to mechanically prove the correctness of the protocol, also detecting a subtle bug [26]. One successor project in the Grand Verification Challenge was the verification of a file system for flash memory, initiated by Jim Woodcock and NASA motivated by a failure of the flash file system on the Mars rover Spirit [36,63]. The case study required the development of a new theory for the modular specification of such a large software system [16], including crash safety, concurrent components [74], and caches [7,60]. With a code size of about 20,000 lines of code, the case study is one of the largest ever conducted worldwide. Only Wolfgang's group has succeeded in developing a fully functional and correct version [6].

PhD Theses. We list all theses supervised by Wolfgang in the domain of formal methods with their defense dates in ascending order. All theses have been awarded when Wolfgang had his chair in Augsburg, with Gerhard being the notable exception already defending his thesis in Ulm.

Gerhard SCHELLHORN Verifikation abstrakter Zustandsmaschinen (9.6.1999)

Kurt STENZEL Verification of Java Card Programs (30.5.2005)

Michael BALSER Verifying Concurrent Systems with Symbolic Execution (12.7.2005)

Andriy DUNETS Automatische Fehlersuche in Algebraischen Spezifikationen (11.2.2010)

Simon BÄUMLER Modulares Beweisen temporallogischer Eigenschaften paralleler Programme (15.7.2010)

Bogdan TOFAN Compositional Concurrent Program Verification with RGITL (11.7.2014)

Gidon ERNST A Verified POSIX-Compliant Flash File System (28.11.2016)

Jörg PFÄHLER A Modular Verification Methodology for Caching and Lock-Based Concurrency in File Systems (9.7.2018)

Stefan BODENMÜLLER Caching, Crashing & Concurrency — Verification Under Adverse Conditions (27.7.2022)

Projects. Wolfgang successfully acquired and completed the following projects in the domain of formal methods, most of them funded by the German Research Council (DFG) highlighting their fundamental nature.

KORSO — Korrekte Software (1991–1994, BMFT-Verbundprojekt)

VSE — Verification Support Environment (c. 1990–1995, BSI)

Integration interaktiven und automatischen Beweisens (c. 1995–2000, DFG-PP Deduktion)

INOPSYS Interoperability of Calculi for System Modelling (2001–2009, DFG)

FLASHIX — Construction and Verification of a State-of-the-Art File System for Flash Memory (2009–2018, DFG)

VERICAS — Verification of Fine-Grained Concurrent Algorithms (2010–2016, DFG)

VERICODE — Correct Translation of Abstract Specifications to C-Code (2022–today, DFG)

3   Security and Safety

Already while in Ulm but in particular at his chair on Software Engineering and Programming Languages in Augsburg, Wolfgang first added the topics of security and safety to his research domain; in both areas software and model engineering were combined with formal methods.

In security, Wolfgang and his group worked on mandatory access policies, model-driven development of secure systems, and the integration of information flow. In an early application in 2002, the group certified the security model for a smart card developed by IBM [75]. Its model-driven security approach [9] has been prominently applied to Woodcock's abstract Mondex protocol, resulting in a fully verified JavaCard implementation [25]. The model-based approach has subsequently been extended in the IFLOW projects (cooperating with Heiko Mantel and Georg Snelting) to include information flow analysis [8,37,89,90].

The major contributions of Wolfgang and his group to the development of safety-critical applications have been a formal foundation for fault tree analysis (FTA [59]), its extension into a model checking-based analysis (DCCA [56]) including qualitative and quantitative (probabilities of failure) properties [22], and a model-based analysis framework [23,49]. Important case studies were the radio-based railroad crossing control, which was part of the German Research Programme "Integration von Techniken der Softwarespezifikation für ingenieurwissenschaftliche Anwendungen", led by Hartmut Ehrig and Martin Große-Rhode [55,58], and the safety analysis of the height control system of the Hamburg Elbtunnel, initiated by Siemens [57]. This line of research is still continued by Wolfgang's former PhD student Frank Ortmeier at Otto-Guericke-Universität Magdeburg.

PhD Theses. As both security and safety involve formal methods, there is no crisp border between the fields. Interestingly, most theses in these fields have been written in German (where, by the way, "Sicherheit" means both safety and security...).

Andreas THUMS Formale Fehlerbaumanalyse (13.12.2004)

Frank ORTMEIER Formale Sicherheitsanalyse (11.11.2005)

Dominik HANEBERG Sicherheit von Smart-Card-Anwendungen (9.11.2006)

Holger GRANDY Formale Verifikation der Korrektheit sicherheitskritischer
Java Anwendungen (2.6.2008)

Jonathan SCHMITT Modellierung und Verifikation medizinischer Leitlinien
(25.11.2008)

Nina MOEBIUS Modellgetriebene Entwicklung sicherer Smart
Card-Anwendungen(5.3.2013)

Axel HABERMAIER Design Time and Run Time Formal Safety Analysis using
Executable Models (21.11.2016)

Marian BOREK Modellgetriebene Entwicklung sicherer Web
Service-Anwendungen (27.3.2017)

Kuzman KATKALOV Ein modellgetriebener Ansatz zur Entwicklung
informationflusssicherer Systeme (3.7.2017)

Johannes LEUPOLZ Probabilistic Safety Analysis of Executable Models
(18.6.2018)

Projects. The third-party funding of the security & safety projects range from
the industrial over the national to the European level:

SMACOS — Secure Multiapplicative SmartCard Operating System
(1995–1998, BSI, in cooperation with IBM)

FORMOSA — Integrating Formal Models and Safety Analysis (1998–2008,
DFG-PP)

GO!CARD — Formal Methods for Secure Java Smart Cards (1999–2009,
DFG-PP)

PROTOCURE Improving medical protocols by formal methods (2001–2006,
European FET project)

SECUREMDD — A Model-Driven Development Method for Secure
Applications (2008–2012, DFG)

IFLOW — Developing Systems with Secure Information Flow (2010–2018,
DFG-PP)

4   Self-organizing Systems

In 2005, Wolfgang initiated the investigation of self-adaptive, self-organizing systems: What can still be *guaranteed* for autonomously deciding systems and how can they be engineered from a software perspective in the first place [52,85]? In fact, in a series of projects, Wolfgang and his team shaped the field of "Organic Computing" [50]. In the SAVE ORCA project, part of the DFG Priority Programme "Organic Computing" (SPP 1183), his group introduced the so-called "restore invariant approach" [21] enabling the system's decisions to be shifted to runtime by providing a corridor of correct behavior that can be verified by model checking or interactive theorem proving in KIV [51]. The developed safety analysis method DCCA (see Sect. 3) was extended to adaptive systems (aDCCA [71]).

The formal methods approach was supplemented by aspects of software engineering to design these systems, such as guidelines for integration into the standard software engineering process and design patterns [84].

In the DFG research group OC-TRUST, Wolfgang's team extended the approach to formal analysis and software architectures for trustworthy self-organizing systems. This includes the challenge of controlling large-scale smart energy systems, which are highly dynamic with energy producers and consumers entering and exiting at any time and, as critical infrastructure, have to guarantee availability, resilience, and performance despite reorganization at all times. In cooperation with Christian Müller-Schloer, Jörg Hähner, and others, the group introduced trust mechanisms for ensuring secure and robust operation [2,39,86,87]. Furthermore, the rigid corridor of correct behavior was relaxed to a corridor for desired behavior by including soft constraints [77,78] which enforces that the system autonomously reorganizes itself when a step outside the corridor of correct behavior is imminent. In a subsequent cooperation with Alexander Knapp in the DFG project TESOS about the quality assurance of such distributed, large-scale systems, Wolfgang's group demonstrated how traditional testing techniques [15] and performance metrics [14] can be adapted to self-organizing systems.

Most recently, in the DFG project COMBO the team led by Wolfgang investigated how self-organization can be employed in swarms of mobile ground and aerial robots. The group considered a combination of conventional planning and truly self-organizing mechanisms for maintaining a high level of autonomy while keeping response times low [45] and introduced semantic description concepts both on the software and hardware level for fostering self-awareness [29,100]. To control these systems and achieve users' goals, the group investigated the possibilities for defining missions in different approaches [41,82]. Therewith, missions can involve swarms of flying systems to find special points of interest and determine and observe relevant parameters, e.g., through distributed search or continuous observation [40]. To simplify the interface to such swarms, the group successfully developed a concept to flexibly change the executed swarm behavior without the need to modify its concrete implementation [38,42].

PhD Theses. Gerrit seems to hold the record of the longest thesis title.

Hella SEEBACH Konstruktion selbst-organisierender Softwaresysteme
    (8.7.2011)
Florian NAFZ Verhaltensgarantien in selbst-organisierenden Systemen
    (22.5.2012)
Jan-Philipp STEGHÖFER Large-Scale Open Self-Organising Systems (6.2.2014)
Gerrit ANDERS Self-Organized Robust Optimization in Open Technical
    Systems: Self-Organization and Computational Trust for Scalable and
    Robust Resource Allocation under Uncertainty (24.4.2016)
Florian SIEFERT Selbst-organisiertes, trust-bewusstes Supply Demand
    Management in Smart Grids (31.5.2017)
Alexander SCHIENDORFER Soft Constraints in MiniBrass: Foundations and
    Applications (12.11.2018)

Benedikt EBERHARDINGER Model-Based Testing for Self-Organization
    Mechanisms (3.12.2018)
Oliver KOSAK Mission programming for flying ensembles: combining planning
    with self-organization (2.11.2021)

Projects. Having all been funded by the German Research Council, the
projects underline the disruptive, fundamental nature of self-adaptation and
self-organization.

SAVE ORCA — Formal Modeling, Safety Analysis, and Verification of
    Organic Computing Applications (2005–2014, DFG-PP)
OC-TRUST — Trustworthy Organic Computing Systems (2009–2017,
    coordination of DFG research group)
FORSA@OC-TRUST — Formal Analysis and Software Architectures for
    Trustworthy Organic Computing (2009–2017, inside DFG research group
    OC-TRUST)
TeSOS — Testing Self-Organizing, Adaptive Systems (2016–2020, DFG)
COMBO — Combination of Planning, Self-Organization, and Reconfiguration
    in a Robot Ensemble for Handling ScORe Missions (2019–2022, DFG)

5   Robotics

In Augsburg, Wolfgang established a long-standing cooperation with KUKA,
one of the world-leading industrial robot manufacturers. In this context, he has
led the development of a comprehensive software engineering framework for
programming cooperative robot systems with hard real-time guarantees in the
SOFTROBOT project [3,31,99]. The goal was to base robot programming on best
practices in software engineering [31,32]. This was achieved by separating an
object-oriented interface [3,4] from a real-time kernel [97,98] using a flexible
mapping between both layers [79]. The approach had a significant influence on
KUKA's own design of future robot programming [33]. The object-oriented interface also allows for describing the devices and workpieces present in the robot
cell [34] facilitating the programming and planning of complex robot tasks. In
cooperation with the DLR, Wolfgang's group has contributed to the offline programming of complex robot tasks [5,54] as well as to planning algorithms for
cooperating industrial robots [53]. Moreover, the SOFTROBOT approach was successfully adapted to teams of mobile robots allowing to model uncertainty [80].

   More recently, an approach has been developed to establish distributed control services that communicate with real-time guarantees via OPC UA over
TSN [18]. This approach was then used to allow for a further step in applying
industrial robots and improving the production process was the idea of "plug-
&-work for material processing" addressing the shift from mass production to
customized manufacturing, where each product can be unique. Wolfgang's group

developed a service architecture that enables the seamless deployment of modular software components across multiple control devices at runtime [17,19]. As part of the WiR AUGSBURG project, the group also successfully developed a flexible, programmable robotic test bench for destructive component testing. This includes a planning and execution framework for various test movements, as well as a unified data platform for sensor-based motion control and test data recording [27,28].

Besides industrial robots, Wolfgang and his team also contributed to the field of mobile robotics and human-robot interaction. The SINA project, a cooperation with the Karlsruhe Institute for Technology, focused on mobile assistance robots. The group achieved a fast estimation of the distance between capacitive sensors on the robot structure and a human hand while the robot is in motion [62] ensuring both safe and efficient cooperation in close proximity [94]. In the KOAROB project, Wolfgang's group further investigated the topic of human-robot interaction to deal with the demographic shift in Germany. There, assistance robots can help mitigate labor shortages and reduce the physical strain on older workers. The group aimed at ensuring safety through sensors that define dynamic protection zones around robots, slowing or stopping them when a person enters [61]. A key aspect was the dynamic definition of the robot's workspace based on human positioning and predicting human movements.

Research by Wolfgang and his group also covers flying mobile robotics and the numerous possibilities they offer in various disciplines. First investigations successfully focused on the flexibility and autonomy of swarms of small-sized quad-rotor drones for meteorologic research outdoors [43,101]. In his own in-house, more than $750\,\text{m}^3$ covering flying arena, Wolfgang further expanded the idea of flying swarms with currently over 50 drones that can serve as test-bed for formation-flight [44], swarm-behavior [42], or the in-door inspection of large size components in production-settings [81,83].

PhD Theses. There are ten theses in the field of robotics, which is equal to the number of those in the field of security & safety ...

Andreas ANGERER Object-oriented Software for Industrial Robots (28.5.2014)

Alwin HOFFMANN Serviceorientierte Automatisierung von Roboterzellen: Modularität und Wiederverwendbarkeit von Software in der Robotik (27.4.2015)

Michael VISTEIN Embedding Real-Time Critical Robotics Applications in an Object-Oriented Language (21.5.2015)

Andreas SCHIERL Object-Oriented Modeling and Coordination of Mobile Robots (30.5.2016)

Ludwig NÄGELE PaRTs: Automatische Programmierung in der robotergestützten Fertigung (8.9.2020)

Constantin WANNINGER Semantic Plug & Play – Selbstbeschreibende Hardware für modulare Robotersysteme (25.7.2022)

Alexander POEPPEL SensorClouds—A framework for real-time processing of multi-modal sensor data for human-robot-collaboration (23.10.2023)

Christian EYMÜLLER RealCaPP—Real-time capable plug & produce for
   distributed robot-based automation (28.11.2023)

Julian HANKE Robot-based component testing on software-defined test benches
   (19.12.2023)

Matthias STÜBEN Constraint-based specification, planning and control for
   mobile manipulators in dynamic environments (27.5.2024)

Projects. ... but robotics has the highest number of third-party projects.

CFK-TEX — Automated handling and placement of large dry carbon fiber
   textiles (2007–2010, StMWiVT/EU)

SOFTROBOT — A new software architecture for controlling industrial robots,
   combining modern software paradigms and hard real-time (2007–2012,
   StMWiVT)

CFK-OPP — Development of an offline programming platform aimed for
   automizing specification of robot-based CFRP-manufacturing processes
   (2013–2014, BMWi)

SAFEASSISTANCE — Intelligent obstacle detection using capacitive sensors for
   safe human robot interaction (2013–2016, BMBF)

SINA — Reliable Perception for Flexible Assistance in Dynamic and
   Unstructured Environments (2017–2020, BMBF)

TEAMBOTS — A tool-based methodology for developing software for dynamic
   robot teams (2017–2023, DFG with DLR)

WIR AUGSBURG — Robotic component inspection and Plug-&-Work for
   composite materials processing in concert with Industry 4.0 technologies
   (2018–2022, BMBF and others)

BILDUNG 4.0 FÜR KMU — Digital learning as a competitive advantage in
   light-weight production (2018–2021, BMBF, ESF)

KOAROB — Combining assistive and sensory systems for safe
   Human-Robot-Collaboration (2020–2024, StMWi)

PIA — Potentiale heben in Augsburg (2020–2024, BMWi)

6 Artificial Intelligence Methods

The most recent addition to Wolfgang's research topics are artificial intelligence
methods, bringing the advances achieved in deep learning to challenging industrial use cases. This transition is a natural continuation of his involvement in
self-organizing systems, organic computing, and multi-agent systems where core
AI methods such as constraint solving and machine learning have been investigated and applied with a particular focus on large-scale distributed systems.

   The KOGNIA project developed a recommendation system for assembly modeling in computer-aided design to transfer known, insightful part recommendations from one application to another in a meaningful way [48]. The AI team led

by Wolfgang showed that a self-supervised approach with pre-trained component embeddings on graph neural networks achieves particularly high predictive performance [20], reducing the cognitive load of choosing between 2,000 and 3,000 components by recommending the ten most likely components with 82–92% accuracy. In the CoSiMo project, the flow front characteristics of a resin transfer molding process in the context of carbon-fiber-reinforced polymers were reconstructed from in-situ sensor data using deep convolutional networks trained with transfer learning from simulation to reality. The goals were to analyze and simulate the process [93], to determine material properties from observations [92], and finally to optimize the process by applying deep reinforcement learning [91]. In the LufPro4.0 project, machine learning methods were used to control and optimize an adaptive spreading process for carbon fibers utilizing reinforcement learning or neuro-evolution [46]. And finally, in the project AICut neural networks facilitated high-quality process monitoring in the context of metal-cutting using strain gauge sensors.

PhD Theses. Carola is currently the most recent descendant of Wolfgang's academic family.

Simon STIEBER Machine learning for carbon fiber reinforced polymer production (1.12.2023)

Carola LENZEN Machine learning for assembly modeling in computer-aided design (9.12.2024)

Projects. The goal of the AI projects has been and still is to learn, to analyze, and to optimize industrial manufacturing and design processes.

LufPro4.0 — Lösungen zur Vernetzung/Analyse einer digitalisierten Produktion für die Luftfahrt (2018–2020, StMWi)

CoSiMo — Leveraging machine learning for the efficient high-volume production of carbon-fiber reinforced lightweight plastic components (2018–2022, StMWi)

KOGNIA — Machine-learning-based Recommender System for Mechanical Design (2020–2023, StMWi)

AICut — Automated detection of process disturbances and quality deviations in machining production with machine learning (2020–2023, StMWi)

FORinFPRO — Intelligent manufacturing processes & Closed-loop production (2024–today, BMWi)

7   Contributions to the Community

As the founding dean of the Faculty of Applied Computer Science at the University of Augsburg from 2004 to 2011, Wolfgang shaped the main course of studies in computer science in Augsburg. He is now the scientific director of the Institute for Software & Systems Engineering which he established in 2008. From 2011 to 2017 he served as vice-president for research at the University of Augsburg.

In 2006, Wolfgang, together with Manfred Broy and Martin Wirsing, initiated the Elite Master study programme "Software Engineering" as a cooperation of the University of Augsburg, the Technical University of Munich, and the Ludwig-Maximilians-University of Munich. Being part of the Elite Network of Bavaria, this programme keeps on attracting highly gifted and highly motivated software engineering students from all over the world to study at all three universities in Augsburg and Munich. Furthermore, in 2012, Wolfgang initiated a unique study programme in "computer science in engineering" ("Ingenieurinformatik") that combines mechanical, electrical, and production engineering with computer science.

Wolfgang coordinated the research group "OC-Trust" from 2009 to 2017 that formed a cooperation between Augsburg and Hannover (see Sect. 4). As part of this effort, he co-edited the "Autonomic and Trusted Computing" conference editions of 2009 and 2010 with Springer. In 2016 his group organized the 10th SASO conference on self-organization in Augsburg where he took on the responsibility of General Chair. From 2017 to 2021 he served as a member of the steering committee of the SASO / ACSOS conference series.

As a service to all interest groups and fostering applications, Wolfgang is very active in disseminating his scientific results to society and industry. In cooperation with the AMU ("Anwenderzentrum Material- und Umweltforschung") and the ACE ("Augsburg Center for Entrepreneurship") within the project PiA ("Potentiale heben in Augsburg"), Wolfgang's group developed a makerspace including a high-tech workspace to transfer results from its robotic projects into practical applications (see Sect. 5). In the WIR AUGSBURG project ("Wissenstransfer Region Augsburg"), Wolfgang together with the Institute for Material Research Management (MRM) aimed to enhance knowledge transfer between the University of Augsburg and regional mid-sized enterprises, addressing the key challenges posed by advancing digitalization in production.

8   There Has Been More and There Will be a Lot More

In his research, Wolfgang not always only delved into the details of modal logic or inverse kinematics. In particular, in 2001, he started a cooperation with Friedrich Pukelsheim from the Institute of Mathematics at the University of Augsburg and Günther Hägele from the university's library on creating a website supporting the analysis and discussion of three manuscripts of Ramon Llull (c. 1232–1315/6) on electoral systems [12]. With a resulting publication in ≪Le médiéviste et l'ordinateur≫ [13], Wolfgang made himself a name also as a medievalist.

And Wolfgang is still active and full of ideas. One of his ambitious and intriguing goals addresses the latest digitization processes in industry and the possibilities of self-organized production systems supported by artificial intelligence [47]. This is part of the "KI-Produktionsnetzwerk Augsburg", where

Wolfgang is playing an indispensable role within the leading board that navigates a cooperation-network of 27 professors and 287 employees in the two main research areas "Materials & Production Technology" and "Digitalization & Self-Organization" through the rough sea of digital transformation.

Concluding, we want to thank Wolfgang for his inspiring research and challenging, always fruitful collaborations over the years insisting on to "go where the bugs are"—and we want to congratulate him on his 65th birthday: Ad multos annos!

References

1. Ahrendt, W., et al.: Integrating automated and interactive theorem proving. In: Automated Deduction – A Basis for Applications, vol. II: Systems and Implementation Techniques, pp. 97–116. Kluwer Academic Publishers (1998)
2. Anders, G., Steghöfer, J.P., Siefert, F., Reif, W.: Patterns to measure and utilize trust in multi-agent systems. In: Proceedings of 5th IEEE Conference on Self-Adaptive and Self-Organizing Systems Workshops, pp. 35–40. IEEE (2011). https://doi.org/10.1109/sasow.2011.21
3. Angerer, A., Hoffmann, A., Schierl, A., Vistein, M., Reif, W.: The robotics API: an object-oriented framework for modeling industrial robotics applications. In: Proceedings of IEEE/RSJ International Conference Intelligent Robots and Systems (2010)
4. Angerer, A., Hoffmann, A., Schierl, A., Vistein, M., Reif, W.: Robotics API: object-oriented software development for industrial robots. J. Softw. Eng. Robotics 4(1), 1–22 (2013)
5. Angerer, A., Vistein, M., Hoffmann, A., Reif, W., Krebs, F., Schnheits, M.: Towards multi-functional robot-based automation systems. In: Proceedings of 12th International Conference Informatics in Control, Automation and Robotics (ICINCO), vol. 2. pp. 438–443. IEEE (2015)
6. Bodenmüller, S., Schellhorn, G., Bitterlich, M., Reif, W.: Flashix: modular verification of a concurrent and crash-safe flash file system. In: Raschke, A., Riccobene, E., Schewe, K.-D. (eds.) Logic, Computation and Rigorous Methods. LNCS, vol. 12750, pp. 239–265. Springer, Cham (2021). https://doi.org/10.1007/978-3-030-76020-5_14
7. Bodenmüller, S., Schellhorn, G., Reif, W.: Verification of crashsafe caching in a virtual file system switch. Formal Aspects Comp. 34, 1–33 (2022)
8. Borek, M., Stenzel, K., Katkalov, K., Reif, W.: Secure integration of third party components in a model-driven approach. In: Hameurlain, A., Küng, J., Wagner, R., Schewe, K.-D., Bosa, K. (eds.) Transactions on Large-Scale Data- and Knowledge-Centered Systems XXX. LNCS, vol. 10130, pp. 66–86. Springer, Heidelberg (2016). https://doi.org/10.1007/978-3-662-54054-1_3
9. Borek, M., Moebius, N., Stenzel, K., Reif, W.: Model checking of security-critical applications in a model-driven approach. In: Hierons, R.M., Merayo, M.G., Bravetti, M. (eds.) SEFM 2013. LNCS, vol. 8137, pp. 76–90. Springer, Heidelberg (2013). https://doi.org/10.1007/978-3-642-40561-7_6
10. Broy, M., Jähnichen, S. (eds.): KORSO — Methods, Languages, and Tools for the Construction of Correct Software, LNCS, vol. 1009. Springer (1995). https://doi.org/10.1007/BFB0015452

11. Bäumler, S., Balser, M., Nafz, F., Reif, W., Schellhorn, G.: Interactive verification of concurrent systems using symbolic execution. AI Comm. **23**(2–3), 285–307 (2010)
12. Drton, M., Hägele, G., Haneberg, D., Pukelsheim, F., Reif, W.: The augsburg web edition of Llull's electoral writings (2001). www.uni-augsburg.de/llull/
13. Drton, M., Hägele, G., Haneberg, D., Pukelsheim, F., Reif, W.: A rediscovered Llull tract and the Augsburg web edition of Llull's electoral writings. Le médiéviste et l'ordinateur, vol. 43 (2004)
14. Eberhardinger, B., Anders, G., Seebach, H., Siefert, F., Reif, W.: A research overview and evaluation of performance metrics for self-organization algorithms. In: Proceedings of IEEE International Conference on Self-Adaptive and Self-Organizing Systems Workshops, pp. 122–127. IEEE (2015). https://doi.org/10.1109/sasow.2015.25
15. Eberhardinger, B., Seebach, H., Knapp, A., Reif, W.: Towards testing self-organizing, adaptive systems. In: Merayo, M.G., de Oca, E.M. (eds.) ICTSS 2014. LNCS, vol. 8763, pp. 180–185. Springer, Heidelberg (2014). https://doi.org/10.1007/978-3-662-44857-1_13
16. Ernst, G., Pfähler, J., Schellhorn, G., Reif, W.: Modular, crash-safe refinement for asms with submachines. Sci. Comp. Program. **131**, 3–21 (2016)
17. Eymüller, C.: RealCaPP — Real-time capable plug & produce for distributed robot-based automation. Dissertation, Universität Augsburg (2024)
18. Eymüller, C., Hanke, J., Hoffmann, A., Reif, M.K.W.: Real-time capable opc-ua programs over tsn for distributed industrial control. In: Proceedings of 25th IEEE International Conference Emerging Technologies and Factory Automation (ETFA), vol. 1. pp. 278–285. IEEE (2020)
19. Eymüller, C., Hanke, J., Poeppel, A., Wanninger, C., Reif, W.: RealCaPP: real-time capable plug & produce service architecture for distributed robot control. In: Proceedings of 7th IEEE International Conference Robotic Computing (IRC), pp. 352–355 (2023). https://doi.org/10.1109/IRC59093.2023.00063
20. Gajek, C., Schiendorfer, A., Reif, W.: A recommendation system for cad assembly modeling based on graph neural networks. In: Proceedings of Joint European Conference Machine Learning and Knowledge Discovery in Databases (ECML/PKDD), Part I, LNCS, vol. 13713, pp. 457–473. Springer (2022)
21. Güdemann, M., Ortmeier, F., Reif, W.: Formal modeling and verification of systems with self-x properties. In: Yang, L.T., Jin, H., Ma, J., Ungerer, T. (eds.) ATC 2006. LNCS, vol. 4158, pp. 38–47. Springer, Heidelberg (2006). https://doi.org/10.1007/11839569_4
22. Saglietti, F., Oster, N. (eds.): SAFECOMP 2007. LNCS, vol. 4680. Springer, Heidelberg (2007). https://doi.org/10.1007/978-3-540-75101-4
23. Habermaier, A., Knapp, A., Leupolz, J., Reif, W.: Fault-aware modeling and specification for efficient formal safety analysis. In: ter Beek, M.H., Gnesi, S., Knapp, A. (eds.) FMICS/AVoCS -2016. LNCS, vol. 9933, pp. 97–114. Springer, Cham (2016). https://doi.org/10.1007/978-3-319-45943-1_7
24. Hähnle, R., Heisel, M., Reif, W., Stephan, W.: An interactive verification system based on dynamic logic. In: Siekmann, J.H. (ed.) CADE 1986. LNCS, vol. 230, pp. 306–315. Springer, Heidelberg (1986). https://doi.org/10.1007/3-540-16780-3_99
25. Haneberg, D., Moebius, N., Reif, W., Schellhorn, G., Stenzel, K.: Mondex: engineering a provable secure electronic purse. Intl. J. Softw. Inform. **5**(1), 159–184 (2011)

26. Haneberg, D., Schellhorn, G., Grandy, H., Reif, W.: Verification of mondex electronic purses with KIV: from transactions to a security protocol. Formal Aspects Comp. **20**(1) (2008)

27. Hanke, J.: Robot-based component testing on software-defined test benches. Dissertation, Universität Augsburg (2024)

28. Hanke, J., Eymüller, C., Reichmann, J., Trauth, A., Sause, M., Reif, W.: Software-defined testing facility for component testing with industrial robots. In: Proceedings of 27th IEEE International Conference Emerging Technologies and Factory Automation (ETFA), pp. 1–8 (2022).https://doi.org/10.1109/ETFA52439.2022.9921625

29. Hanke, J., Kosak, O., Schiendorfer, A., Reif, W.: Self-organized resource allocation for reconfigurable robot ensembles. In: Proceedings of 12th IEEE International Conference Self-Adaptive and Self-Organizing Systems (SASO), IEEE (2018). https://doi.org/10.1109/saso.2018.00022

30. Heisel, M., Reif, W., Stephan, W.: Program verification using dynamic logic. In: Börger, E., Büning, H.K., Richter, M.M. (eds.) CSL 1987. LNCS, vol. 329, pp. 102–117. Springer, Heidelberg (1988). https://doi.org/10.1007/3-540-50241-6_32

31. Hoffmann, A., Angerer, A., Ortmeier, F., Vistein, M., Reif, W.: Hiding real-time: a new approach for the software development of industrial robots. In: Proceedings of IEEE/RSJ International Conference Intelligent Robots and Systems (2009)

32. Hoffmann, A., Angerer, A., Schierl, A., Vistein, M., Reif, W.: Managing extensibility and maintainability of industrial robotics software. In: Proceedings of 16th International Conference Advanced Robotics (ICAR) (2013)

33. Hoffmann, A., Nägele, L., Angerer, A., Schierl, A., Reif, W.: Using object-oriented development for planning and controlling industrial robot systems. In: Proceedings of Workshop Recent Advances in Planning and Manipulation for Industrial Robots (2016)

34. Hoffmann, A., Angerer, A., Schierl, A., Vistein, M., Reif, W.: Service-oriented robotics manufacturing by reasoning about the scene graph of a robotics cell. In: Proceedings of 41st International Symposium on Robotics (ISR/Robotik), pp. 1–8. VDE (2014)

35. Hußmann, H., Loeckx, J., Reif, W.: KORSO: das Verbundprojekt "Korrekte Software". In: Proceedings of 23. GI-Jahrestagung: Informatik — Wirtschaft — Gesellschaft, pp. 266–271. Informatik Aktuell, Springer (1993). https://doi.org/10.1007/978-3-642-78486-6_42

36. Joshi, R., Holzmann, G.: A mini challenge: Build a verifiable filesystem. Formal Aspects Comput. **19**(2) (2007)

37. Katkalov, K., Stenzel, K., Reif, W.: Code abstractions for automatic information flow control in a model-driven approach. In: Wang, G., Atiquzzaman, M., Yan, Z., Choo, K.-K.R. (eds.) SpaCCS 2017. LNCS, vol. 10658, pp. 209–218. Springer, Cham (2017). https://doi.org/10.1007/978-3-319-72395-2_20

38. Kosak, O.: Mission programming for flying ensembles: combining planning with self-organization. Dissertation, Universität Augsburg (2021)

39. Kosak, O., Anders, G., Siefert, F., Reif, W.: An approach to robust resource allocation in large-scale systems of systems. In: Proceedings of 9th IEEE International Conference Self-Adaptive and Self-Organizing Systems (SASO), pp. 1–10 (2015). https://doi.org/10.1109/SASO.2015.8

40. Kosak, O., et al.: Swarm and collective capabilities for multipotent robot ensembles. In: Margaria, T., Steffen, B. (eds.) ISoLA 2020. LNCS, vol. 12477, pp. 525–540. Springer, Cham (2020). https://doi.org/10.1007/978-3-030-61470-6_31

41. Kosak, O., Huhn, L., Bohn, F., Wanninger, C., Hoffmann, A., Reif, W.: Maple-swarm: programming collective behavior for ensembles by extending HTN-planning. In: Margaria, T., Steffen, B. (eds.) ISoLA 2020. LNCS, vol. 12477, pp. 507–524. Springer, Cham (2020). https://doi.org/10.1007/978-3-030-61470-6_30

42. Kosak, O., Kastenmüller, P., Wanninger, C., Reif, W.: An approach for extended swarm formation flight with drones: PROTEASE$^{2.0}$. In: Proceedings of 12th International Symposium on Leveraging Applications of Formal Methods, Verification and Validation, Part II, LNCS, vol. 15220, pp. 263–280. Springer (2024). https://doi.org/10.1007/978-3-031-75107-3_16

43. Kosak, O., Wanninger, C., Angerer, A., Hoffmann, A., Schiendorfer, A., Seebach, H.: Towards self-organizing swarms of reconfigurable self-aware robots. In: Proceedings of 1st IEEE International Workshops on Foundations and Applications of Self* Systems (FAS*W), pp. 204–209 (2016). https://doi.org/10.1109/FAS-W.2016.52

44. Kosak, O., Wanninger, C., Angerer, A., Hoffmann, A., Schierl, A., Seebach, H.: Decentralized coordination of heterogeneous ensembles using jadex. In: Proceedings of 1st IEEE International Workshops on Foundations and Applications of Self* Systems (FAS*W), pp. 271–272 (2016). https://doi.org/10.1109/FAS-W.2016.65

45. Kosak, O., Wanninger, C., Hoffmann, A., Ponsar, H., Reif, W.: Multipotent systems: combining planning, self-organization, and reconfiguration in modular robot ensembles. Sensors **19**(1) (2019). https://doi.org/10.3390/s19010017

46. Krützmann, J., Schiendorfer, A., Beratz, S., Moosburger-Will, J., Reif, W., Horn, S.: Learning controllers for adaptive spreading of carbon fiber tows. In: Nicosia, G., et al. (eds.) LOD 2020. LNCS, vol. 12566, pp. 65–77. Springer, Cham (2020). https://doi.org/10.1007/978-3-030-64580-9_6

47. Lehner, C., Kosak, O., Ponsar, H., Reif, W.: Flexible assembly planning for self-organizing production cells. In: Proceedings of 6th International Conference Industry 4.0 and Smart Manufacturing. Procedia Computer Science, vol. 253, pp. 1860–1869 (2025). https://doi.org/10.1016/j.procs.2025.01.248

48. Lenzen, C., Schiendorfer, A., Reif, W.: Graph machine learning for assembly modeling. In: Proceedings of 1st Learning on Graphs Conference (2022)

49. Leupolz, J., Knapp, A., Habermaier, A., Reif, W.: Qualitative and quantitative analysis of safety-critical systems with s#. Int. J. Softw. Tools Technol. Transf. **20**(4), 359–377 (2018). https://doi.org/10.1007/s10009-017-0464-3

50. Müller-Schloer, C., Tomforde, S.: Organic Computing-Technical Systems for Survival in the Real World, vol. 1. Springer (2017)

51. Nafz, F., Ortmeier, F., Seebach, H., Steghöfer, J.-P., Reif, W.: A universal self-organization mechanism for role-based organic computing systems. In: González Nieto, J., Reif, W., Wang, G., Indulska, J. (eds.) ATC 2009. LNCS, vol. 5586, pp. 17–31. Springer, Heidelberg (2009). https://doi.org/10.1007/978-3-642-02704-8_3

52. Nafz, F., Seebach, H., Steghöfer, J.-P., Bäumler, S., Reif, W.: A formal framework for compositional verification of organic computing systems. In: Xie, B., Branke, J., Sadjadi, S.M., Zhang, D., Zhou, X. (eds.) ATC 2010. LNCS, vol. 6407, pp. 17–31. Springer, Heidelberg (2010). https://doi.org/10.1007/978-3-642-16576-4_2

53. Nägele, L., Hoffmann, A., Schierl, A., Reif, W.: Legobot: automated planning for coordinated multi-robot assembly of lego structures. In: Proceedings of IEEE/RSJ International Conference Intelligent Robots and Systems (IROS), pp. 9088–9095. IEEE (2020)

54. Nägele, L., et al.: A backward-oriented approach for offline programming of complex manufacturing tasks. In: Proceedings of 6th International Conference on Automation, Robotics and Applications (ICARA), pp. 124–130. IEEE (2015)

55. Ortmeier, F., Reif, W., Schellhorn, G.: Formal safety analysis of a radio-based railroad crossing using deductive cause-consequence analysis (DCCA). In: Dal Cin, M., Kaâniche, M., Pataricza, A. (eds.) EDCC 2005. LNCS, vol. 3463, pp. 210–224. Springer, Heidelberg (2005). https://doi.org/10.1007/11408901_15

56. Ortmeier, F., Reif, W., Schellhorn, G.: Deductive cause-consequence analysis (DCCA). In: Proceedings of IFAC World Congress. Elsevier (2006)

57. Ortmeier, F., Schellhorn, G., Thums, A., Reif, W., Hering, B., Trappschuh, H.: Safety analysis of the height control system for the Elbtunnel. In: Anderson, S., Felici, M., Bologna, S. (eds.) SAFECOMP 2002. LNCS, vol. 2434, pp. 296–308. Springer, Heidelberg (2002). https://doi.org/10.1007/3-540-45732-1_29

58. Ortmeier, F., Schellhorn, G., Reif, W.: Safety optimization of a radio-based railroad crossing. In: Proceedings of Formal Methods for Automation and Safety in Railway and Automotive Systems (FORMS) (2004)

59. Ehrig, H., et al. (eds.): Integration of Software Specification Techniques for Applications in Engineering. LNCS, vol. 3147. Springer, Heidelberg (2004). https://doi.org/10.1007/b100778

60. Pfähler, J., Ernst, G., Bodenmüller, S., Schellhorn, G., Reif, W.: Modular verification of order-preserving write-back caches. In: Polikarpova, N., Schneider, S. (eds.) IFM 2017. LNCS, vol. 10510, pp. 375–390. Springer, Cham (2017). https://doi.org/10.1007/978-3-319-66845-1_25

61. Poeppel, A., Eymüller, C., Reif, W.: SensorClouds: a framework for real-time processing of multi-modal sensor data for human-robot-collaboration. In: Proceedings of 9th International Conference on Automation, Robotics and Applications (ICARA), pp. 294–298 (2023). https://doi.org/10.1109/ICARA56516.2023.10125740

62. Poeppel, A., Hoffmann, A., Siehler, M., Reif, W.: Robust distance estimation of capacitive proximity sensors in hri using neural networks. In: Proceedings of 4th IEEE International Conference on Robotic Computing (IRC), pp. 344–351 (2020). https://doi.org/10.1109/IRC.2020.00061

63. Reeves, G., Neilson, T.: The mars rover spirit FLASH anomaly. In: Aerospace Conference on, pp. 4186–4199. IEEE (2005)

64. Reif, W.: Vollständigkeit einer modifizierten Goldblatt-Logik und Approximation der Omega-Regel durch Induktion. Diplomarbeit, Universität Karlsruhe, Fakultät für Informatik (1984). (in German)

65. Reif, W.: Korrektheit von Spezifikationen und generischen Moduln. Ph.D. thesis, Karlsruhe Institute of Technology, Germany (1991). https://d-nb.info/931877865

66. Reif, W.: Correctness of full first-order specifications. In: Proceedings of 4th International Conference on Software Engineering and Knowledge Engineering (SEKE), pp. 276–283. IEEE Computer Society (1992). https://doi.org/10.1109/SEKE.1992.227918

67. Reif, W.: Correctness of generic modules. In: Nerode, A., Taitslin, M. (eds.) LFCS 1992. LNCS, vol. 620, pp. 406–417. Springer, Heidelberg (1992). https://doi.org/10.1007/BFb0023893

68. Reif, W.: The KIV system: systematic construction of verified software. In: Kapur, D. (ed.) CADE 1992. LNCS, vol. 607, pp. 753–757. Springer, Heidelberg (1992). https://doi.org/10.1007/3-540-55602-8_218

69. Shyamasundar, R. (ed.): FSTTCS 1992. LNCS, vol. 652. Springer, Heidelberg (1992). https://doi.org/10.1007/3-540-56287-7

70. Reif, W.: The KIV-approach to software verification. In: Broy and Jähnichen [10], pp. 339–370. https://doi.org/10.1007/BFB0015471

71. Reif, W., Ortmeier, F., Güdemann, M.: Safety and dependability analysis of self-adaptive systems. In: Proceedings of 2nd International Symposium on Leveraging Applications of Formal Methods, Verification and Validation (ISoLA), pp. 177–184. IEEE (2006). https://doi.org/10.1109/ISoLA.2006.38

72. Schellhorn, G., Ahrendt, W.: The WAM Case Study: Verifying Compiler Correctness for Prolog with KIV. In: Automated Deduction — A Basis for Applications, vol. III: Applications, pp. 165–194. Kluwer Academic Publishers (1998)

73. Schellhorn, G., Bodenmüller, S., Bitterlich, M., Reif, W.: Software & system verification with KIV. In: The Logic of Software. A Tasting Menu of Formal Methods: Essays Dedicated to Reiner Hähnle on the Occasion of His 60th Birthday, LNCS, vol. 13360, pp. 408–436. Springer (2022)

74. Schellhorn, G., Bodenmüller, S., Pfähler, J., Reif, W.: Adding concurrency to a sequential refinement tower. In: Raschke, A., Méry, D., Houdek, F. (eds.) ABZ 2020. LNCS, vol. 12071, pp. 6–23. Springer, Cham (2020). https://doi.org/10.1007/978-3-030-48077-6_2

75. Schellhorn, G., Reif, W., Schairer, A., Karger, P., Austel, V., Toll, D.: Verified formal security models for Multiapplicative smart cards. J. Comp. Sec. **10**(4), 339–367 (2002)

76. Schellhorn, G., Tofan, B., Ernst, G., Pfähler, J., Reif, W.: RGITL: a temporal logic framework for compositional reasoning about interleaved programs. Ann. Math. Artif. Intell. 131–174 (2014). https://doi.org/10.1007/s10472-013-9389-z

77. Schiendorfer, A., Knapp, A., Anders, G., Reif, W.: MiniBrass: soft constraints for MiniZinc. Constraints **23**(4), 403–450 (2018). https://doi.org/10.1007/s10601-018-9289-2

78. Schiendorfer, A., Steghöfer, J.-P., Knapp, A., Nafz, F., Reif, W.: Constraint Relationships for Soft Constraints. In: Bramer, M., Petridis, M. (eds.) Research and Development in Intelligent Systems XXX, pp. 241–255. Springer, Cham (2013). https://doi.org/10.1007/978-3-319-02621-3_17

79. Schierl, A., Angerer, A., Hoffmann, A., Vistein, M., Reif, W.: From robot command to real-time robot control — transforming high-level robot commands into real-time dataflow graphs. In: Proceedings of 9th International Conference on Informatics in Control, Automation and Robotics, vol. 2 (2012)

80. Schierl, A., Angerer, A., Hoffmann, A., Reif, W.: Consistent world models for cooperating robots: separating logical relationships, sensor interpretation and estimation. In: Proceedings of 1st IEEE International Conference on Robotic Computing (IRC), pp. 101–108. IEEE (2017)

81. Schörner, M., Katschinsky, R., Wanninger, C., Hoffmann, A., Reif, W.: Towards fully automated inspection of large components with UAVs: offline path planning and view angle dependent optimization strategies. In: Gusikhin, O., Madani, K., Zaytoon, J. (eds.) Rev. Sel. Papers 17th International Conference on Informatics in Control, Automation and Robotics (ICINCO), LNEE, vol. 793, pp. 105–123. Springer (2022)

82. Schörner, M., Wanninger, C., Hoffmann, A., Kosak, O., Ponsar, H., Reif, W.: Modeling and execution of coordinated missions in reconfigurable robot ensembles. In: Proceedings of 4th IEEE International Conference on Robotic Computing (IRC), pp. 290–293 (2020). https://doi.org/10.1109/IRC.2020.00053

83. Schörner, M., et al.: UAV inspection of large components: determination of alternative inspection points and online route optimization. In: Proceedings of 5th

IEEE/ACM International Workshops on Robotics Software Engineering (RoSE), pp. 45–52 (2023). https://doi.org/10.1109/RoSE59155.2023.00012

84. Seebach, H., Nafz, F., Steghofer, J.P., Reif, W.: A software engineering guideline for self-organizing resource-flow systems. In: Proceedings of 4th IEEE International Conference on Self-Adaptive and Self-Organizing Systems, pp. 194–203. IEEE (2010). https://doi.org/10.1109/SASO.2010.26

85. Seebach, H., Ortmeier, F., Reif, W.: Design and construction of organic computing systems. In: Proceedings of IEEE Congress on Evolutionary Computation (2007). https://doi.org/10.1109/cec.2007.4425021

86. Steghöfer, J.P., Anders, G., Siefert, F., Reif, W.: A system of systems approach to the evolutionary transformation of power management systems. In: Proceedings of 43. Jahrestagung der Gesellschaft für Informatik e.V. GI (2013). https://dl.gi.de/handle/20.500.12116/20584

87. Steghöfer, J.-P., et al.: Trustworthy organic computing systems: challenges and perspectives. In: Xie, B., Branke, J., Sadjadi, S.M., Zhang, D., Zhou, X. (eds.) ATC 2010. LNCS, vol. 6407, pp. 62–76. Springer, Heidelberg (2010). https://doi.org/10.1007/978-3-642-16576-4_5

88. Rattray, C., Maharaj, S., Shankland, C. (eds.): AMAST 2004. LNCS, vol. 3116. Springer, Heidelberg (2004). https://doi.org/10.1007/b98770

89. Stenzel, K., Katkalov, K., Borek, M., Reif, W.: A model-driven approach to non-interference. J. Wireless Mobile Netw. Ubiq. Comput. Depend. Appl. 5(3), 30–43 (2014)

90. Stenzel, K., Katkalov, K., Borek, M., Reif, W.: Declassification of information with complex filter functions. In: Proceedings of 2nd International Conference on Information Systems Security and Privacy, pp. 490–497 (2016)

91. Stieber, S., Heber, L., Obertscheider, C., Reif, W.: Control of composite manufacturing processes through deep reinforcement learning. In: Proceedings of International Conference on Machine Learning and Applications (ICMLA), pp. 17–22. IEEE (2023)

92. Stieber, S., Schröter, N., Fauster, E., Bender, M., Schiendorfer, A., Reif, W.: Inferring material properties from frp processes via sim-to-real learning. Intl. J. Adv. Manufact. Technol. 1–17 (2023)

93. Stieber, S., Schröter, N., Schiendorfer, A., Hoffmann, A., Reif, W.: Flowfront-net: improving carbon composite manufacturing with cnns. In: Proceedings of Joint European Conference on Machine Learning and Knowledge Discovery in Databases (ECML/PKDD), Part IV, LNCS, vol. 12460, pp. 411–426. Springer (2020)

94. Stüben, M., Hoffmann, A., Reif, W.: Constraint-based whole-body-control of mobile manipulators in human-centered environments. In: Proceedings of 26th IEEE International Conference on Emerging Technologies and Factory Automation (ETFA), pp. 1–8 (2021). https://doi.org/10.1109/ETFA45728.2021.9613281

95. Tofan, B., Bäumler, S., Schellhorn, G., Reif, W.: Temporal logic verification of lock-freedom. In: Bolduc, C., Desharnais, J., Ktari, B. (eds.) MPC 2010. LNCS, vol. 6120, pp. 377–396. Springer, Heidelberg (2010). https://doi.org/10.1007/978-3-642-13321-3_21

96. Tofan, B., Schellhorn, G., Reif, W.: Formal verification of a lock-free stack with hazard pointers. In: Cerone, A., Pihlajasaari, P. (eds.) ICTAC 2011. LNCS, vol. 6916, pp. 239–255. Springer, Heidelberg (2011). https://doi.org/10.1007/978-3-642-23283-1_16

97. Vistein, M., Angerer, A., Hoffmann, A., Schierl, A., Reif, W.: Interfacing industrial robots using realtime primitives. In: Proceedings of IEEE International Conference on Automation and Logistics (2010)

98. Vistein, M., Angerer, A., Hoffmann, A., Schierl, A., Reif, W.: Instantaneous switching between real-time commands for continuous execution of complex robotic tasks. In: Proceedings of IEEE International Conference on Mechatronics and Automation (2012). https://doi.org/10.1109/icma.2012.6284329

99. Vistein, M., Angerer, A., Hoffmann, A., Schierl, A., Reif, W.: Flexible and continuous execution of real-time critical robotic tasks. Intl. J. Mechatron. Automat. 4(1), 27–38 (2014)

100. Wanninger, C., Eymüller, C., Hoffmann, A., Kosak, O., Reif, W.: Synthesizing capabilities for collective adaptive systems from self-descriptive hardware devices bridging the reality gap. In: Proc. 8th Intl. Symp. Leveraging Applications of Formal Methods, Verification and Validation (ISoLA), Part III, Distributed Systems. LNCS, vol. 11246, pp. 94–108 (2018). https://doi.org/10.1007/978-3-030-03424-5_7

101. Wolf, B., et al.: The SCALEX campaign: scale-crossing land surface and boundary layer processes in the TERENO-preAlpine observatory. Bull. Am. Meteor. Soc. 98(6), 1217–1234 (2017). https://doi.org/10.1175/BAMS-D-15-00277.1

Eine REIFe Leistung – Anecdotes from Behind the Scenes

Gidon Ernst[1] , Dominik Haneberg[2] , Matthias Güdemann[3] ,
Oliver Kosak[2] , Florian Nafz[3] , Hella Ponsar[2(✉)] ,
and Gerhard Schellhorn[2]

[1] Ludwig-Maximilians-Universität München, Munich, Germany
gidon.ernst@lmu.de
[2] Universität Augsburg, Augsburg, Germany
{dominik.haneberg,oliver.kosak,hella.ponsar,gerhard.schellhorn}@uni-a.de
[3] Munich University of Applied Sciences HM, Munich, Germany
{matthias.guedemann,florian.nafz}@hm.edu

Abstract. "This is Australia calling..." – that's the sound of Wolfgang Reif. When this tune can be heard echoing through the corridors of the ISSE, you can be sure: the boss is here. Science is not just dry theory, it lives from the people who shape it. And if there's anyone who combines research, teaching and a pinch of lightheartedness and academic humor, it's Wolfgang Reif. A brilliant mind who not only enriched the academic world with ground-breaking ideas, but also with legendary anecdotes. With a wink, but with the utmost respect, we take a closer look at Wolfgang Reif as a person.

1 How It All Began: Karlsruhe and the Years in Ulm

After his diploma studies Wolfgang Reif became a researcher in the group of Prof. Menzel at the Institute for Logic, Complexity and Deduction. Together with Maritta Heisel and Werner Stefan he immediately submitted a successful proposal for a project to the German Research Foundation (DFG). The topic was a new approach to program verification. Core elements were the use of Dynamic Logic in a theorem prover for the first time, and the development of a functional "Proof Programming Language" (PPL), that used proof trees as its core data structure instead of "S-expressions" (like in LISP, that was common at that time). The first version of the KIV-system was developed. Initially it did not have a graphical interface, but only shortly after Motif was used to have a convincing interface that shows proof trees graphically. In order to develop a compiler for the new programming language (based on byte code of the SECD machine) and to implement a proof calculus soon half a dozen students were employed who had weekly meetings with the researchers to discuss results and next steps. The author of these lines remembers that there were big discussions, why a chair for theoretical computer science needed the biggest hard disks in the

---

German "reif" is "ripe" or "mature" and a "reife Leistung" is a "great performance".

© The Author(s), under exclusive license to Springer Nature Switzerland AG 2025
G. Ernst et al. (Eds.): Wolfgang Reif Festschrift, LNCS 15765, pp. 20–35, 2025.
https://doi.org/10.1007/978-3-031-92196-4_2

department (at that time work was on Sun Sparc computers). The reason was simply that a saved proof tree could consume several hundred kilobytes, which was a gigantic amount of memory at that time.

Fig. 1. Wolfgang Reif's group in Ulm

Wolfgang Reif obtained his PhD 1991 in Karlsruhe, and the projects KORSO ("Korrekte Software", sponsored by the German Ministry of Research) and VSE (Verification Support Environment) started, that allowed to employ new researchers, so the team expanded. Shortly after in 1994 he accepted a professorship in Ulm. There he soon had a productive working group (see Fig. 1).

An important annual workshop at that time was the "Deduktionstreffen", where many rivalries between advocates of interactive and fully automated provers were fought out. The slogan at that time was "Demo or Die". So it happened, that as a gag at the end of a talk we often showed a proof tree with 5000 nodes, that was presented on many pages in DIN-A4-format taped together to a size of ca. $2\,\mathrm{m}^2$.

Having a graphical interface for a tool also had its risks: our first demo in Berlin required a long telephone discussion with Ulm (the internet was only just in its beginning) to get a patch, since on a brand new color monitor the edges of the proof tree suddenly had ten times the width they had on the monochrome display that we still used. And then there were lots of heated discussions what colors to use for the KIV GUI.

Worth remembering are also the two FM-TOOLS-Workshops in 2000 and 2002 on the Reisensburg. The author still remembers today that he learned at the workshop, that lunch is not after, but before dinner, and that organizing a workshop sometimes results in unexpected problems: one participant arrived *without* any money, but *with* his family. His talk was with pages from the submission just copied to his presentation slides (which were of course unreadable). Nevertheless his talk was much too long, and the session chair almost had to resort to physical violence to end it.

In 2000, Wolfgang Reif was offered a professorship in Augsburg and the next move followed. Various things accompanied us to the new home: an ancient Apple Macintosh, offprints from KIT and the never-ending discussion how workstation computers should be named. This was one of the few instances where Prof. Reif's opinion did not prevail. His proposal, to name them after deadly diseases was not accepted, instead they were still named after brands of Scottish whisky (glenfiddich, bruichladdich etc.).

2   Computer Science in Augsburg

When Prof. Reif began his work in Augsburg, computer science still stood in the shadow of the Faculty of Mathematics. But then, the computer science study programme was introduced – and Prof. Reif had the honor of welcoming the first full-time computer science students to Augsburg in his very first lecture for "Informatik 1". However, it was not enough for Prof. Reif to simply teach students. He fought for a true "student association" that would advocate computer science, and was instrumental in founding the computer science student body, which finally gave students their own voice in the academic councils.

Just a few years later, in 2004, he followed in the footsteps of Augsburg's great visionaries by becoming the founding dean of the Faculty of Applied Computer Science (FAI). This marked the beginning of the new computer science building, into which so many colleagues soon moved that it almost felt like a software update – more space, more people, and ever-improving performance. But that was only the beginning, for Prof. Reif it was very clear: Computer science in Augsburg still had much more potential. Through various offers for appointments from other universities, he made it clear to his own university how important the field of software and systems engineering is, and thus founded the ISSE (Institute for Software and Systems Engineering) within the FAI in 2008. Under his leadership, at its peak, up to 23 scientific employees and 31 research assistants (see Fig. 2) were dedicated to developing new, sometimes crazy, but certainly brilliant ideas for the digital world – a true research co-living space for innovative minds. Three secretaries also tirelessly helped organize conference trips, festive occasions, various meetings, and much more.

Fig. 2. Wolfgang Reif's SWT group in Augsburg

As if that were not enough, in 2011, Prof. Reif became Vice President of the University of Augsburg for "Technologie und Innovation" – and suddenly computer science in Augsburg truly began to flourish. For instance, he captivated local medical professionals with its possibilities, who, up until then, might have thought that computer science was only for people spending their entire day sitting in dark rooms staring at screens. Through Prof. Reif's efforts, a "Medical Informatics" study programme was established when the university finally received its own medical faculty. Another challenge from the increasing attractiveness of computer science – which Prof. Reif gladly accepted – was that the department had gained so many students, colleagues, and staff that even the new building soon began to burst at the seams. Since also the old home in Eichleitner Straße no longer sufficed for the constant growth, more space was required for expansion. During the construction of the MRM building, Prof. Reif was once again able to put his evidently existing qualifications as an architect to good use. Of course, he also took the opportunity to organize a massive research arena for the ISSE, which would likely remind any visitor more of the factory halls of tech giants than typical university spaces.

Wolfgang Reif's influence extends far beyond the confines of his chair, the ISSE, the computer science faculty, and the University of Augsburg. His engagement in many other important aspects of Augsburg's society is a testimony for this. In particular, he is deeply involved in the regional and even supra-regional development of the location: He has been participating in various technology alliances (advisory boards of the aiti-Park, Innovation Park, FZG, the Chamber of Commerce and Industry, the Mechatronics & Automation Cluster, and many more) to offer his expertise to local companies. In various transfer projects, he and his team share important insights into computer science with the business and education sectors. In addition to specifically established laboratories and provided equipment, numerous delegations and visitor groups regularly enjoy the exchange on-site – from countless school classes to numerous business partners and interested university colleagues, as well as visitors from around the world. Most recently, since 2020, he has been serving as the director of the "KI Produktionsnetzwerk Augsburg", further demonstrating his commitment to the development of both the academic and economic landscape in Augsburg. Through numerous projects, discussion panels, lectures, and individual conversations, he conveys his belief in the importance of computer science as a driving force for regional development.

3   Building Up the Elites of Tomorrow – Master Programme Software Engineering

"What's software engineering to you?" With this seemingly simple question, Prof. Reif regularly encourages new students in the Master's programme in Software Engineering in the Elite-Network Bavaria right at the beginning of the titular lecture to engage with their future professional and research field. With his enthusiasm for the subject and the discipline, he fosters discussion and exchange among the students. Over the course of two intense and challenging years, the group grows together, forms networks, and builds friendships. This is where the experts and leaders of tomorrow are shaped.

Prof. Reif played a leading role in developing this Elite programme as a collaboration between the University of Augsburg and the Munich universities TUM and LMU. Prof. Reif's initiative combined the software competence from three universities, dedicated professors who all are highly renowned in their respective fields, to create an innovative and academically excellent environment. Within the programme, "Pushing the Limits" projects are developed, international exchanges with partner institutions take place, and thesis work leads to scientific publications. Prof. Reif also took the idea of excellence to the heart and made sure to be as well prepared as possible for his lectures, what often led to research assistants perfecting the lecture slides minutes before the lecture started.

The results are impressive: many students pursue doctorates and embark on academic or research-oriented careers, leading to a new generation of Software Engineering professors who actually studied Software Engineering. But graduates also become multipliers in professional environments, transferring the state of the art in research to industrial practice, sometimes in important roles in large companies and sometimes by founding their own companies. The elite programme in Software Engineering was launched in 2006. We now look back on 19 successful years, with more than 260 graduates. A highlight is the annual graduation celebration: Even though the (admittedly excellent) traditional pasta with

Parmigiano may tend to take center stage, the focus is always on the achievements of the students. Prof. Reif could not always attend the celebration, his passion for Australia interfering, but he never missed sending a video greeting from "down under".

4   Inventor of Anything

Prof. Reif underwent an impressive evolution during his academic career, ranging from the abstract heights of formal methods to the more grounded realm of software engineering and finally to tangible hardware, which, if necessary, could even be adjusted with a screwdriver. In Augsburg, he not only focused on educating students but also created optimal learning conditions, enabling both young talents and experienced researchers to experiment with controlling hardware-intensive systems with scientific precision, great curiosity, and the occasional realization: "This really shouldn't have crashed. . . ".

Fig. 3. Before the metamorphose . . .   **Fig. 4.** . . . and now: the ISSE research hall.

4.1   A New Programme: Computer Science for Engineers

In the course of numerous research projects with industry, Wolfgang Reif and his group made a crucial discovery: Communication between those who understand how machines, robots, and processes work and those who specialize in computer science and software development is often quite complicated. This led him to the insight that there is a lack of true "tech interpreters" – individuals who feel at home and are competent in both worlds. It was precisely this missing type of professional that he aimed to train, leading to the thought: "If such an educational programme doesn't yet exist in Augsburg, then I'll just create one myself."

And so, in 2013, he launched the "Ingenieurinformatik" study programmes in Augsburg. With persuasive skills (and probably a good amount of persistence), he convinced the university and faculty to establish three professorships for engineers who teamed up with computer scientists to educate the next generation of high-tech all-rounders. Today, the graduates of this program are in high demand in the industry. No surprise – after all, they are the rare blend of machine whisperers and code wizards.

To ensure that the new "Ingenieurinformatik" students were accommodated appropriately, Prof. Reif and his colleagues promptly arranged for a dedicated building – Building W, complete with its own research hall (Figs. 3 and 4). But not just any hall – of course, it had to be a true technological mecca. The result was a high-tech palace featuring a massive research facility, where robots now buzz through the air or diligently assemble objects.

Ingenieurinformatik
Computer Science in Engineering

But what would a new degree program be without a stylish logo? Exactly – unfinished. However, the search for the perfect logo turned out to be a true test of patience. A professional design firm was commissioned to create a suitable emblem. Yet, when the first drafts arrived, Prof. Reif was dismayed – all the proposed logos consisted solely of zeros and ones. But computer science is so much more – why always fall back on the binary cliché unwilling to settle for this, he refused to accept fate and instead launched a never-ending brainstorming offensive at the institute. After countless heated discussions, the idea arose. And true to Prof. Reif's motto, "you just have to take matters into your own hands", we now only needed the external support of the design firm to transform our brilliant in-house concept into a polished design.

4.2   We Are the Robots – The Credo of Demonstrators

"Every dissertation needs a demonstrator" – this guiding principle has been consistently upheld by Prof. Reif in supervising his doctoral students. His students, in turn, have embraced this credo – sometimes even in their sleep – leading to numerous "aha moments" (which, indeed, often occur while sleeping, showering, commuting, driving on the highway, or simply reflecting in the office or laboratory). Visitors to the ISSE frequently experience similar "aha moments", as demonstrators effectively showcase the scientific achievements of Prof. Reif's research group in an accessible and engaging manner. These demonstrators have been highlights at numerous events, including Open-Lab days, Rotary Club Augsburg gatherings, and public exhibitions such as the "Lange Nacht der Wissenschaft" in the "Goldener Saal" of Augsburg's Town Hall. They have also gained recognition at trade fairs from Augsburg to Friedrichshafen, Hannover, and even Rimini, making them a true attraction. Prof. Reif places particular emphasis on ensuring that the demonstrators not only convey paradigm shifts in scientific research but also provide sufficient depth for expert discussions. At first glance, they should be easy to understand, offer palpable added value,

Fig. 5. The "Energietisch"

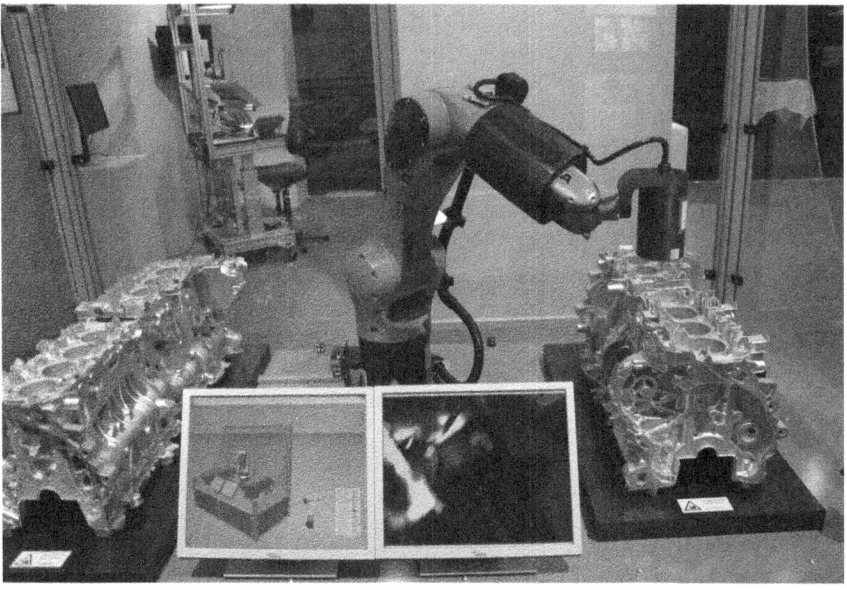

Fig. 6. Automated engine inspection

and at the same time, serve as a foundation for in-depth debates among professionals. This approach drives the motivation of all students involved in their
development.

Who would immediately realize that the working force behind the two robots – produced by different manufacturers – that skillfully maneuver an uncooked spaghetti noodle at impressive speeds without breaking it, is the real-time capable, manufacturer-independent, and open-access RoboticsAPI with its RPI networks, developed by Prof. Reif's team? Or that this very software serves as the backbone for the flying "serving robot", which skillfully balances drinks in mid-air within the lab's flight arena? Interested visitors quickly grasp that there is more than meets the eye – sparking fascinating discussions about the presented concepts, underlying details, and scientific insights. These insights also address pressing contemporary issues, such as solutions for the challenges of the energy transition. For instance, the "Energietisch" miniature model (cf. Fig. 5) visualizes nature-inspired strategies for addressing future scalability and energy security concerns. Alternatively, a self-organizing ensemble – featuring an off-road capable mobile robot (internally called "Gulaschkanone"[1] for its form) paired with quadcopters – assists in validating climate models. Yet, there is also room for "La Dolce Vita" at Prof. Reif's institute (cf. Fig. 7). Even the preparation of "authentic Italian coffee" using an intelligently programmed industrial robot poses no major challenge for him and his research team. While it may have seemed ambitious to present such an automated coffee service at a trade fair in Rimini – home to the quasi-inventors of this culinary specialty – the resounding success proved them right.

Listing all demonstrators developed under Prof. Reif's supervision would be nearly impossible, and the available space here is far from sufficient to describe them all. From a portrait-drawing robot, to a simulated traffic control system for the "Elb-Tunnel", a gesture-controlled lightweight robot using Kinect image analysis, a mysterious dual-arm platform, a capacitive sensor-based system for secure engine block analysis (cf. Fig. 6), a quadcopter-based component inspection system, self-organizing swarm flight, robots playing either Lego or Poker, and an unbeatable AI-powered "Rock-Paper-Scissors" machine – the pool of innovative ideas seems limitless.

And one thing is certain: the journey is far from over. New ideas and concepts are already emerging, such as a self-organizing, fully automated, and highly flexible production system of the future – one that could even be deployed on Mars to autonomously manufacture rovers. Of course, this would require additional research funding. But if anyone could secure these funds – naturally just before the end of the fiscal year – it would undoubtedly be Prof. Reif.

5   Community Feeling or "Life During the Doctoral Phase"

"Have you ever thought about a doctorate? You will rarely find the luxury of being paid for thinking elsewhere." This motivation comes with an implicit praise, with which Prof. Reif wants to win over the best of the best for himself. The principle is to find and promote talents. Of course, pure thinking is rather metaphorical, because although research is theoretically based, it comes to life

---

[1] Field kitchen.

Fig. 7. La Dolce Vita at ISSE – Italian Coffee brewed by the Robarista

through practical implementation. When you join, you immediately become part of a lively community at the chair. And, of course, your own ambitious research takes center stage. However – "have you got a minute?" – important issues are best discussed in person. While in one moment you are busy and desperate working on a KIV proof, developing the energy architecture of the future, or letting the quadcopters dash. At another moment, you are deeply immersed in a discussion with Prof. Reif, who drops by the office. Sometimes it's about teaching, sometimes about project applications, sometimes about university matters and sometimes about configuring his e-mail client.[2]

There are many aspects to the academic world. One of the most important of these is personal dialogue. There are many opportunities for this at Prof. Reif's department, which we would like to highlight below.

5.1 FMSOFTies – Meetings in the Mountains

Once a year, it's time to get down to the nitty gritty. When the entire department sets off together to the idyllic scenery of the (foothills of the) alps to present and discuss the previous year's scientific results at FMSOFT, the annual department meeting in a mountain hut, the anticipation of the ambiance and the activities is sometimes mixed with some quiet doubt: Did I prepare enough slides for my

---

[2] We are proud to report that Prof. Reif, as a long-time user of `mutt` successfully switched to modern alternatives at an unspecified point in time.

	Locations of FMSOFT
2002	Barbara-Hütte
2003	Achensee
2004f.	Breitenbach
2006	Schönbach
2007	Kaunerberg
2008	Ramsau
2009f.	Radstadt
2011ff.	Bezau

Fig. 8. The boss lends a hand, of course!

presentation? How is my dissertation progressing and what feedback might I get? Do I have enough issues for an extended discussion of technical details?

Some of these discussions are characterized by their depth, some by their length, the intersection by both; sometimes it became foundational: "Does a corridor only have top and bottom or also other sides?", sometimes the foundations were missing: "Where does that leave the *trust*?" Naturally, the overflowing lecture schedule often necessitated the implementation of a timekeeper, a role traditionally handled diligently but rarely with sufficient impact. This resulted in a common occurrence: the allure of the available food served as a siren call, ultimately silencing the discourse.

For this reason alone, great importance was and is attached to communal cooking at the huts (Fig. 8). Agreeing on a selection of foods was supported by a preference and soft-constraints-based tool, which, in addition to roasts and sausage salads, took into account a vegetarian Bolognese just as fairly as the apodictic veto against the use of peas. The chair's collected barbecue skills were also regularly put to the test, the results of which only had to be further praised ("Schweinebauch!"[3]) when everyone was already comfortably full. Scientific interests and culinary delights also came together in "Egg Engineering", where the "Master of Egg Technology" had to find out how to prepare the eggs soft-boiled at the various heights of the different mountain huts on the very first day of each stay.

Joint undertakings and activities – for team-building – were an essential part of the chair huts. Not everyone was able to join the "Badebus" (drive to the nearby lake) early in the morning, but Prof. Reif was always there. In addition

---

[3] Pork belly.

 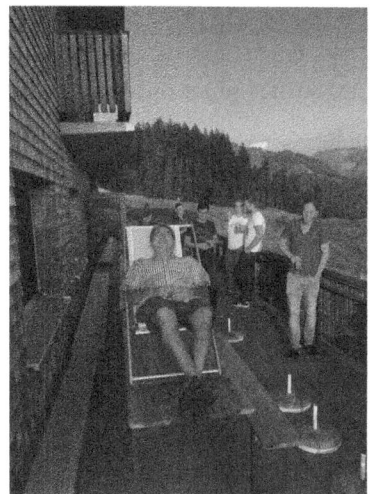

Fig. 9. Well-deserved break after reaching the top

Fig. 10. Time for relaxation: a break after intense debates

to hikes (Fig. 9), be it to the nearby "Buttermilch"[4] hut or to more distant peaks, be it at "touring" or "wuss" level, there were scavenger hunts, geocaching, capture the flag and much more; Prof. Reif showed great commitment to his team in each of these competitions (Figs. 11, 12, 13 and 14). After the day's discussions in the large group (Fig. 10), most people ended the evening with classic games or table tennis; for Prof. Reif and a small group, on the other hand, dissertation discussions were on the agenda, sometimes late into the night. But when Prof. Reif had time for a game, he played with full commitment. In the "Werewolf" game, for example, he first skillfully threw all his fellow "werewolves" under the bus to finally win the gruesome victory. The cozy "Wertschätzungsabend" (evening of appreciation) was a completely different experience, where people were eager to show their appreciation for one another over beer, wine and perhaps even a schnapps.

5.2   ISSE-Sational – Summer Parties and Christmas Celebrations

In addition to research-intensive stays at FMSOFT, social field research was not to be missed. Summer parties, Christmas celebrations, and the legendary "Hiwi" (student research assistant) barbecue have always provided ample opportunity to study social dynamics in the wild. Don't worry, at the Hiwi barbecue, the Hiwis have so far only ended up *at* the table, rarely *under* it, and never *on* it or even on the grill. The summer parties took us to picturesque locations such as the Wertach or the Kälberhalle, where current employees and alumni and their families perfected the fine art of barbecuing and talking shop over chilled drinks.

---

[4] Buttermilk.

At Christmas time, we often swapped the lab for the pub to get in the mood for the festive season with mulled wine or mulled beer - and to discover that the best theories often only emerge after the second cup.

5.3   Viva Voce, No Thesis Defense! – The Doctors

"Did we just get lucky with this?" – You should be prepared for this question from Prof. Reif as soon as it comes to the crowning conclusion of the doctorate. Of course, the expected answer is: no! Success is based on scientific excellence, current and challenging research topics, but also cleverly and systematically chosen approaches.

Prof. Reif's department has a tradition of choosing the viva voce examination ("Rigorosum") instead of the disputation. As of today, only one of 39 successful doctoral students has broken this tradition (nothing too bad is said to have happened). As a result, each subsequent generation has always been attentive and diligent in writing down the questions asked in the debates and oral examinations of the viva voce, and we have all become experts in topics such as shading methods in computer graphics, geographical routing, or virtual function tables with multiple inheritance.

Excellent preparation for this performance is a matter of course. Prof. Reif makes candidates reflect on the far-reaching implications of their own research in advance. Not only does the initial question prompt reflection on this point, but also the question of translating results into practice.

6   Networking in the World of Science

Prof. Reif had a clear plan from the very beginning: not just to wait for the future but shape it together with other scientists. Already early on (see the section "How it all began...."), he showed a talent for getting DFG projects approved – and, what is even more challenging, successfully renew them. He also enthusiastically organized meetings to exchange ideas and build communities.

When it came to digital transformation and controlling large, emergent, distributed systems, Prof. Reif was not only active in the scientific world – he was also frequently on an expedition "down under". Between 2009 and 2023, he traveled to Australia no fewer than five times, and another trip is already planned. It is no coincidence that his unmistakable phone ringtone is "This is Australia calling . . . ". But he did not just share the insights of his research group in Australia, he did so worldwide at conferences, from Japan and India to Russia and the USA.

Particularly close to his heart was SASO (Self-Adaptive and Self-Organizing Systems), one of the leading conferences on self-organizing systems. Through workshops, he introduced critical aspects such as Trust (TSOS) and quality assurance (QA4SASO) for these distributed autonomous systems, that meanwhile are integral parts of the main conference. A true highlight was SASO

Fig. 11. Didn't I already lose the exhaust last year?

Fig. 12. How do I get rid of this beeping sound when I open the roof window?

Fig. 13. Only with paper and tape: the egg must not break if it falls from the 2nd floor

Fig. 14. The winning design (before the test flight).

2016, which he chaired as General Chair. Beyond a week packed with technical expertise, there was an equally full entertainment programme: a city tour by tram, a reception in the magnificent "Goldener Saal" of Augsburg's Town Hall, and a legendary Bavarian social event at the "Kurhaus Göggingen". Here, traditional Bavarian hospitality was on full display, featuring a "Schuhplattler" dance group and real cowboys from Arizona (hosting SASO 2017).

As the host, Prof. Reif wanted to be fully prepared and decided that he must learn Schuhplattler himself. He turned to online tutorial videos and started secret late-night training sessions. Naturally, these activities did not remain a secret for long, causing some amusement. Much to everyone's regret, the anticipated performance never took place. As rumor has it Prof. Reif still laments this missed opportunity and considers founding the "Rotary Schuhplattler Club", provided that his other dream (owning a ride-on lawnmower) has not yet been realized as a compensatory measure by then.

7 Our Wishes for Your Future, Prof. Reif

On the occasion of your 65th birthday, we would like to extend our heartfelt congratulations to you, Professor Reif! To us, you are not only an outstanding professor, but also a source of inspiration, with your energy, humor, and your enthusiasm for everything you do. You always motivated us even through scientific dry spells and encouraged perseverance. We wish you unvarying creativity and innovation, that you keep on generating new and unexpected ideas, planting them in minds, and nurturing them with knowledge. We hope your coming years are as exciting, adventurous, and joyful as those you have already shared with us and that you keep finding time to ride your bike, take on the waves, and, most importantly, indulge in your passion for Australia. Merrily celebrate this special day with us and your colleagues and friends, sharing perhaps a story or two from more than 40 years in academia.

Observation and Control of Hybrid Organic Computing Systems – Centralised Planning Combined with Autonomous Entities

Sven Tomforde[1] , Jonas Lange[1] , Pia Schweizer[2] ,
and Christian Krupitzer[2(✉)]

[1] Intelligent Systems, Christian-Albrechts-Universität zu Kiel, Kiel, Germany
{st,jla}@informatik.uni-kiel.de
[2] Food Informatics, Universität Hohenheim, Stuttgart, Germany
{pia.schweizer,christian.krupitzer}@uni-hohenheim.de

Abstract. In response to the ever-increasing complexity of technical systems and the corresponding challenges in their controllability, robustness, and safety, initiatives such as Autonomic Computing, Organic Computing, and Pro-Active Computing emerged within the last two decades. In the core, they all propose to change the way systems are developed by adding a self-adaptation unit on top of productive units in order to adapt the runtime behaviour to changing conditions and unforeseen events. This resulted in various design patterns and architectural concepts that mostly assume some kind of control loop. Those include steps such as (i) perceiving the environmental and internal status by sensors, (ii) analysing this information towards a situation-awareness model, (iii) using these models as a basis for planning the next adaptation steps, and (iv) finally executing these plans, maybe in a coordinated fashion within collectives of systems. In this chapter, we summarise these efforts with an emphasis on the Organic Computing and Autonomic Computing domains. We highlight that the control of large-scale systems that are made of a potentially large set of fully autonomous entities is limited when using these approaches and propose a hybrid control model. This combines system-wide and centralised planning capabilities with fully autonomous local behaviour of entities. The main impact is on how systems decide on their current behaviour and how this impacts their long-term learning behaviour. We use platooning as a running example and mention further applications sharing the same characteristics.

Keywords: Organic Computing · Autonomic Computing · MAPE-K loop · Observer/Controller paradigm · Hybrid control

1 Introduction

With roots in initiatives such as cybernetics and complex adaptive systems, various fields of research have emerged since the turn of the millennium that

G. Ernst et al. (Eds.): Wolfgang Reif Festschrift, LNCS 15765, pp. 36–55, 2025.
https://doi.org/10.1007/978-3-031-92196-4_3

deal with the design and realisation of self-adaptive systems to solve technical problems. On the one hand, these were already established areas of research such as multi-agent systems and, on the other, new areas such as proactive computing [37], autonomic computing [16] and organic computing [27]. These differ in their focus, applications and the techniques used, but share the same motivation and basic approach.

The advancing mechanisation and capabilities of technical systems have led to a sharp increase in complexity that is assumed to be no longer mastered using conventional approaches, which is reflected in increasing failures and malfunctions. To counteract this trend, there is a paradigm shift in system development: instead of complete reaction planning by the developer at design time, the system itself is enabled to find solutions for previously unknown situations at runtime. This is based on a system-inherent distribution of tasks: in addition to the productive part, an adaptation mechanism becomes part of the system, which observes the current status and adapts the productive behaviour on this basis.

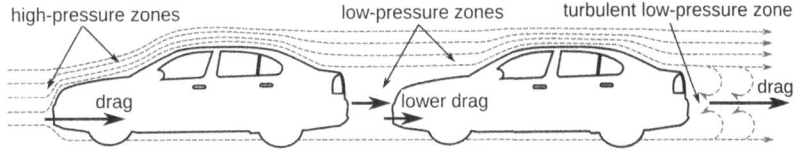

Fig. 1. Effect of the slipstream in platoons (adapted from [33])

In this chapter, we consider platooning as a running example for self-adaptive decisions. The term describes the behaviour of several vehicles that coordinate in a row to form a large group of vehicles, establishing minimum distances between them. This results in reduced fuel consumption through slipstream effects and reduced emissions [43]. As shown in Fig. 1, the leading and last vehicles also benefit from reduced slipstream effects; however, the effects are reduced compared to those in-between. These effects are particularly relevant on highways due to the higher driving speeds. The coordinated vehicles communicate constantly, enabling them to accelerate and break in unison. In addition to the potential benefits for the individual participants, the motorway system as a whole benefits as well, as the throughput of vehicles increases [30]. Obviously, such an approach requires automated and context-dependent behaviour, as manual influences from human drivers do not have the necessary reaction times. Consequently, platooning has often been investigated with self-adaptive system approaches, although there are a variety of implementation options in terms of control responsibility or coordination procedures. The productive part from above refers to the autonomous driving unit, while the adaptation mechanism decides about possible platooning manoeuvres such as joining, leaving, dissolving a platoon, or changing the position within a platoon.

From the perspective of SASO systems, in the platooning scenario the aforementioned design-related division between the production and adaptation unit is even found conceptually in two places: In the vehicle and in a possible central coordination system. At the same time, the eponymous steps for certain design patterns for controlling adaptation (i.e., monitor-analyse-plan-execute of autonomic computing) can be found in both places. However, a conflict that has so far received less attention in such approaches is already becoming clear: What are the implications for the design of self-adaptive systems when fully autonomous local units (i.e., the vehicles in the platooning scenario) have to interact with additional centralised control mechanisms (i.e., the centralised coordination mechanism with planning function) with limited intervention capabilities?

In this chapter, we summarise the development of design approaches of the corresponding initiatives and show that very strong commonalities can be derived across the corresponding initiatives. We also provide a summary of existing deployment scenarios, which we use to highlight potential challenges. On this basis, we present an adapted design approach for hybrid self-adaptive systems, which is tailored to the challenges mentioned and is examined using the platooning scenario, among others.

2   Design of Self-adaptive Systems

Back in 2000, Tennenhouse postulated a paradigm shift in system engineering, putting an emphasis on closing the gap between computer science and control theory to allow for continuous responses to external stimuli at faster-than-human speeds [37]. He called such systems "proactive" computers or systems, already sketching a separation of parts: at a basic level, the proactive system is connected to the physical world via sensors and actuators; at a proactive level, the system adapts its behaviour to perceived stimuli; and at an external level, it accepts user intervention and provides reporting.

Even if this has not yet been further specified, the basic division and idea in the design of SASO systems is already recognisable: the actual system behaviour is configurable and adapted to external observations and evaluations by an additional adaptation mechanism at runtime. Based on this basic idea, various initiatives have emerged that deal with the preferably autonomous, self-adaptive, and self-organised behaviour of technical systems.

Central to SASO systems are the Autonomic Computing (AC) [16] and Organic Computing (OC) [27] initiatives, which were dedicated to the topic relatively in parallel with a slightly different focus (industry vs. academia driven, applications, technology, local roots). In both initiatives, an AC or OC system – we summarise both as SASO systems in the following – adapt their behaviour at runtime as a reaction to observed environmental changes or the system itself [19]. Therefore, these systems are composed of an *adaptation part*—also called, i.a., adaptation manager, autonomic manager, or adaptation part—that controls the *productive part* [45] (see Fig. 2). To deal with dynamic environments and unanticipated events, the adaptation part triggers the adaptation of parameters or

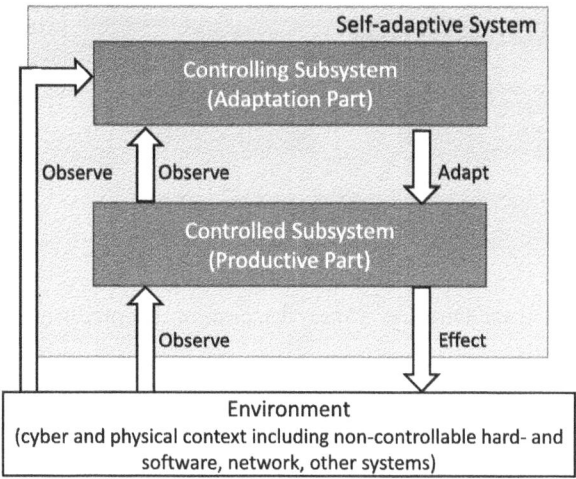

Fig. 2. Basic concept of a self-adaptive system, according to [45].

components of the controlled subsystem. Weyns at al. [45] explained that similar concepts can all be reduced to the basic concept shown in Fig. 2, which then serves as a root to define variants of patterns for SASO systems.

In the following, we present in Sect. 2.1 the basic realisations of this pattern from the OC and AC initiatives: Initially, we discuss the Observer/Controller (O/C) pattern [29] with its subsequent variants as a system model for individual agents covering the OC perspective [27]. The focus here is on the structure and design of the eponymous basic structuring of the adaptation mechanism. Afterwards, we compare this to the Monitor-Analyse-Plan-Execute-Knowledge (MAPE-K) cycle as established for the AC perspective [16] and show that both can be mapped onto each other. Finally, as SASO systems are often large-scale systems-of-systems [26], we also describe distribution patterns and implications for larger system contexts.

2.1 Individual Agent

For the OC initiative, the basic division into a 'system under observation and control' (SuOC, for the productive part) and an 'observer and a controller' unit (O/C, for the adaptation part) was initially created. In addition, it was specified that the observer's perception - i.e., the selected and processed environmental information in the form of features and aggregated indicators - is configured by an 'observation model', which in turn can be modified by the controller depending on the situation. The controller, in turn, works on the basis of external goal functions that are specified by the user and can be changed during operation. Figure 3 illustrates this basic concept.

Afterwards, this abstract pattern was refined towards a concretisation of observer and controller components, first described in [29] and then extended

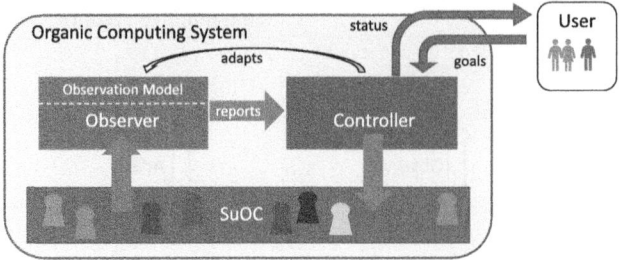

Fig. 3. Basic Observer/Controller pattern, adapted from [27].

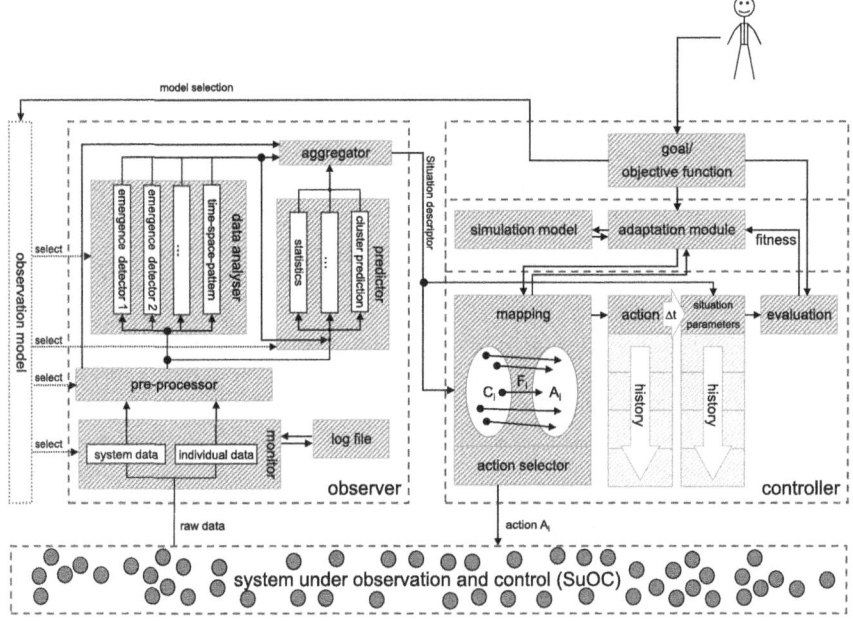

Fig. 4. Generic Observer Controller model [27].

in [40]. Figure 4 illustrates the outcome: the generic observer/controller pattern. Conceptually, the SuOC already consists of one or more productive subsystems and is controlled by the overlying O/C unit. For this purpose, the raw data is first recorded in the observer and divided into individual and system data. This corresponds to the division between the local system and the overall system context, which was later further concretised, for example, as part of the 'self-improving system integration' initiative [2]. After processing the data in the preprocessor, the data is analysed and augmented by predictions. Characterised by the focus of OC in terms of emergent system behaviour in self-organised systems, the main focus of the analysis is on the detection of emergent behaviour using appropriate metrics. The values aggregated into a situation description are then passed on to the controller, which in turn selects an appropriate response based on an existing set of actions (the action selector mapping context C_i via a function F_i to possible actions A_i) that adapts the SuOC to the current circumstances. 'Appropriate' is determined by the user's target and requires corresponding metrics [29]. The action set can be adapted dynamically using components such as the simulator and adaptation module, albeit typically with a time horizon that extends far beyond the spontaneous reaction. Usually, a reinforcement learner serves as a basis for the selection [38], enriched with a separation between the exploitation of current knowledge and the exploration of new, suitable behaviour using encapsulated environments.

In the platooning example, the observer generates a situation description that includes, for example, the availability of other vehicles with desired characteristics and the route context, while the controller adapts the autonomous driving behaviour on this basis using trajectories of speed and direction. The SuOC therefore corresponds to an autonomous driving mechanism that accepts control commands in the form of modifications to current target values.

In parallel to the developments for the OC perspective, a specific feedback loop structure has been developed for the AC domain: the MAPE cycle by Kephart and Chess [16] (see Fig. 5). The MAPE cycle is named after its four main functions: monitor, analyse, plan, and execute. The *monitor* function collects data of the system and the environment and pre-processes it, e.g., through filtering, reliability analysis, or categorisation. Afterwards, the *analyser* decides if adaptation is needed by comparing the monitored data and system objectives. The *planner* computes the adaptation plan based on the monitored information and interpretations of the analyser. If more than one possible adaptation is found, the planner has to decide which one matches the given system objectives best. Finally, the *executor* allocates the planning instructions and sends them to the corresponding productive part. This way, the adaptation is triggered in the productive part. Afterwards, the process starts over again. Sensors and effectors are interfaces that gain data from productive parts or send instructions.

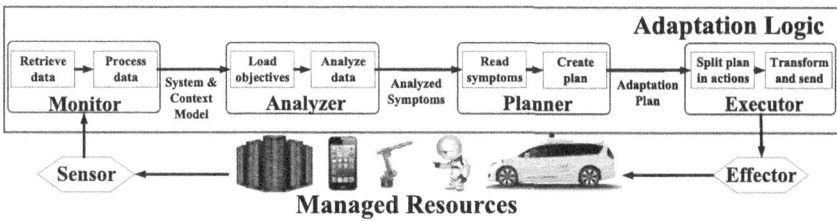

Fig. 5. MAPE Cycle according to Kephart and Chess [16]. Own visualisation based on [18]. The knowledge base is omitted here.

The MAPE- and O/C-concepts structure the adaptation process similarly, just with a varying emphasis. Conceptually, the 'M' part belongs to the

'Observer' part, while 'P' and 'E' are covered by the 'Controller'. However, aspects of the 'A' part are found in both components: Most of the tasks are observer tasks, but, e.g., success estimation, feedback generation, credit assignment, or updating existing knowledge are part of the controller.

Even the basic O/C structure above distinguishes between reaction times in the controller: an immediate reaction to observed stimuli and a longer-term adaptation and optimisation of the knowledge base through simulations. This was subsequently schematised and organised into layers based on reaction times, inspired by approaches from real-time systems and robotics (such as the sense-plan-act scheme [5] and the subsumption architecture [4]). The result was the multi-level observer/controller architecture (MLOC) as illustrated in Fig. 6.

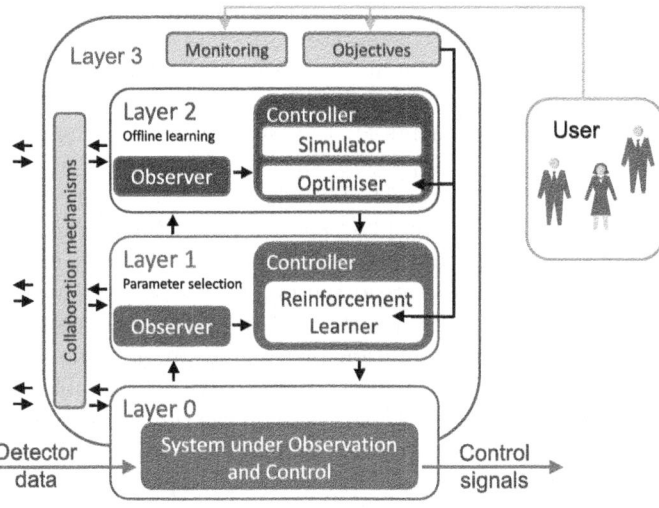

Fig. 6. The Multi-Level Observer/Controller architecture as proposed in [40].

The lowest layer of the MLOC approach encapsulates the SuOC (as in the O/C architecture) and works in real time. The system remains functional even if higher layers fail–but then in a static mode without context adjustments. Layer 1 ('adaptation layer') continuously monitors the behaviour of layer 0 and decides on any necessary reparameterisation. This typically takes place at a certain sampling rate and is, therefore, slightly slower than productive operation. In contrast, layer 2 ('offline learning') works with a considerably longer time horizon: it observes the behaviour of layer 1 and evaluates the extent to which the knowledge base of the layer is sufficient. Accordingly, rules are generated on the basis of simulations using digital twins [8] and the use of optimisation heuristics. Such a simulation-coupled approach is associated with considerable computing effort, which can lead to reaction times in the range of minutes or hours, depending on the application. MLOC also provides for explicit cooperation with similar systems on all three levels ('collaboration mechanisms'), e.g.

to expand the sensor horizon (layer 0), to coordinate parameterisations (layer 1, e.g. to establish green waves in traffic systems [41]) and to exchange rules or knowledge (layer 2).

2.2 Distribution Patterns

Often, SASO systems are systems-of-systems. SASO systems can be found in many different domains, such as transportation systems [32], robotics [9,17], smart grids [11], or production facility control [12,13]. Central questions are, whether the adaptation part should be decentralised or centralised and how the control functionality should be distributed. In [1], the authors define self-adaptation as a top-down approach to system control. This means that a central unit has to control the system. Self-organisation is seen as the opposite: a bottom-up approach. This work does not strongly differentiate self-organisation and self-adaptation as both require mechanisms to adapt the productive part. However, we can distinguish the locus of control. In *centralised* approaches (see Fig. 7a), a central instance controls the productive part; this is equal to the definition of self-adaptation in [1].

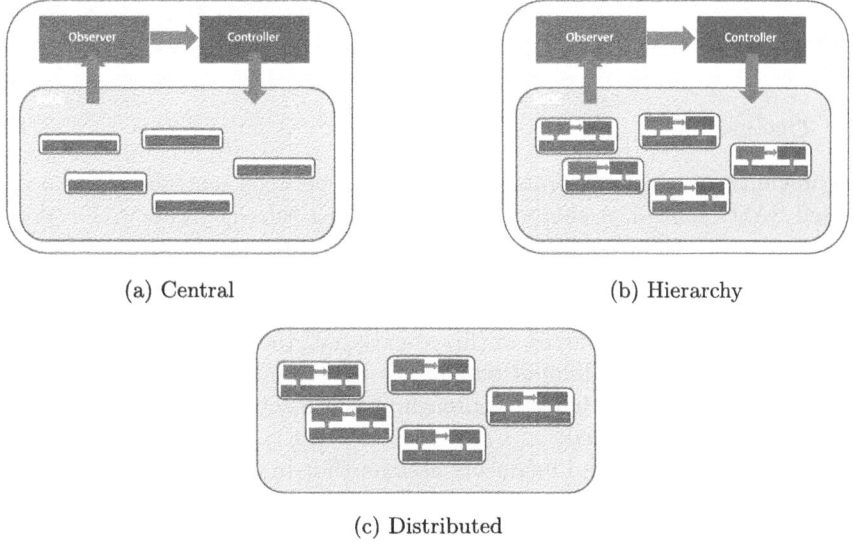

(a) Central   (b) Hierarchy

(c) Distributed

Fig. 7. Variants for the realisation of O/C patterns, adapted from [27].

Whereas a global maximum can be achieved because the central instance is aware of all information, this approach is not suitable for large systems with a high amount of information or resource-poor devices because of the calculation power needed [44,45]. For large systems, a central approach is hard to achieve because of the size of the system and real-time constraints.

Alternatively, in *distributed* approaches (see Fig. 7c), developers define an interaction pattern that describes which sub-systems control functionality and adaptation. The different adaptation part elements can communicate and coordinate to achieve global goals. In the IBM reference model for Autonomic Computing [16], additional orchestrating autonomic managers improve the coordination, introducing a *hierarchical approach* (see Fig. 7b). Weyns *et al.* describe patterns for how to distribute the control functionality on various sub-systems with different levels of interaction and decentralisation [45]: (i) *Coordinated Control Pattern*, (ii) *Information Sharing Pattern*, (iii) *Master/Slave Pattern*, (iv) *Regional Planning Pattern*, and (v) *Hierarchical Control Pattern*. Patterns (i) and (ii) are fully decentralised with interaction, whereas patterns (iii) - (v) have centralised elements and, therefore, are hybrid approaches.

According to [42], all of these patterns require an infrastructure provided externally or by the systems themselves, which is responsible for key tasks such as system discovery and authentication, communication, trust management, or basic negotiation schemes. For the OC domain, the middleware OCμ has been discussed, covering some of these tasks in a structured manner [31]. Designers of SASO systems must be aware of the implications resulting from the distribution of their applications: Suitable patterns depend on the number of sub-systems, the coordination needed between them for adaptation decision-making, or the relevance of achieving global goals. Especially if the local instances take non-coordinated decisions, this might result in conflicts with global goals.

2.3   Discussion

The previously described literature distinguishes between centralised and decentralised SASO control. A single central controlling subsystem allows for global, optimal planning. The contrary is a fully decentralised approach: each productive part has its dedicated controlling subsystem. In these scenarios, the entities might collaborate fully autonomously: They (i) cooperate for fulfilling common tasks, (ii) compete for resources, or (iii) co-exist, i.e., their autonomously planned interactions are not coordinated but do not have conflicting goals. In between these two extremes lies a hybrid approach in which some aspects of the control functionality are centralised.

Usually, researchers and engineers assume that in the case of central planning, the autonomous entities adapt according to the plan. However, with higher degrees of autonomy, these entities act independently, i.e., potentially ignore the planning [6]. Hence, conflicts arise between central planning and local decision-making, especially in cases where autonomous decisions are based on runtime learning that may explore behaviours other than those proposed by the controlling subsystem. Next, we present a system model to integrate those perspectives.

3   Hybrid Observation and Control

As outlined above, SASO systems define a broad field of research. To focus on the research efforts, we restrict our work to coordinated SASO systems (see Fig. 8). Therefore, we discuss several assumptions in the following section.

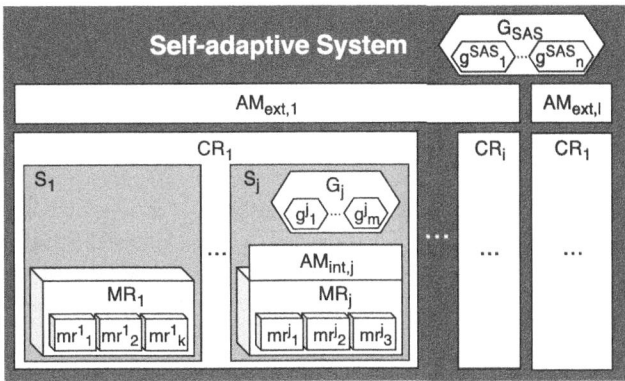

Fig. 8. System model for hybrid coordinating SASO systems (from [25]).

An adaptation manager $AM_{SASO} = (G_{SASO}, AM_{ext,1}, \ldots, AM_{ext,l})$ modelling the adaptation part has a set of certain goals $G_{SASO} = (g_1^{SASO}, \ldots, g_n^{SASO})$ which may contain several goals g_i^{SASO} and (depending on the level of control) is split into several elements $AM_{ext,l}$. The productive part of the SASO system $MR_j = (mr_1^j, \ldots, mr_k^j)$ is composed of hardware and software elements mr_k^j belonging to subsystems $S_j = (MR_j, G_j, AM_{int,j})$. In the case of trying to accomplish certain goals $G_j = (g_1^j, \ldots, g_m^j)$ at the subsystem level, an internal adaptation manager $AM_{int,j}$ coordinates the MR_j. Several subsystems can be grouped into coordinated resource groups $CR_i = (S_1, \ldots, S_j)$. The position of each subsystem S_j, i.e., the position of each MR_j within the CR_i group and the common properties of this group have a direct impact on the goal achievement of each subsystem and the overall SASO system.

The individual part $AM_{ext,l}$ of one subsystem might control one or several CR_i, hence, $AM_{ext,l} = (CR_1, \ldots, CR_i)$. If the MR_j are fully controlled by the AM_{ext} (e.g., in a centralised approach) the AM_{int} just forwards the control signals. The AM_{ext} and the AM_{int} can cooperatively coordinate the SASO system in hybrid approaches. To this end, the different distributed elements of the AM_{ext} also communicate for finding a consensus in the adaptation decision. In fully decentralised approaches, i.e., the AM_{int} are fully responsible for adaptation decisions, the AM_{ext} can be neglected. For hybrid and (in particular) decentralised system settings, conflicts can arise between the local autonomous subsystems S_j within one group.

We use the term *coordination scheme* to describe the act of participation and order of entities as part of adaptation decisions within a coordination process between different $AM_{ext,l}$ and/or AM_{int}. Additionally, processes describe the formation (`initialise`, `join`, and `leave`) and reorganisation (`change position` or `group properties`) of such a CR. Performing these processes can either be done externally (AM_{ext}) or internally (via negotiations between the AM_{int}), both resulting in an optimisation problem.

The targeted category of coordinated systems, such as in platooning, requires decentralised, local decision-making to optimise the goals of individual drivers. However, platooning also requires central coordination to assign vehicles to platoons. Accordingly, we integrate central adaptation recommendations (rather than strict centrally controlled optimisation) with decentralised optimisation techniques. Further, based on our observations in [10], we propose to include situation-aware optimisation schemes to handle the trade-off between local and fast, but potentially conflicting, optimisation versus time-consuming but global optimisation. On the other hand, we compare the suitability of different techniques based on whether they deliver a usable result when the optimisation procedure is stopped, i.e., they allow for anytime learning.

Regarding the execution of adaptation, research in SASO systems assumes (implicitly) that autonomous entities follow the adaptation plan of a central instance, i.e., the autonomous entities behave cooperatively to coordinate adaptation actions. This includes that entities might behave altruistically [24] in case the global optimal adaptation plan decreases their individual utility. However, autonomous entities often follow own goals. In current work, we study the resulting issues from such situations, including identifying (i) mechanisms to reward entities for decreased utility and (ii) interaction mechanisms to control adaptation execution. This is also highly relevant to the platooning example. Here, vehicles have their own goals, e.g., optimising their individual fuel consumption (which conflicts with the role of the platoon leader). Current approaches ignore those individual preferences/constraints (cf. [23]). Also, multi-objective decision-making for optimising the composition of platoons based on several factors is often ignored [35]. The proposed concept extends the state-of-the-art as participants receive recommendations and choose which platoon to join individually, obeying local preferences, instead of receiving a fixed instruction on which platoon to join or homogenising parameters (e.g., velocity) with static values for all platoons. Further, we shift platooning coordination from a single objective towards a multi-level optimisation problem with a many-objective (different objectives from different vehicles) and multi-objective (integration of various global objectives) solution space, which better reflects the complex nature and can increase the acceptance of platooning as individual preferences are addressed.

4   Open Research Challenges

In the previous Sect. 3, we described a hybrid approach in which control is neither fully centralised nor fully decentralised. The goal is to design a system that combines the advantages of both centralised and decentralised systems, such as central knowledge distribution, agent autonomy, robustness, and fast adaptation. This could be achieved by implementing a central entity that does not have full control over the agents. Instead, it just provides recommendations in the form of plans with corresponding incentives to them. Each agent can then decide how to implement a given plan or even disregard it entirely. In exchange, the agents provide the central planner with observations gathered from their environment, such as their current position and speed. Figure 9 illustrates this flow of information and control in the hybrid system. In the following, we describe research challenges for implementing such an architecture.

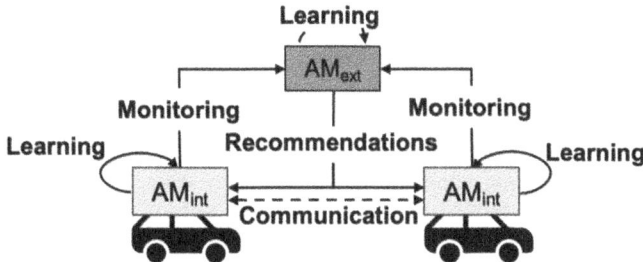

Fig. 9. The hybrid approach to adaptive multi-agent coordination. In the platooning example, each vehicle represents an agent.

Challenge 1 – Evaluation of the Systems' Performance

The performance of SASO systems can be quantified by applying various different evaluation metrics (e.g., [39] or [7]), while the properties self-adaptation and self-organisation are thereby often assessed in isolation. Metrics relevant to the self-adaptation property include but are not limited to runtime metrics, which, e.g., enable the assessment of the response time; the situation performance metric, for the quality assessment of the final adaptation outcome; as well as the decision benefit metric, to measure the final advantage gained from the adaptation, e.g., [36]. In this context, Kaddoum et al. [14] propose several measures for evaluating the adaptive properties of self-*systems, introducing metrics that focus on methodological, intrinsic, architectural, and runtime evaluation criteria. To evaluate the self-organising aspect of the systems, interaction metrics, that e.g., quantify the amount and effectiveness of interactions, but also graph-theoretic metrics and the metric of entropy are relevant, e.g., [3]. Eberhardinger et al. [7] further present key metrics for assessing the performance of self-organization algorithms, with particular emphasis on time and solution quality. The high diversity of evaluation metrics leads to the following research questions (RQ):

RQ 1 Which metrics best suit the evaluation of the described class of cooperating SASO systems?

RQ 2 How can the system performance be mapped to application-specific metrics and system situations?

The first application to which the developed centralised, decentralised, and hybrid systems are mapped represents the use case of platooning coordination. This includes the organisation of the platoon composition, the assignment of vehicles to appropriate platoons, as well as the management of the inter- and intra-platoon interactions, which comprises platoons overtaking other platoons and vehicles changing their inner-platoon-position [22]. Metrics, such as the average travel time of all vehicles or the difference between an individual's current and desired speed, serve to quantify the capabilities of the applied system's architecture. The relevance of a specific metric is thereby expected to be situation-dependent. Time, for instance, might be a more critical factor during rush hour, when the drivers' objective will most probably be to find a fast way to work or home. Fuel consumption, on the other hand, might be more relevant during off-peak traffic situations, where it is less hard to stay within a specific travel time frame. In their study, Sturm et al. [35] presented a comprehensive spectrum of metrics relevant to platooning coordination. This first research challenge will, accordingly, comprise an empirical creation of a taxonomy for when which metric is more relevant, depending on the underlying traffic scenario. Such a scenario is thereby not solely restricted to the traffic volume but also includes factors like the number of lanes, the vehicle type, as well as the car-to-truck ratio, and the authorised maximum speed. The taxonomy will then be tested by applying a static central planner and a static local decision-maker, respectively. Static implies that both systems operate according to one platooning coordination strategy, such as assigning a vehicle to its closest platoon.

Challenge 2 – Central Coordination vs. Local Decision-Making

So far, both system architectures, centralised and decentralised, are static. This means that they face every possible scenario similarly by adhering to one strategy with fixed parameters. However, a changing environment might require a different strategy or adjusted parameters, i.e., integrated situation-awareness. In the context of platooning coordination, for instance, it might suit to search for platoons in the vicinity of the respective vehicle, as long as there is a high traffic volume, e.g., during rush hour. Yet, with a lower vehicle density, it could be advantageous to expand this search radius. Such a demand requires systems that can adapt their behaviour at runtime to changes in the environment. To enable such an adaptive system's behaviour, the central planner will be supplied with an additional optimiser that either adjusts the strategy's parameters, such as the maximum search distance for a possible platoon, depending on the situation, or directly optimises a platoon's composition by applying a genetic algorithm. The decentralised system transitions to be adaptive through the application of Reinforcement Learning (RL). The shift from static to adaptive system behaviour thus results in four different architectures: two fully centralised architectures, one static and one adaptive, focusing on global optima, and on the contrary, two fully decentralised architectures, one static and one adaptive, aiming for individual local optima. Subsequently, the following two research questions emerge:

RQ 3 In which situations (i.e., evaluation settings) are which architecture properties (i.e., static vs. adaptive, global vs. local) preferable?

RQ 4 How can the specifics of the four different architectures be consolidated into a hybrid architecture?

Each mentioned system approach has its strengths and weaknesses. While a fully centralised system can actively strive for global optima, it ignores local interests. Conversely, in a fully decentralised system, each local entity is responsible for fulfilling its own goals but remains unaware of global optima. For both approaches, it is expected that the shift from static to adaptive behaviour will result in an increase in the system's complexity with a simultaneous enhancement of its efficiency. Therefore, this second challenge aims at a comprehensive comparison of these systems (cf. Table 1), in dependence on the underlying situation. The insights derived from this comparison will then be utilised to bundle the situation-dependent individual strengths into a hybrid SASO system.

Table 1. Relevant system setups that will be compared.

	global perspective	local perspective
static	central planner with fixed strategy	local decision-maker with fixed strategy
adaptive	central planner with optimiser	reinforcement learner

Challenge 3 – Merge Towards the Hybrid SASO System

This research challenge merges the top-down approach with a global adaptation manager with the RL agents from the bottom-up approach. The aim is to create a hybrid SASO setup that integrates system-wide macro-level planning based on optimisation techniques with micro-level autonomous decision-making of entities based on autonomous learning. The following research questions need to be considered to enable a successful integration of the two opposing systems:

RQ 5 How to encourage compliance of the autonomous adaptation decisions with the central plan?

RQ 6 How to consider local preferences for generating the central plan?

The coordination of the individuals inevitably provokes an unfair distribution of their investments. For instance, being a platoon leader always results in higher costs due to a higher drag, which leads to a proportionally higher fuel consumption compared to an inner platoon position. To counteract such negative effects and to avoid possible conflicts, mechanisms for compensation will be investigated. The central planner will directly integrate compensation into the design of the optimisation algorithms, thereby differentiating between direct (e.g., exchanging money) and indirect (e.g., frequently changing the inner-platoon positions) compensation [23]. Whereas the local entities will apply a decentralised incentivisation scheme to be robust against a single point of failure [15]. Furthermore, they will learn to differentiate between utility-based solutions (e.g., the fastest or most cost-efficient) and incentive-based solutions to enable a decision about their participation based on a trade-off analysis.

Currently, central coordination in SASO systems does not consider local preferences. Instead, it gives strict adaptation commands. The adaptation plan will be further equipped with degrees of freedom to liberate the individual entities from complying with the instructions. This enables the individuals to select from numerous different planning constellations depending on the one that best fits their interests. This could be achieved by integrating solvers for constraint satisfaction problems to identify suitable solutions for the entities within the degrees of freedom, relying on weighted utility functions.

Challenge 4 – From Static to Dynamic Environments

For the sake of simplification, we plan to study the aforementioned challenges in stable system settings. As the dynamics in the environment are the main driver for a SASO system, we now relax the static properties and integrate dynamics. The resultant research questions are the following:

RQ 7 How robust are the established systems?
RQ 8 In which settings is the hybrid system superior to the purely centralised and the decentralised ones?

Congestion, road accidents, and construction sites are common in real highway traffic. All these occurrences represent disruptions to which the platooning coordination system has to adapt at runtime. More specifically, such disturbance factors can be separated, depending on their origin, into *Environment* or *System*. Furthermore, depending on their predictability, into *anticipated* or *spontaneous* factors. While highway entries and exits, lane reductions, and public events that cause higher traffic during certain periods of time are considered as predictable and are therefore classified as anticipated disturbances, the occurrence of a traffic accident or a wrong-way driver is not predictable and is, hence, categorized as a spontaneous disturbance. All five examples are of an environment-origin. Anticipated disturbances that the system itself could cause include sensor problems due to bad weather, which reduces the sensor's reliability, as well as local entities that make egoistic decisions and do not adhere to the global plan (which is specifically relevant to the central planner). System disturbances that could happen spontaneously include a sudden failure of the central planner, which leaves the local entities on their own, as well as sensor defects. The mentioned relevant disturbance factors that will be considered are summarized in Table 2.

Table 2. Classification of possible disturbances in highway traffic scenarios.

	Environment	System
anticipated	highway entries and exits, rush hour, lane reduction	local entities disobeying the global plan, sensor problems (bad weather)
spontaneous	traffic accident, wrong-way driver	sensor problems (defect), failure of central planner

5   Application Scenarios

SASO systems can be found in a large spectrum of applications. The presented system model can be applied in various application domains. We describe several applications that might benefit from our approach in the following. Besides platooning, this includes coordination of robotics and smart grids.

5.1   Platooning Coordination

As already mentioned, platooning needs to integrate central decision-making for assigning vehicles to platoons fulfilling global goals with autonomic decision-making of individual drivers/vehicles [20,22]. Hence, a system for platooning coordination needs to balance several objectives, such as optimisation of (i) the global traffic flow, (ii) interactions between platoons, (iii) the vehicle order within a platoon. Some of them conflict with the objectives of individual agents, while others complement them. Sturm et al. [35] devise a taxonomy of optimizing factors for platooning on the levels of individual vehicles, platoons, and global traffic optimization. Not only do the authors provide insight into the optimization factors on different levels, but also on the arising conflicts and synergies between them.

The execution of the adaptation plans as the output of the central decision-making approach is enforced on autonomously deciding vehicles. We assume fully autonomous driving; however, since vehicles act on behalf of their drivers', transparency of plans is required, and autonomous behaviour may ignore the plan. Accordingly, the execution requires a mechanism with incentives to balance individual goals and preferences - as described above. Additionally, uncertainty is added through non-controllable vehicles that cannot platoon. This introduces further constraints compared to usual approaches for coordinating unmanned autonomous vehicles, such as [32].

5.2   Coordination of Robots

In Industry 4.0, autonomous robots and smart factories aim to increase flexibility, efficiency, and adaptability by connecting machines, devices, and systems, enabling them to communicate and make decisions without human intervention.

In such modern production facilities, often self-driving vehicles/robots can transport items or goods within the facility. As the systems share routes, coordination is necessary. Further, the flexibility demand for Industry 4.0 production processes also requires flexibility in planning the routes. This is a multi-dimensional optimisation problem as it has to integrate different aspects besides finding the shortest path [13], e.g., prioritising items for premium customers. As this requires the integration of various data sources, the coordination of the vehicles is often done centrally, i.e., central decision-making is applied.

However, solely central decision-making can be problematic in several cases. First, the amount of information that must be processed can be large, and the central system would be a potential single point of failure. Second, the dynamics in such manufacturing environments do not necessarily fit with a central coordinator. Hence, local decision-making is required to find the path according to the local conditions, such as pedestrians or objects blocking the path. Third, coordination with the other processes might be necessary, e.g., in scenarios with autonomous production systems [9,28]. In this context, mobile robots and autonomous forklifts could leverage a central planner that integrates real-time inventory tracking, prioritization of urgent or high-value orders, and shortest path optimization, while still pursuing their own objectives, such as meeting time constraints.

5.3   Smart Grids

Smart grids are advanced electrical grids integrating digital communication, automation, and control technologies to efficiently manage electricity flows from multiple sources, including renewable energy. These grids improve reliability and enable real-time monitoring, allowing for better demand-response management and integrating distributed energy resources. SASO systems play a key role [21] by dynamically adjusting to fluctuations in energy supply and demand, optimising storage use, and enhancing grid stability, thereby ensuring efficient and resilient grid operations despite the inherent variability of renewable energy sources. One important aspect of successful implementations of smart grids is the photovoltaic power forecast [11].

Smart grids are hierarchical systems because they involve both centralised and decentralised components, each responsible for different levels of control. Centralised systems manage large-scale power generation, distribution, and overall grid stability, while decentralised, local systems handle real-time adjustments like integrating renewable energy sources and responding to local demand changes. This structure ensures both efficiency and adaptability, allowing the grid to respond quickly to local fluctuations while maintaining global stability. One of the main challenges is to align the goals of the local systems with the global goals. Steghfer et al. propose a market-based system of systems approach to tackle this challenge [34]. This solution would involve a multilevel system of virtual power plants which can buy and sell electricity to and from agents on the same level. To deter malicious actors, the authors propose a centrally controlled reputation system, which allows agents to evaluate potential trading partners beforehand. Hence, in the smart grid domain, this system of systems approach could serve as a starting point to move toward the proposed hybrid system.

6 Conclusion

The coordination of autonomous agents is a central challenge in self-organised traffic systems. This work argues that conventional centralised and decentralised SASO architectures are unsuitable for combining local decision-making with global optimisation, e.g., as required in applications like platooning coordination. While centralised systems limit the agents' individual degrees of freedom, decentralised systems suffer from worse coordination between agents. In response to that, this work proposes a hybrid architecture that results from a merge of two approaches: The top-down approach takes a centralised view, and the bottom-up approach takes a decentralised view. Starting from the isolated, static agents, the systems initially become adaptive before they are merged into the hybrid system. This work identifies four research challenges that need to be tackled to build hybrid Organic Computing systems: (i) definition of relevant evaluation metrics and scenarios, (ii) comparison of centralised and decentralised architectures, (iii) building a hybrid SASO system, and (iv) analysing the hybrid SASO system in settings with dynamic disturbances. Additionally, we described how to transfer the system model to additional application scenarios besides platooning: coordination of robotics, and smart grids.

Based on the current efforts on implementing and evaluating our system model for hybrid OC systems, we will propose metrics and scenarios to evaluate the performance of such hybrid OC systems. Second, we will evaluate the performance of centralised and decentralised architectures in different conditions. Third, we will merge the centralised and decentralised systems to build a hybrid SASO system and also analyse the system in settings with dynamic disturbances. We will focus on the use case of platooning coordination but also plan to generalise our work in further application domains.

References

1. Babaoglu, O., Shrobe, H.E.: Foreword from the general co-chairs. In: Proceedings of SASO, pp. ix–x (2007)
2. Bellman, K.L., et al.: Self-improving system integration: mastering continuous change. Future Gener. Comput. Syst. **117**, 29–46 (2021)
3. Birdsey, L., Szabo, C., Falkner, K.: Identifying self-organization and adaptability in complex adaptive systems. In: Proceedings of SASO, pp. 131–140 (2017)
4. Brooks, R.: A robust layered control system for a mobile robot. IEEE J. Robot. Autom. **2**(1), 14–23 (1986)
5. Choset, H., Lynch, K.M., Hutchinson, S., Kantor, G.A., Burgard, W.: Principles of Robot Motion: Theory, Algorithms, and Implementations. MIT Press (2005)
6. Diaconescu, A., et al.: Architectures for Collective Self-aware Computing Systems, pp. 191–235. Springer (2017)

7. Eberhardinger, B., Anders, G., Seebach, H., Siefert, F., Reif, W.: A research overview and evaluation of performance metrics for self-organization algorithms. In: 2015 IEEE International Conference on Self-Adaptive and Self-Organizing Systems Workshops, pp. 122–127 (2015). https://doi.org/10.1109/SASOW.2015.25

8. Esterle, L., Gomes, C., Frasheri, M., Ejersbo, H., Tomforde, S., Larsen, P.G.: Digital twins for collaboration and self-integration. In: Proceedings of ACSOS, pp. 172–177 (2021)

9. Eymüller, C., Hanke, J., Poeppel, A., Reif, W.: Towards self-configuring plug & produce robot systems based on ontologies. In: Proceedings of ICARA (2023)

10. Fredericks, E.M., Gerostathopoulos, I., Krupitzer, C., Vogel, T.: Planning as optimization: dynamically discovering optimal configurations for runtime situations. In: Proceedings of SASO (2019)

11. Gajek, C., Schiendorfer, A., Reif, W.: A chained neural network model for photovoltaic power forecast. LNCS, vol. 11943, pp. 566–578 (2020)

12. Hirsch, J., Neumayer, M., Ponsar, H., Kosak, O., Reif, W.: Deadlock avoidance for multiple tasks in a self-organizing production cell. In: Proceedings of ACSOS, pp. 178–187 (2020)

13. Hirsch, J., Neumayer, M., Ponsar, H., Kosak, O., Reif, W.: Distributed constraint optimization for task allocation in self-adaptive manufacturing systems. In: Proceedings of ACSOS, pp. 62–67 (2021)

14. Kaddoum, E., Raibulet, C., Georgé, J.P., Picard, G., Gleizes, M.P.: Criteria for the evaluation of self-* systems. In: Proceedings of SEAMS, pp. 29–38 (2010)

15. Kantert, J., Tomforde, S., Diaconescu, A., Mueller-Schloer, C.: Incentive-oriented task assignment in holonic organic systems. In: Proceedings of ARCS, pp. 1–8 (2017)

16. Kephart, J.O., Chess, D.M.: The vision of autonomic computing. IEEE Comput. **36**(1), 41–50 (2003)

17. Kosak, O., Wanninger, C., Hoffmann, A., Ponsar, H., Reif, W.: Multipotent systems: combining planning, self-organization, and reconfiguration in modular robot ensembles. Sensors **19**(1), 17 (2019)

18. Krupitzer, C., Breitbach, M., Saal, J., Becker, C., Segata, M., Lo Cigno, R.: RoCoSys: a framework for coordination of mobile IoT devices. In: Proceedings of PerComW, pp. 485–490 (2017)

19. Krupitzer, C., Roth, F.M., VanSyckel, S., Schiele, G., Becker, C.: A survey on engineering approaches for self-adaptive systems. Pervasive Mob. Comput. **17**, 184–206 (2015)

20. Krupitzer, C., Segata, M., Breitbach, M., El-Tawab, S., Tomforde, S., Becker, C.: Towards infrastructure-aided self-organized hybrid platooning. In: Proceedings of GCIoT, pp. 1–6 (2018)

21. Lehna, M., Holzhüter, C., Tomforde, S., Scholz, C.: Hugo - highlighting unseen grid options: combining deep reinforcement learning with a heuristic target topology approach. Sustain. Energy Grids Netw. **39**, 101510 (2024)

22. Lesch, V., Breitbach, M., Segata, M., Becker, C., Kounev, S., Krupitzer, C.: An overview on approaches for coordination of platoons. IEEE T-ITS **23**(8), 10049–10065 (2022)

23. Lesch, V., et al.: A comparison of mechanisms for compensating negative impacts of system integration. Futur. Gener. Comput. Syst. **116**, 117–131 (2021)

24. Lesch, V., Krupitzer, C., Tomforde, S.: Emerging self-integration through coordination of autonomous adaptive systems. In: Proceedings of FAS*W (2019)

25. Lesch, V., Krupitzer, C., Tomforde, S.: Multi-objective optimisation in hybrid collaborating adaptive systems. In: Proceedings of ARCS, pp. 1–8 (2019)

26. Maier, M.W.: Architecting principles for systems-of-systems. Syst. Eng. J. Int. Council Syst. Eng. **1**(4), 267–284 (1998)
27. Müller-Schloer, C., Tomforde, S.: Organic Computing – Technical Systems for Survival in the Real World. Autonomic Systems, Birkhäuser Verlag (2017)
28. Nägele, L., Schierl, A., Hoffmann, A., Reif, W.: Modular and domain-guided multi-robot planning for assembly processes. In: Proceedings of ICINCO, pp. 595 – 604 (2019)
29. Richter, U., Mnif, M., Branke, J., Müller-Schloer, C., Schmeck, H.: Towards a generic observer/controller architecture for organic computing. In: Proceedings of INFORMATIK, pp. 112–119 (2006)
30. Robinson, T., Chan, E., Coelingh, E.: Operating platoons on public motorways: an introduction to the sartre platooning programme. In: ITS World Congress, vol. 1, p. 12 (2010)
31. Roth, M., Schmitt, J., Kiefhaber, R., Kluge, F., Ungerer, T.: Organic computing middleware for ubiquitous environments. In: Müller-Schloer, C., Schmeck, H., Ungerer, T. (eds.) Organic Computing - A Paradigm Shift for Complex Systems, pp. 339–351. Springer (2011)
32. Schörner, M., Wanninger, C., Hoffmann, A., Kosak, O., Reif, W.: Architecture for emergency control of autonomous UAV ensembles. In: Proceedings of ICSEW, pp. 41–46 (2021)
33. Segata, M.: Safe and Efficient Communication Protocols for Platooning Control. Ph.d. thesis, University of Innsbruck (2016)
34. Steghöfer, J.P., Anders, G., Siefert, F., Reif, W.: A system of systems approach to the evolutionary transformation of power management systems. In: INFORMATIK 2013 - Informatik angepasst an Mensch. Organisation und Umwelt, pp. 1500–1515. Gesellschaft für Informatik e.V, Bonn (2013)
35. Sturm, T., Krupitzer, C., Segata, M., Becker, C.: A taxonomy of optimization factors for platooning. IEEE T-ITS **22**(10), 6097–6114 (2021)
36. Taranu, S., Tiemann, J.: On assessing self-adaptive systems. In: IEEE PERCOM Workshops, pp. 214–219 (2010). https://doi.org/10.1109/PERCOMW.2010.5470667
37. Tennenhouse, D.: Proactive computing. Commun. ACM **43**(5), 43–50 (2000)
38. Tomforde, S., Brameshuber, A., Hähner, J., Müller-Schloer, C.: Restricted on-line learning in real-world systems. In: Proceedings of CEC, pp. 1628–1635 (2011)
39. Tomforde, S., Goller, M.: To adapt or not to adapt: a quantification technique for measuring an expected degree of self-adaptation. Computers **9**(1) (2020)
40. Tomforde, S., et al.: Observation and control of organic systems. In: Organic Computing-A Paradigm Shift for Complex Systems, pp. 325–338 (2011)
41. Tomforde, S., et al.: Decentralised progressive signal systems for organic traffic control. In: Proceedings of SASO, pp. 413–422 (2008)
42. Tomforde, S., Rudolph, S., Bellman, K.L., Würtz, R.P.: An organic computing perspective on self-improving system interweaving at runtime. In: Proceedings of ICAC, pp. 276–284 (2016)
43. Tsugawa, S., Jeschke, S., Shladover, S.E.: A review of truck platooning projects for energy savings. IEEE T-IV **1**(1), 68–77 (2016)
44. Weyns, D., Malek, S., Andersson, J.: On decentralized self-adaptation: lessons from the trenches and challenges for the future. In: Proceedings of SEAMS, pp. 84–93 (2010)
45. Weyns, D., et al.: On patterns for decentralized control in self-adaptive systems. In: Software Engineering for Self-Adaptive Systems II, pp. 76–107. Springer (2013)

The Many Uses of Dynamic Logic

Wolfgang Ahrendt[1], Bernhard Beckert[2], Richard Bubel[3],
Reiner Hähnle[3(✉)], and Mattias Ulbrich[2]

[1] Chalmers University of Technology and University of Gothenburg, Gothenburg,
Sweden
ahrendt@chalmers.se
[2] Karlsruhe Institute of Technology, Karlsruhe, Germany
{beckert,ulbrich}@kit.edu
[3] Technische Universität Darmstadt, Darmstadt, Germany
{richard.bubel,reiner.haehnle}@tu-darmstadt.de

Abstract. Dynamic logic is a multi-modal logic for reasoning about
programs. In deductive verification systems, it can be used as a versatile
alternative to the Floyd-Hoare calculus with uniform syntax and semantics. Dynamic logic has not only been used in functional verification,
but one can represent a plethora of verification scenarios in it, including
relational and hyperproperties, program equivalence, information flow,
incorrectness logic. Dynamic logic is the basis for three deductive verification tools that are highly competitive in their application domain. In
this article, we present the foundations of dynamic logic and we review
its many uses in state-of-the-art deductive verification.

Keywords: Dynamic logic · deductive verification · symbolic
execution · weakest preconditions

1 Introduction

Dynamic logic [52,73] is a modal logic for reasoning about programs introduced
by V. Pratt [66] in 1976. The term *dynamic logic* was coined in the paper [42] and
it was initially investigated by D. Harel [41] and R. Goldblatt [36]. Wolfgang's
association with dynamic logic starts in 1984 with his Master's Thesis [67] that
laid some of the theoretical foundations of the calculus used in the KIV deductive
verification system. Until then, excepting a short-lived attempt [57], dynamic
logic had been mainly the object of theoretical investigation and had not been
used as a program logic in an actual verification system. Then (and now) the
Floyd-Hoare calculus [47] was far more popular, despite clear advantages on the
side of dynamic logic including (i) greater expressiveness, (ii) syntactic closure
by first-order connectives and quantifiers, and (iii) the proximity to relational
program semantics.[1]

---

[1] For some reasons how this might have come about, see [37, Sect. 6.2.7].

This article is dedicated to Wolfgang Reif on the occasion of his 65th birthday.

G. Ernst et al. (Eds.): Wolfgang Reif Festschrift, LNCS 15765, pp. 56–82, 2025.
https://doi.org/10.1007/978-3-031-92196-4_4

The work of Wolfgang (together with M. Heisel and W. Stephan) was the first serious attempt to apply the benefits of dynamic logic within the context of a deductive verification system [38, 44]. Specifically, Wolfgang presented in his Ph.D. work [68, 69] an elegant solution to a pivotal problem in program verification: how to realise procedure-modular verification. For this he used abstract data types and step-wise refinement of program modules, its correctness being justified by dynamic logic. This approach within a few years yielded impressive practical success [70]. The KIV system is still developed and maintained in Wolfgang's research group [32] and was used for some of the most intricate and comprehensive formal verification efforts to date [72], see also Sect. 4.1 below.

When some of the present authors began considerations [39] on their deductive verification system KeY [4] for the Java programming language, one decision was obvious: To use dynamic logic as its theoretical foundation. Thanks to Wolfgang's work, its usefulness was abundantly clear and we did not regret the choice for a second. In this article we celebrate the versatility of dynamic logic, not only as a language for expressing and verifying the correctness of *real* programs and case studies, but also as a language to express many other problems such as information flow analysis or relational verification.

In Sect. 2 we review the foundations of dynamic logic in terms of modal logic, relational semantics, and the original, "pure" programming language inspired by regular expressions used by Pratt, Harel et al. In Sect. 3 we highlight the expressive power of dynamic logic in *practical terms*: its ability to characterise or incorporate a wide range of specification approaches and program analysis techniques. In Sect. 4 we collect some of the success stories achieved with deductive verification systems based on dynamic logic, before we briefly conclude in Sect. 5.

2 Foundations of Dynamic Logic

2.1 Logics of Change

Dynamic logic belongs to the family of logics reflecting the fact that the *state of affairs* (of the word, of the mind, of a system, etc.) can change. Such logics can be referred to as *logics of change*. For a long time, philosophers had been thinking about the impact of change on reasoning, for example, Aristotle and William of Ockham. In modern times, logic became the subject of mathematical studies in the form of *mathematical logic* (or *meta mathematics*). Within that discipline, there evolved a branch of logic which focuses on the modelling of change, namely *modal logic*, starting with the work of C. I. Lewis [55]. The first, surprisingly little known, approach to semantics of modal logic was proposed by B. Jónsson and A. Tarski [49]. Further attempts on semantics were contributed in the 1950s, by A. Prior, J. Hintikka, and S. Kripke [53], where the latter had the biggest impact by far.[2] In the early years, motivating application areas of modal logic were largely the philosophy of language, epistemology, and metaphysics. But

---

[2] Kripke was at that point unaware of Tarski's work on modal logic.

over time, *computation* became more and more a prime application of modal
logic, and variations thereof.

In modal logic, we assume a non-modal base logic, for instance propositional
logic or first-order logic. Moreover, we assume a set of *possible worlds*, and a
definition of when a formula of the base logic is valid in any given world. So
far, there is no aspect of change. This comes in if we assume, in addition to
the above, a relation R between worlds, called *accessibility relation*. When two
worlds w and w' are in this relation, i.e., $(w, w') \in R$, it means that we can go
from w to w' in *one* R-step (whatever an R-step is supposed to mean intuitively
in the application at hand). With these ingredients, we obtain what is called
a *Kripke structure*. As an example, let us consider Fig. 1 (where we ignore the
formulas containing $\Box$ or $\Diamond$ for now). In two of the possible worlds, p is true,
whereas p is false in the others. The arrows depict the accessibility relation R. In
our example, R does not enjoy many properties. (In fact, R is neither reflexive,
symmetric, anti-symmetric, transitive, total, nor deterministic.)

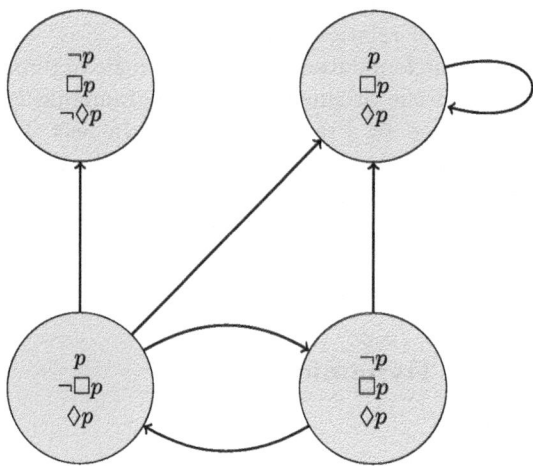

Fig. 1. Kripke Structure: possible worlds with (one-step) accessibility relation

A given base logic, such as propositional or first-order logic, can be extended
to a modal logic by adding two logical operators, $\Box$ and $\Diamond$. A formula $\Box\phi$ is
valid in a world w iff ϕ is valid in *all* worlds accessible from w via R (in one
step). And a formula $\Diamond\phi$ is valid in a world w iff ϕ is valid in *some* world
accessible from w via R (in one step). The reader may check that all formulas in
Fig. 1 are valid in the worlds they are depicted in. Assume the validity of the
propositional literals, the modal formulas follow then from this assumption and
the chosen transition relation R. A perhaps unintuitive corner case is the upper
left world, from where no world is accessible via R (not even that world itself).
Therefore, $\Box p$ and $\neg\Diamond p$ are vacuously true that world.

Modal logic is syntactically closed under all propositional operators, the ones from the base logic as well as $\square$ and $\diamond$. For instance, $\diamond p \wedge \neg \square p$ and $\square\square p$ are formulas of modal logic (both are valid in the lower left world). In general, modal logics discussed in the literature come in many flavours, called K, T, S4, S5, D, among others, which vary in their properties of the accessibility relation.

The reader may have noticed the syntactic similarity of modal logic to temporal logic as introduced by A. Pnueli [65] (even if he wrote G for $\square$ and F for $\diamond$). But in temporal logic, $\square p$ and $\neg\diamond p$ could not be true at the same time (see Fig. 1 upper left world), neither could $\neg\square p$ and $\square\square p$ be true at the same time (see Fig. 1 lower left world). Moreover, in temporal logic, $\square$ and $\diamond$ refer to arbitrarily many steps, not just one as in modal logic. Still, quoting Pnueli, his temporal logic *"is completely isomorphic to the modal logic system S4"* [65].[3] The reason is that, in temporal logic, the accessibility relation is reflexive and transitive, such that arbitrarily many steps forward are at the same time a single step in the modal reading.

2.2 Dynamic Logic as a Multi-Modal Logic

Dynamic logic was introduced in 1976 by V.R. Pratt in *"Semantical considerations on Floyd-Hoare Logic"* [66]. It extends modal logic by a language of *actions*. Actions can moreover be composed to new actions, such that they are reminiscent of *programs*. Every (elementary or composed) action gives rise to its own accessibility relation. In that sense, dynamic logic is a *multi-modal logic*.

In formulas of dynamic logic, the modalities are "parameterised" by actions. Syntactically, we write the actions inside the modalities. The syntax looks as follows, given here together with its intuitive meaning:

– $\langle\alpha\rangle\phi$ means "ϕ holds in *some* state we can *reach by executing* α"
– $[\alpha]\phi$ means "ϕ holds in *all* states we can *reach by executing* α"

Here, "reach by executing α" refers to one step in the α-accessibility relation. Please note that the worlds of modal logic are called "states" here and in the following.

2.3 Propositional Dynamic Logic

Propositional Dynamic Logic (PDL) [33] is typically based on *non-deterministic* actions. In applications of dynamic logic, the non-determinism serves mainly two purposes: abstraction, and the modelling of an uncontrollable environment. The following notation and definitions are inspired by [52,64].

Definition 1 (Propositional Dynamic Logic: Formulas and Actions).
We assume a set of atomic formulas and a set of atomic actions. If ϕ, ψ are formulas, and α, β are actions, then

---

[3] Temporal logic as introduced by Pnueli [65] did not have a "next" operator.

- $\neg\phi$
- $\phi \vee \psi$
- $\langle\alpha\rangle\phi$ *("some execution of α leads to a state where ϕ holds")*

are also formulas, and

- $\alpha;\beta$ *(sequence)*
- $\alpha \cup \beta$ *(non-deterministic choice)*
- α^* *(execute α a finite, non-deterministic number of times)*
- $?\phi$ *(proceed if ϕ, otherwise fail)*

are also actions.

The above action language is also called *"regular programs"* [33]. Further formulas ($\phi \wedge \psi$, $\phi \rightarrow \psi$,$[\alpha]\phi$) and actions (see below) can be derived from the ones above. In particular, just like modal logic and temporal logic have dual modalities $\square$ and $\lozenge$, dynamic logic has, for every action α, a modality $[\alpha]$ which is dual to $\langle\alpha\rangle$. The former can be defined as follows:

$$[\alpha]\phi \equiv \neg\langle\alpha\rangle\neg\phi \tag{1}$$

The intuitive meaning of $[\alpha]\phi$ is that *"all executions of α lead to a state where ϕ holds"*. Let us provide some intuition by means of an example.

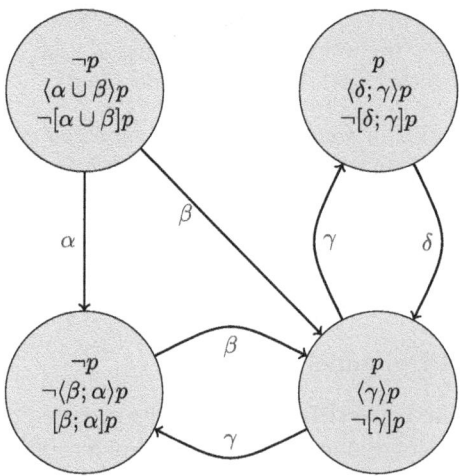

Fig. 2. Kripke structure with multiple atomic action relations

Figure 2 depicts, for each state, a few dynamic logic formulas that are valid there. The example highlights the difference between box and diamond modalities. For instance, in the upper left state, there exists a state accessible by non-deterministic choice of α or β where p holds. But in the same state, it is not the case that p is true in *all* states accessible by $\alpha \cup \beta$. A similar situation occurs when executing δ and γ in sequence in the upper right state. The dynamic logic formulas given in the lower left state talk about the action $\beta; \alpha$, which is, however, not feasible from that state (because, after executing β, there is no α that can be executed). We will come back to this phenomenon.

To further explain PDL, we provide semi-formal definitions of its semantics (for a formal account, we refer to [52]). In particular, we build on notions not fully defined in this chapter. In the following, we assume a set of states S, a meaning function $\mathcal{M}$ of atomic formulas $\phi^{\mathcal{M}} \subseteq S$ (assigning to each formula the set of states in which it is true)[4], and the (overloaded) meaning function $\mathcal{M}$ of atomic actions $\alpha^{\mathcal{M}} \subseteq S \times S$ (such that each action denotes a relation between states). This relation does not have to be deterministic, neither does it have to be total. For instance, $\gamma^{\mathcal{M}}$ in Fig. 2 is non-deterministic on the lower right state, and undefined on all other states. The following definition extends the meaning function $\mathcal{M}$ from atomic to non-atomic formulas and actions.

Definition 2 (Relational Semantics of PDL Formulas). *Meaning of formulas* $\phi^{\mathcal{M}} \subseteq S$, *meaning of actions* $\alpha^{\mathcal{M}} \subseteq S \times S$:

- $(\neg\phi)^{\mathcal{M}} = S - \phi^{\mathcal{M}}$
- $(\phi \vee \psi)^{\mathcal{M}} = \phi^{\mathcal{M}} \cup \psi^{\mathcal{M}}$
- $(\langle\alpha\rangle\phi)^{\mathcal{M}} = \{u \mid \exists v.\ (u, v) \in \alpha^{\mathcal{M}} \ and \ v \in \phi^{\mathcal{M}}\}$
- $(\alpha; \beta)^{\mathcal{M}} = \{(u, v) \mid \exists w.\ (u, w) \in \alpha^{\mathcal{M}} \ and \ (w, v) \in \beta^{\mathcal{M}}\}$
- $(\alpha \cup \beta)^{\mathcal{M}} = \alpha^{\mathcal{M}} \cup \beta^{\mathcal{M}}$
- $(\alpha^*)^{\mathcal{M}} = \bigcup_{n \in \mathbb{N}}(\alpha^n)^{\mathcal{M}}$, *where* $\alpha^{n+1} \equiv \alpha; \alpha^n$ *and* $\alpha^0 \equiv \{(u, u) \mid u \in S\}$
- $(?\phi)^{\mathcal{M}} = \{(u, u) \mid u \in \phi^{\mathcal{M}}\}$

The meaning of non-deterministically choosing between α and β is the union of the behaviours of α and β. The meaning of α^* is all behaviours resulting from choosing some $n \in \mathbb{N}$, non-deterministically, and repeating α n times. This includes zero repetitions of α, which is a *skip* operation in all states. Note that α^* denotes *finite* iterations only. The *test* action $?\phi$ deserves special attention. In states where ϕ holds, it behaves like *skip*, i.e., we stay in the same state. But in states where ϕ does *not* hold, there is *no* state to go to with this action, not even the same state. This is similar to applying a partial function outside the domain, where it is defined. The result would be undefined, not the identity. Intuitively, we can see a computation where this happens as a *failed* computation.[5]

According to Definition 2, $\alpha; \beta$ fails as soon as any of α or β fail. We illustrate this with Fig. 3. The crossed out β arrow from state s_1 illustrates that atomic action β fails on s_1, i.e., there is no state s' such that $(s_1, s') \in \beta^{\mathcal{M}}$. Therefore, by Definition 2, there is also no state s' such that $(s_0, s') \in (\alpha; \beta)^{\mathcal{M}}$, hence $\alpha; \beta$ fails on s_0. Therefore, in state s_0, the choice $\alpha; \beta \cup \gamma; \delta$ collapses to $\gamma; \delta$.

This is an important point in the interplay of choice and failure as defined by the relational semantics. In cases where non-determinism allows numerous

---

[4] Note that this semantic modelling avoids truth values.
[5] A failed computation is sometimes referred to as *abort* in the DL literature.

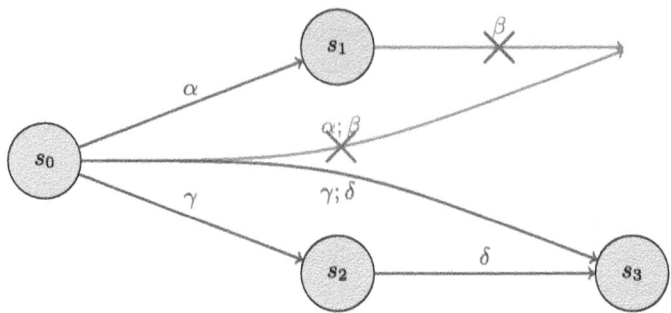

Fig. 3. Propagation of failure over sequence and choice

computation paths, any path p where a failure occurs is taken out from the set of possible behaviours, even if the failure occurs arbitrarily late in p. Accordingly, it is a consequence of Definition 2 that failed computations of α are not included in the set of behaviours which are quantified over in the modalities $\langle\alpha\rangle\phi$ and $[\alpha]\phi$. In other words, we can think of $\langle\alpha\rangle\phi$ as "some non-failing computation of α leads to ϕ", and $[\alpha]\phi$ as "all non-failing computations of α lead to ϕ".

To support a more common notion of "programs", well-known programming constructs are definable in PDL's action language:

- **skip** $\equiv$ *?true*
- **fail** $\equiv$ *?false*
- **if** ϕ **then** α **else** β **fi** $\equiv (?\phi; \alpha) \cup (?\neg\phi; \beta)$
- **while** ϕ **do** α **od** $\equiv (?\phi; \alpha)^* ; ?\neg\phi$

skip was discussed informally already; **fail** is a computation that always fails: $(?false)^{\mathcal{M}} = \{(u, u) \mid u \in false^{\mathcal{M}}\} = \{(u, u) \mid u \in \emptyset\} = \emptyset$. In the definition of the conditional, one of the choices $(?\phi; \alpha)$ and $(?\neg\phi; \beta)$ necessarily fails and is discarded from the possible behaviours. The definition of loops is concise but intricate. The * operator permits an *arbitrary* number of iterations, whereas the α in the **while** loop is supposed to be repeated *exactly* as long as ϕ is true, not more, not less. The resolution of this seeming conundrum is illustrated in Fig. 4, where we depict a scenario where ϕ becomes false after three iterations of α. The key insight is that most choices of the number n of iterations lead to failure and are therefore discarded. If n is chosen too small, we exit the loop at a point where ϕ is still true and the following action $?\neg\phi$ will fail. If, on the other hand, n is chosen too big, we stay in the loop even when ϕ became false, and the following iteration $?\phi; \alpha$ will fail. In our example, all choices of n other than 3 lead to an execution that is discarded, only the choice $n = 3$ leads to an execution which is kept in $((?\phi; \alpha)^* ; ?\neg\phi)^{\mathcal{M}}$.

How is non-termination handled, given that the * operator permits only a *finite* number of iterations? To make the point, we consider the extreme case of the program **while** *true* **do** α **od**, which is defined as $(?true; \alpha)^* ; ?\neg true$. This can simplified to $\alpha^* ; ?false$. Regardless of which number n of iterations is chosen (non-deterministically), the execution will fail because of the last

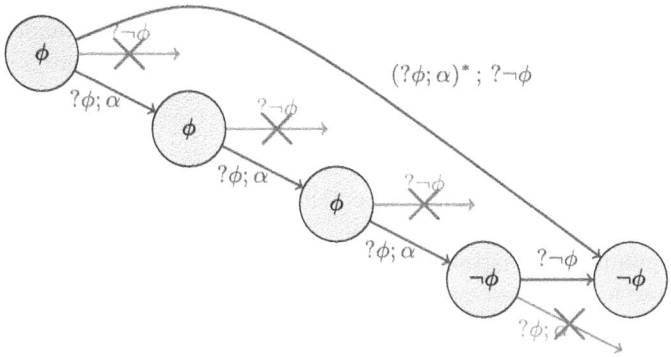

Fig. 4. Illustration of loop definition in dynamic logic

action $?false$. Instead of semantically modelling infinite executions, the relational semantics models *infinitely* many attempts on *finite* executions, all of which fail. The resulting relation is empty, i.e., the final state of the program is not defined. (In general, a program's final state may be defined for certain initial states, but not for others.)

2.4   Deterministic PDL

In case the action language has a deterministic semantics, we have the special case of *deterministic PDL*. We assume the atomic actions to be deterministic: For all atomic actions α, if $\{(s, s'), (s, s'')\} \subseteq \alpha^{\mathcal{M}}$, then $s' = s''$. We further assume that the non-deterministic constructs $\cup$ and $*$ appear *only* to abbreviate **if** and **while** (where all but one of the non-deterministic choices fail).

In this setting, we can move from relations to partial functions as meaning of (atomic or composed) actions: $\alpha^{\mathcal{M}} \in S \rightharpoonup S$. Note that the behaviour of an action (program) can still be undefined. But there is at most one defined behaviour for every initial state.

In deterministic PDL (and all other deterministic versions of DL), $[\alpha]\phi$ denotes *partial correctness*, whereas $\langle\alpha\rangle\phi$ denotes *total correctness*. Moreover, partial correctness implies total correctness, i.e., we have $\langle\alpha\rangle\phi \rightarrow [\alpha]\phi$. (This is not the case in general PDL, see $\langle\gamma\rangle\phi$ in Fig. 2.)

2.5   First-Order Dynamic Logic

So far, there is no notion of program variables, which is, however, essential for imperative programs. To address this, D. Harel introduced *first-order dynamic logic* (FDL) [41], adding variables to programs, and quantification to formulas. In FDL, atomic programs have the following form:

- $v := t$ (deterministic assignment)
- $v := *$ (non-deterministic assignment)

where v is a program variable, and t is an expression in an underlying side effect-free expression language. Atomic formulas are of the forms:

- $p(t_1, \ldots, t_n)$
- $t_1 \doteq t_2$

Regarding composite formulas, we extend the operators from PDL with quantification. If ϕ is an FDL formula, then $\exists x.\phi$ and $\forall x.\phi$ are also FDL formulas. Otherwise, composite programs and formulas are formed exactly as in PDL.[6] FDL is syntactically closed under all its logical operators, the quantifiers as well as the logical operators from PDL. For instance, all of the following are formulas of FDL:

1. $\forall x.\,(\langle t := a; a := b; b := t\rangle b = x \leftrightarrow \langle a := a + b; b := a - b; a := a - b\rangle b = x)$
2. $\langle \alpha \rangle \exists x.\phi(x)$
3. $\exists x.\langle \alpha \rangle \phi(x)$

We see that operators from propositional and first-order logic can appear inside or outside the scope of modalities. The first formula states equivalence of two programs (relative to the final value of b)—a property most non-dynamic logic program logics cannot express. The other two formulas are equivalent for a simple programming language as the one given here. But in richer programming languages, where resources can be generated by programming constructs (such as object creation in object-oriented languages), these formulas can have a different meaning when α extends the domain we quantify over (see [5] for an in-depth discussion).

The non-deterministic assignment $v := *$ is a tool for abstracting away from irrelevant details, or for receiving input from an unknown environment. Moreover, the combination of non-deterministic assignment and test actions allows for a kind of declarative programming when desired. Concretely, the program fragment $v := *; ?\phi$ expresses the following command: "choose v such that $\phi(v)$ is true". This is used when modelling assumptions on the environment during verification of controllers, as in the KeYmaera X theorem prover, see Sect. 4.3.

3   Expressive Power of Dynamic Logic

As seen in Sect. 2, in dynamic logic, programs as the constituents of modalities are first-class citizens of its syntax, which makes dynamic logic closed under syntactic composition. This property sets dynamic logic apart from most other program logic frameworks that limit the formulation of program properties to pre-defined patterns (like Hoare calculus), or that see programs as operators on formulas (like Dijkstra's weakest precondition calculus). These formulate *functional* properties corresponding to the canonical pattern $\psi \to [\alpha]\phi$ in dynamic logic which asserts that a postcondition ϕ holds in some post-state of a program α under the assumption of a precondition ψ in the pre-state. Dynamic logic is

---

[6] Usually, FDL allows no quantifiers in $?\phi$.

not restricted to this pattern, and without the need to define syntactic or semantic extensions, or introducing *ad-hoc* notation, one can use the well-studied and semantically clear composition operations of propositional and first-order logic to obtain formalisations of program properties beyond the ones mentioned. For instance, one may

– quantify over dynamic logic formulas containing modalities,
– nest dynamic logic operators so that the formula in the scope of a modality may again contain modalities, or
– use modalities inside test operators within programs

to name only a few of the possibilities. While the functional property pattern is prominent in functional verification and covers many application scenarios, there are interesting properties that heavily benefit from the expressive power of dynamic logic.

In this section, we show how the core concepts of some important program analysis frameworks can be expressed within dynamic logic. Moreover, we look at a number of interesting program properties beyond the functional pattern. The advantage of the expressive power of dynamic logic is that one can rely on a sound basis in a well-understood program logic when addressing novel verification questions and approaches.

3.1 Hoare Calculus

The Hoare calculus [47] allows us to specify a program's behaviour in terms of the eponymous Hoare triple

$$\{pre\} \; \alpha \; \{post\} \; ,$$

where *pre*, *post* are first-order logic formulas representing the precondition and postcondition, respectively, of program α.[7]

The meaning of the triple is that if program α is executed in a state in which formula *pre* is satisfied *and* program α terminates, then in the[8] reached final state formula *post* holds. In other words, a Hoare triple states the partial correctness of a program. Manna and Pnueli [58] propose an extension for reasoning about total correctness and built-in support to access the pre-state value of variables in the postcondition. For that, they require the provision of a *convergence function* whose value must be shown to decrease by each iteration of the loop. For a general overview of the Hoare calculus and its influence, we refer to [7].

Hoare's objective was to provide a logical framework to prove the behavioural correctness of programs. However, this framework does not define a *logic* as it is not closed under the logical connectives and, hence, not a universal algebra. This omission results in missing compositionality and limitations of properties that are

---

[7] Originally, the syntax was *"pre {α} post"*, but at some point the braces were moved.
[8] The Hoare calculus was formulated for deterministic programs which we assume here as well.

directly expressible within the framework itself. For instance, non-interference is not expressible within Hoare logic without self-composition [26].

Dynamic logic [41, 66] addresses this shortcoming, being a program logic, where programs are first-class citizens (as in the Hoare calculus) of its formulas but which is closed under its operators (logical connectives). As mentioned above, Hoare triples can then be expressed as $pre \rightarrow [\alpha]post$, total correctness is expressible as $pre \rightarrow \neg[\alpha]\neg post$, or simply, $pre \rightarrow \langle\alpha\rangle post$ (for deterministic programs).

3.2   Flexible Verification Patterns

A common scenario where the syntactic flexibility of dynamic logic is useful occurs when a specification contains logic symbols whose definition is given in terms of code (for instance, as a pure function). Consider the proof obligation `valid(x)` $\rightarrow$ $[p]post$, where `valid` references a function in the programming language. Dynamic logic allows one to formulate this *as a formula*

$$[r = \mathtt{valid(x)}]r \not\doteq \mathrm{true} \rightarrow [p]post$$

that refers to `valid` as a program function. Consequently, there is more than one piece of code in the proof goals which is perfectly admissible in dynamic logic.

In a further scenario, dynamic logic is used to specify reachability. Consider the singly linked list Java implementation in Listing 1.1, where we navigate from one node of the list to the next via attribute `next`. The final element of the list is reached when its attribute `next` is `null`. Assume we want to specify that method `contains(int)` returns `true` if the list contains a node with the content passed as parameter `e`. To specify the intended behaviour, we use that the precondition (and postcondition) of a dynamic logic specification in turn can contain programs. Consider program `p` which assigns program variable `l` the value of the `idx`-th element in the list (assuming program variable `idx` is initially non-negative). Then the following dynamic logic formula specifies the intended behaviour of `contains`:

```
class ListNode {
  ListNode next;
  int cnt;

  boolean contains(int e) { ... }
}
```

```
// program p for specification
ListNode l = this;
while (l != null && idx >= 0) {
  l = l.next;
  idx--;
}
```

Listing 1.1. Singly linked list and program used for specification

$$\Big(\exists k.(k \geq 0 \ \wedge \ \langle \mathtt{idx} \not\doteq \mathtt{k}; \mathtt{p}\rangle(\mathtt{l} \not\doteq \mathtt{null} \wedge \mathtt{l.cnt} \doteq \mathtt{e}))\Big)$$
$$\leftrightarrow \langle \mathtt{res} \not\doteq \mathtt{contains(e)};\rangle(\mathtt{res} \doteq \mathit{TRUE})$$

3.3   Dijkstra's Weakest Precondition Calculus

Dijkstra's weakest precondition calculus [29] defines a predicate transformer wp
that computes the *weakest precondition* from a given postcondition. To prove
correctness of a program, it remains to be shown that the given precondition
implies the weakest precondition. As wp is a predicate transformer and thus a
meta construct, which transforms formulas, it is not a first-class citizen of the
logic.

The original wp-operator includes to prove termination of the given program;
one variant, the *weakest liberal precondition* wlp formalizes partial correctness.
Although the wp-calculus semantics aligns nicely with the "diamond" semantics
of dynamic logic, there is a crucial difference when the target programming language is non-deterministic, as is the guarded command language Dijkstra introduced alongside the wp-calculus. Given a non-deterministic program α that can
choose non-deterministically between continuations $\alpha_1, \ldots, \alpha_n$, then $\mathsf{wp}(\alpha, post)$
requires that (i) program α can make a non-deterministic choice (i.e., it does
not abort or is blocked), and (ii) that for any executable choice α_i, $\mathsf{wp}(\alpha_i, post)$
holds. In contrast, the diamond formula $\langle \alpha \rangle post$ holds if (i) program α can make
a non-deterministic choice (i.e., it does not abort or is blocked), and (ii) that
there is one executable choice α_i for which $\langle \alpha_i \rangle post$ holds.

This difference makes a canonical emulation of the wp-calculus in dynamic
logic impossible. Of course, extending dynamic logic with a dedicated modality
matching wp's semantics is one possibility. This has been realized in KIV [32] by
introducing the strong diamond modality $\langle\!| \cdot |\!\rangle \cdot$. However, wp and wlp share not
the duality of the diamond and box operator in dynamic logic but differ only in
the termination requirement as such. This means wlp is a natural fit for dynamic
logic's box operator and thus can be canonically emulated by inference rules for
dynamic logic.

We provide some examples to showcase how wlp-rules can be directly cast
into dynamic logic axioms (and dynamic logic inference rules):

– It is easy to see that $\mathsf{wlp}(\alpha, post_1 \wedge post_2) = \mathsf{wlp}(\alpha, post_1) \wedge \mathsf{wlp}(\alpha, post_2)$ corresponds to the valid dynamic logic formula $[\alpha](post_1 \wedge post_2) \leftrightarrow ([\alpha]post_1) \wedge ([\alpha]post_2)$.
– The relation between the two systems becomes even clearer when considering
that the equation $\mathsf{wlp}(\alpha_1; \alpha_2, post) = \mathsf{wlp}(\alpha_1, \mathsf{wlp}(\alpha_2, post))$ corresponds to
$[\alpha_1; \alpha_2]post \leftrightarrow [\alpha_1][\alpha_2]post$ thanks to the possibility of nested modalities.

A further significant difference is that the wp-calculus analyses the program
in a backward direction from the end to the start while transforming the postcondition into its weakest precondition. Dynamic logic provides more flexibility
and permits the implementation of both directions of analysis. In particular,
dynamic logic allows one to analyse a program in the forward direction (see
Sect. 3.5), while still computing the weakest precondition (and not the strongest
postcondition).

For a thorough review of Dijkstra's weakest precondition calculus, as well
as a comparison to dynamic logic, we refer to [37]. The parallels between the

wp-calculus and dynamic logic were already observed by Harel & Pratt in 1978 [43] who reproved formally many of Dijkstra's results [30].

3.4   Relational Properties

Relational and Hyperproperties. Certain relevant program properties concern not only one program but express a relationship between the effects and results of one program relative to the effects and results of another program. Such properties are called *relational*.[9] A relational property may put two (or more) different programs into relation, thus comparing the effects of two different, independent realisations. But relational properties can also be about *the same* program in which case two different independent runs of a program α are compared. This is called a *hyperproperty* of α [21]. Originally, hyperproperties are for traces but can also be formulated for pre- and post-states. The general pattern for formalising a relational property in dynamic logic is as follows:

$$pre \rightarrow \langle\!|\alpha_1|\!\rangle \langle\!|\alpha_2|\!\rangle ... \langle\!|\alpha_n|\!\rangle \, post$$

Here, $\langle\!|\alpha|\!\rangle$ is either $[\alpha]$ or $\langle\alpha\rangle$ and the α_i have disjoint program variable sets. For a hyperproperty, one can take identical copies of the same program with renamed variables. The relational precondition *pre* and the relational postcondition *post* connect the state spaces of the different programs in the pre- and post-state. Most relational properties relate two programs ($n = 2$), but there are scenarios relating more than two program executions. Functional properties as defined previously are relational properties for the (corner) case of $n = 1$.

Program Equivalence. The arguably most prominent relational property formulates in its purest form that two programs α and β presented with equal input yield equal output (provided they both terminate):

$$\overline{in}_\alpha = \overline{in}_\beta \;\rightarrow\; [\alpha][\beta](res_\alpha = res_\beta) \tag{2}$$

Here, the equality $\overline{in}_\alpha = \overline{in}_\beta$ correlates the corresponding input variables of the two disjoint name spaces of α and β. An important use case of program equivalence checking is *regression verification*, where a more recent program version replaces a less efficient or buggy older version. In case of a bug fix, one evidently does not want the new version to be equivalent to the old one. Hence, it makes sense to amend the premise of the implication with a relational precondition (excluding the buggy cases) under which equivalence has to hold (*conditional regression verification*). In Eq. (2), we referred to the corresponding inputs in the two variable sets. It is not always and necessarily clear how the state spaces correspond. It may very well be that the value of one variable in α is represented by two variables in β. Hence, equality (2) can take the relaxed form of a one-to-one relationship between state spaces (*relational regression verification*).

---

[9] While sometimes one encounters the term *relational verification*, being relational is not a characteristic of the analysis technique, but one of the property itself. Relational properties can sometimes be encoded as functional properties.

Property (2) encodes "partial equivalence", leaving termination aside. But *mutual termination* [35] can also be formulated in dynamic logic (at least for deterministic languages). A program α terminates under a precondition *pre* in precisely the same cases as the program β iff the following formula is valid:

$$pre \rightarrow \left(\langle\alpha\rangle true \leftrightarrow \langle\beta\rangle true\right)$$

Refinement. A very successful concept [1,2,48] in formal design is the notion of *refinement*: An abstract, often nondeterministic state transition description is enriched with details about the realisation and formally shown to be compatible with the abstract description. One program γ *refines* another program α if any effects observable under γ can also be observed under α: No new behaviour is introduced by γ, so all invariants of α hold for γ, too.

$$in_\gamma = in_\alpha \quad \rightarrow \quad [\gamma]\langle\alpha\rangle(res_\gamma = res_\alpha) \tag{3}$$

Formula (3) closely resembles (2) with the difference that opposing modalities are in use. This is to accommodate the situation that for all observable concrete behaviours there must be an abstract action. To deal with the nondeterminism, this alternation of modalities is necessary. As in the case of program equivalence, data representation is not necessarily identical in α and γ. Hence, the equalities in (3) are usually generalised to *coupling predicates* describing the relation between the abstract and concrete state space.

Secure Information Flow. The best-studied hyperproperty of programs is *non-interference*. Non-interference is an information-flow security property ensuring a program does not leak confidential information. To prove this property, one has to show that the value of program variables holding confidential information does not influence the value of a non-confidential program variable. For example, the following two programs do not satisfy the non-interference property (l is a non-confidential, "low" program variable, while h denotes a confidential, "high" one):

– l = h; is insecure, as the secret h is directly leaked to l
– if (h > 0) l = 0; else l = 1; is insecure, as it leaks the sign of h to l

As shown in [26], it is possible to express non-interference for a program α directly in dynamic logic as

$$\forall \bar{l}.\exists \bar{r}.\forall \bar{h}. \left(\langle\alpha\rangle(\bar{l} \doteq \bar{r})\right) \tag{4}$$

which expresses that the final values $\bar{r}$ of non-confidential program variables $\bar{l}$ only depend on the initial values of the non-confidential variables, but not on the value of the confidential variables $\bar{h}$. This formalisation uses alternating quantification over variables (which is possible in dynamic logic, but not in the Hoare calculus).

Other techniques, such as *self-composition* [8, 25], can be used to express non-interference in Hoare calculus as well as dynamic logic, but require additional encoding effort to allow for independent execution of the same program. A compact representation of this approach in dynamic logic is

$$\bar{l} = \bar{l'} \;\rightarrow\; [\alpha][\alpha'](r = r') \;,$$

where program $\bar{\alpha}$ is identical to α, except that all program variables v in α have been renamed to v'.

The formalisation shown in Eq. (4) requires providing a witness for the result value in contrast to self-composition. This can be mitigated by delaying the provision using free variables. In both cases, dynamic logic permits modelling more complex information-flow security properties, such as declassification, in a natural manner. For example, some information may be intentionally leaked, such as the average salary of all employees, while the salary of a specific employee is confidential.

Relational Properties of Algorithms. Hyperproperties may formally relate *more than two* program runs. We show two examples and their encoding in dynamic logic:

When determining the winner of an election, different vote counting schemes can be applied. One desirable feature of a voting scheme ω is *separability*, requiring that if a candidate *res* wins in two independent elections, then they also win if the ballot boxes *in* and *in'* are joint (*in''*) [9]:

$$in'' = in \cup in' \rightarrow [\omega][\omega'][\omega''](res = res' \rightarrow res = res'')$$

At times it is important to know whether an implemented routine possesses desirable mathematical properties. For example, when implementing distributed routines using the map-reduce paradigm, the independence of the result from the distribution of the input values over computing nodes depends on the fact that the reducers computing intermediate results are commutative and associative, i.e. that reordering and regrouping partial results does not modify the final result. Associativity of α with two inputs a and b and output *res* can be formulated in dynamic logic as a condition relating four program runs:

$$[\alpha][\alpha'][\alpha''][\alpha'''](a = a''' \wedge b = res' \wedge a' = b''' \wedge b' = b'' \wedge a'' = res''' \rightarrow res = res'')$$

Relational Reasoning. For the verification of relational properties, dedicated inference rules tailored to this scenario can be used to increase efficiency. One elegant inference rule for loop induction in dynamic logic is

$$\frac{\phi \rightarrow [\alpha]\phi}{\phi \rightarrow [\alpha^*]\phi}$$

formalising that when ϕ is maintained by a single execution of α, then it is maintained by arbitrarily many iterations of α. This induction rule can be generalised to reason about synchronised loops in two related program runs. The formula ϕ serves as *coupling invariant* describing the relationship between the state spaces of the two program runs under verification. In relational refinement proofs, for example, the following rule can be applied

$$\frac{\phi \rightarrow [\alpha\,]\langle\gamma\,\rangle\phi}{\phi \rightarrow [\alpha^*]\langle\gamma^*\rangle\phi} \ .$$

If a single iteration of γ refines a single iteration of α (modulo ϕ), then the nondeterministic iteration γ^* refines α^*.

This relational reasoning allows us, in the case of programs with (lock-step) synchronised loops, to reason with local relational knowledge: It does not matter what the program computes; it is merely interesting how the state relates to the state of the concurrent execution of the other programs. We can focus on the differences between the states rather than describing the states explicitly.

3.5 Symbolic Execution

Symbolic execution [50] is a general program analysis that can be used to prove programs correct. In contrast to Dijkstra's wp-calculus, it executes a given program in a forward-directed manner (applying forward substitutions) to compute the weakest precondition. Besides deductive verification [20,44], symbolic execution has shown to be advantageous for test generation [18,50,51] and debugging [45,51].

Dynamic logic is a suitable candidate to realise symbolic execution for deductive verification, as well as for test generation and debugging. Dynamic logic permits representing the symbolic states generated during symbolic execution naturally, as well as recording path conditions as preconditions. The nature of logic proof systems, such as sequent or tableaux calculi, is to represent proofs as trees. This renders them ideal to represent the unrolled control-flow (and data) graph obtained by symbolic execution.

The semantics of symbolic execution then serves as a design guideline for developing a deduction system. This becomes obvious when looking at the dynamic logic sequent calculus rule for the conditional statement:

$$\frac{\Gamma, e \Longrightarrow \langle\alpha_1;\, \beta\rangle post, \Delta \qquad \Gamma, \neg e \Longrightarrow \langle\alpha_2;\, \beta\rangle post, \Delta}{\Gamma \Longrightarrow \langle\texttt{if } (e) \texttt{ then } \alpha_1 \texttt{ else } \alpha_2;\, \beta\rangle post, \Delta}$$

Here, the proof goal in the conclusion splits into two subgoals, one for the case where the guard of the conditional statement evaluates to true and one where it evaluates to false. This maps branching of the control-flow into branching of the proof tree at the current proof goal.

An advantage of using dynamic logic as a framework to implement symbolic execution is that it can use the information provided by the specified precondition and the accumulated path condition to simplify the state representation and exclude unreachable code paths early on. In semi-automated deductive verification [4], the close relation between symbolic execution trees and proof trees is helpful for proof comprehension because it facilitates navigation in a proof and understanding the proof situation.

3.6 Incorrectness Logic

Recently, P. O'Hearn proposed *incorrectness logic* [60] as a program logic that does *not* formalise that for all states satisfying the precondition, the postcondition holds in the poststate of the execution. Instead, it formalises that *there exists* a state satisfying the precondition such that after execution the postcondition is satisfied. The intended scenario for this specialised logic is to (automatically) reason during bug finding that a potential bug is not spurious, but indeed reachable by at least one program execution.

Incorrectness logic reuses the notation of Hoare triples (with square brackets instead of curly braces), but a triple in it is interpreted as $[presumption] \, \alpha \, [result]$ indicating that the post-relation *result* can be an *under-approximation* of the states reachable via α, starting from states satisfying the *presumption*. In contrast, postconditions in valid Hoare triples are *over-approximations*.

Dynamic logic is expressive enough to model incorrectness logic if one adds the *converse program* α^- inverting a program α to dynamic logic.[10] This operator is well-studied in dynamic logic [74] and has the semantics that pre- and post-states exchange places: $(\alpha^-)^{\mathcal{M}} = (\alpha^{\mathcal{M}})^{-1}$. The inverted program α^- can be viewed as executing α backwards. The incorrectness triple $[\phi] \, \alpha \, [\psi]$ holds if the dynamic logic formula $\psi \rightarrow \langle \alpha^- \rangle \phi$ is valid.

Interestingly, the *Hoare triple* $\{\phi\} \, \alpha \, \{\psi\}$ corresponds to the validity of the reversed dynamic logic formula $\langle \alpha^- \rangle \phi \rightarrow \psi$. This can be read as: The strongest postcondition of α with respect to the precondition ϕ of the Hoare triple implies the postcondition ψ of the Hoare triple.

3.7 Abstract Interpretation

Abstract interpretation [24] is a static analysis technique based on a lattice model for approximation. The rough idea is to use *abstract domains* to approximate the values of (or relations between) program variables for a given program. One can derive an abstract program semantics from these abstract domains for the target programming language. One can then adapt standard analyses like control- or data-flow analyses or deductive reasoning techniques to the abstract semantics. The intention is that with dedicated abstract domains, one can automatically derive program properties for a given program with sufficient precision to avoid false positives.

In principle, one can define a dynamic logic calculus variant for each abstract program semantics and thus use deductive reasoning in combination with

---

[10] This is already pointed out in the original publication [60].

abstract interpretation. A complementary approach for value abstraction of program variables while reusing the fully precise dynamic logic calculus for the programming language has been developed in [19]. The idea is to use partially interpreted constants that use underspecification to act as representatives for an abstract domain element.

4 Success Story Systems

4.1 The KIV Verification System

Overview. The KIV system[11] [32], in whose design Wolfgang Reif was crucially involved, is a powerful formal verification tool designed to ensure the correctness of software systems.

For specification, KIV combines algebraic data type specifications with different formalisms for describing system behaviour. For the definition of abstract systems, KIV supports Abstract State Machines (ASM) [17]. ASMs consist of abstract programs (rules) that implement the steps of a transition system. Concrete systems to be verified can also be written in Java.

The software development process of KIV is step-wise refinement from abstract specifications to implementations. One can start by verifying an abstract model and then refine it into more concrete implementations, while ensuring that correctness is preserved. KIV supports data refinement as well as system refinement. It offers a graphical user interface for interactive proof development.

Dynamic Logic in KIV. KIV uses higher-order dynamic logic [32], a dynamic logic where the base logic is typed higher-order logic and the modalities can contain abstract imperative programs as well as sequential Java.

The system uses a sequent calculus for its logic that implements wp-style reasoning as well as symbolic execution. It features a powerful simplifier that automatically reduces formulas, and an extensive library of data types with numerous pre-proven theorems.

KIV's Success Stories. KIV was used in various case studies and applications to prove the correctness of real-world systems:

A Prolog compiler that compiles from Prolog to the Warren Abstract Machine was fully verified with KIV using a hierarchy of a dozen refinements [71].

KIV was applied in the area of electronic payment systems to verify a protocol for secure money transfer between Mondex smartcards [40].

KIV was used to verify part of Flashix, a file system for flash memory that had been proposed as a case study for Hoare's Grand Challenge [31]. The file system was decomposed into a hierarchy of components with eleven levels of refinement.

---

[11] www.uni-augsburg.de/de/fakultaet/fai/isse/software/kiv.

4.2   The KeY System: Deductive Verification of Java Programs

Overview. The KeY System[12] [3,4,10,11] is a state-of-the-art program verification tool for one of the most widely used programming languages: Java. Its capabilities enable the formal specification and verification of unmodified industrial Java code at source-code level.

In addition to its role as a program verifier, KeY serves as a versatile research platform for implementing various formal methods for Java using the symbolic execution engine of KeY. For instance, KeY has been used to facilitate the generation of test cases with high code coverage [6] and to implement a symbolic-state debugger [45].

The roots of the KeY project trace back to 1999, when the continuous development and refinement of KeY and its verification methodology were started.

Dynamic Logic in KeY. JavaDL is a dynamic logic, where the programs in the modalities are Java code, i.e., executable fragments of Java programs. It is based on a *typed* first-order logic whose type system includes all Java primitive and reference types that are equipped with Java's typing rules. For example, there is a non-rigid function that returns the length of an array a in the current execution state as $length(a)$. JavaDL also permits Java-style syntax like a.length.

KeY's deductive verification engine is based on a sequent calculus for JavaDL [13]. The calculus rules perform forward symbolic execution whereby all symbolic paths through a program are explored. Method contracts make verification scalable because one can prove one method at a time to be correct relative to its contract. Contracts do not need to be expressed in dynamic logic, but can be given at the source code level as *Java Modeling Language* annotations [54].

KeY's sequent calculus has rules for the features of Java, including operations on data types, heap operations, object creation, inheritance, polymorphism, method invocation, loops, abrupt termination etc. Since a large number of rules is needed, KeY features a domain-specific textual language (called *taclets*) to add axioms of theories and lemmas, and to define proof rules.

KeY allows both semi-automated (user-guided) and automatic (system-driven) verification of program properties. The tool supports modular reasoning, which means it can verify parts of a program in isolation, making the verification process scalable to larger systems.

KeY's Success Stories. Over the years, a plethora of case studies were conducted, where KeY was used to verify real-world algorithms and data structures; a comprehensive list is on the KeY project website.

A verification case study that received much attention is TimSort, an algorithm combining merge and insertion sort. It is prominently used as Java's default for sorting collections of objects. However, that implementation had a bug

---

[12] www.key-project.org.

and crashed for certain large collections. This issue was detected and explained in [28], a fixed version has been presented and verified with KeY in [27].

While the JDK uses TimSort to sort collections of objects, collections of primitive types are sorted using Dual Pivot Quicksort, which is a standard quicksort that partitions into three instead of into two parts. The implementation provided by the JDK has been proven correct in [15], which includes the sortedness property, the permutation property, and the absence of integer overflows.

In [16], the core of the JDK's Identity Hash Map was specified and verified. For that, KeY was used in combination with other JML tools: the bounded model checker JJBMC [12] and OpenJML [23], to exploit the strengths of each of them and jointly verify a large project.

Researchers at CWI showed that Java's LinkedList implementation breaks when lists with more than 2^{31} elements are created [46]. They propose a fixed version and verified it successfully with KeY. This case study shows the capability of KeY to reason about bounded integer data types and handle overflows.

The most recent large case study performed with KeY is the verification of the sorting algorithm in-place super scalar sample sort [14]. This algorithm is efficient on modern machines, as it avoids branch mispredictions, allows high instruction parallelism by reducing data dependencies in the innermost loops, and it is very cache-efficient. This case study shows that with KeY it is possible to verify state-of-the-art sorting algorithms of considerable size (in this case about 900 lines of Java) and complexity *without having to modify the source code*.

4.3   KeYmaera X: A Theorem Prover for Hybrid Systems

Overview. KeYmaera X[13] [34] is a formal verification tool for hybrid systems that combine continuous dynamics (for example, physical processes) and discrete transitions (for example, digital control). Its verification engine is based on differential dynamic logic (d$\mathcal{L}$) [61,63] to model and verify safety properties of these systems. KeYmaera X is particularly useful for verifying cyber-physical systems, such as autonomous vehicles or medical devices.

The tool is built on top of a small, soundness-critical logic kernel, ensuring the reliability of its proofs, and it is extensible for advanced applications, such as differential game logic [62].

KeYmaera X's proof search is partially automated, it offers tight integration with solvers like Z3 and Wolfram Mathematica for handling real arithmetic. Users can guide proof search interactively by manually applying proof rules, selecting tactics, and providing input such as loop invariants.

Dynamic Logic in KeYmaera X. In differential dynamic logic (d$\mathcal{L}$) the modalities contain hybrid programs. These programs model hybrid systems and can describe both continuous changes (using differential equations) and discrete actions (using traditional program constructs like assignments and conditionals). A d$\mathcal{L}$ formula might express that after following the trajectory governed

---

[13] www.keymaerax.org.

by a system of differential equations, a certain safety condition is guaranteed. For example, consider the d$\mathcal{L}$ formula describing a safety property for a car model [34]:

$$v \geq 0 \wedge A > 0 \;\rightarrow\; [((a := A \cup a := 0); \{v' = a\})^*]\, v \geq 0$$

It expresses that a car, when started with non-negative velocity $v \geq 0$ and positive acceleration $A > 0$ (left-hand side of the implication), will always drive forward ($v \geq 0$) after executing $a := A \cup a := 0$ followed by the differential equation $v' = a$ arbitrarily often.

KeYmaera X uses uniform substitution [63] to automatically or interactively prove d$\mathcal{L}$ formulas and, thus, the correctness of hybrid systems. In the uniform substitution framework, variables and formulas can be substituted uniformly within logical rules.

KeYmaera X's Success Stories. Common use cases include autonomous vehicles, aircraft control systems, and robotics.

KeYmaera X has been used to verify collision avoidance systems for autonomous cars. One notable success is the formal verification of the correctness of adaptive cruise control, where KeYmaera X was used to prove that the system would maintain a safe distance from other vehicles under all operating conditions [56]. Another significant application involved automated lane-changing manoeuvres in autonomous cars. The tool verified that under appropriate conditions, the vehicle would change lanes safely without violating safety constraints.

KeYmaera X was involved in the verification of safety properties in robotic systems, especially those operating in dynamic environments. For instance, it has helped ensure that robots navigate safely around obstacles and other moving agents by verifying the correctness of their control algorithms under different scenarios [59].

KeYmaera X has been successfully used in verifying air traffic management systems, particularly the Airborne Collision Avoidance System (ACAS-X), used in airplanes to prevent mid-air collisions [22]. KeYmaera X helped prove that the collision avoidance algorithms work reliably under a wide range of flight scenarios, accounting for the continuous dynamics of aircraft motion and discrete control decisions. Additionally, the tool has been applied in unmanned aerial vehicles (UAVs) for ensuring safe flight path planning.

5   Conclusion

While dynamic logic is well established in theoretical circles as an object of investigation[14], it leads a niche existence in the area of deductive verification: The three tools reported in Sect. 4 represent all the major verification system implementations using dynamic logic we are aware of.

This is a pity, because, as we show in the present paper, dynamic logic is extremely versatile:

---

[14] See, for example, the Dalí workshop series (dblp.org/db/conf/dali/index.html).

- It works with different base logics: Propositional, first-order or higher-order, typed or untyped.
- It can be instantiated to a wide range of modelling and real-world programming languages, including hybrid programs.
- It can express proof obligations deriving from different specification paradigms: Refinement, abstract data types, contract-based.
- It can represent different styles of verification: Symbolic execution (unbounded, bounded, concolic), weakest preconditions, abstract interpretation.
- It can express a wide variety of verification scenarios (see Sect. 3): Relational and hyperproperties, program equivalence, information flow, incorrectness logic.

All of this is possible in the same syntactic (multi-modal) and semantic (relational) framework, without the need of *ad hoc* and *meta* constructs like Hoare quadruples, program composition operators, etc. This makes dynamic logic also a good choice to improve interoperability among verification tools. As witnessed by the success stories mentioned in Sect. 4, the generality and versatility of dynamic logic comes without a performance overhead: Deductive verification tools based on dynamic logic are competitive.

For all these reasons, we believe that dynamic logic deserves to be better known and more widely used than it is. We hope that the present article can serve as an inspiration.

Acknowledgements. This work was supported by the DFG projects BE 2334/9-1, BU 2924/3-1, HA 2617/9-1, and UL 433/3-1, the Helmholtz topic Engineering Secure Systems (KASTEL), the Helmholtz pilot program KiKIT, the ATHENE project "Model-centric Deductive Verification of Smart Contracts', and the Swedish Research Council project "Smart Contract Verification".

References

1. Abrial, J.: Modeling in Event-B - System and Software Engineering. Cambridge Univ. Press (2010). https://doi.org/10.1017/CBO9781139195881
2. Abrial, J., Schuman, S.A., Meyer, B.: Specification language. In: McKeag, R.M., Macnaghten, A.M. (eds.) On the Construction of Programs, pp. 343–410. Cambridge University Press (1980)
3. Ahrendt, W., et al.: The KeY tool. Softw. Syst. Model. **4**(1), 32–54 (2004). https://doi.org/10.1007/s10270-004-0058-x
4. Ahrendt, W., Beckert, B., Bubel, R., Hähnle, R., Schmitt, P.H., Ulbrich, M. (eds.): Deductive Software Verification – The KeY Book: From Theory to Practice. No. 10001 in LNCS, Springer (2016). https://doi.org/10.1007/978-3-319-49812-6
5. Ahrendt, W., de Boer, F.S., Grabe, I.: Abstract object creation in dynamic logic. In: Cavalcanti, A., Dams, D.R. (eds.) FM 2009. LNCS, vol. 5850, pp. 612–627. Springer, Heidelberg (2009). https://doi.org/10.1007/978-3-642-05089-3_39
6. Ahrendt, W., Gladisch, C., Herda, M.: Proof-based test case generation. In: Deductive Software Verification – The KeY Book. LNCS, vol. 10001, pp. 415–451. Springer, Cham (2016). https://doi.org/10.1007/978-3-319-49812-6_12

7. Apt, K.R., Olderog, E.: Fifty years of hoare's logic. Formal Aspects Comput. **31**(6), 751–807 (2019). https://doi.org/10.1007/S00165-019-00501-3
8. Barthe, G., D'Argenio, P.R., Rezk, T.: Secure information flow by self-composition. In: 17th IEEE Computer Security Foundations Workshop, CSFW-17,Pacific Grove, CA, USA, pp. 100–114. IEEE Computer Society (2004)
9. Beckert, B., Bormer, T., Kirsten, M., Neuber, T., Ulbrich, M.: Automated verification for functional and relational properties of voting rules. In: Grandi, U., Rosenschein, J.S. (eds.) Sixth International Workshop on Computational Social Choice (COMSOC 2016), June 2016. https://www.irit.fr/COMSOC-2016/proceedings/BeckertEtAlCOMSOC2016.pdf
10. Beckert, B., et al.: The Java verification tool KeY: A tutorial. In: Platzer, A., Rozier, K.Y., Pradella, M., Rossi, M. (eds.) Formal Methods, 26th International Symposium, FM 2024, pp. 597–623. Springer (2024)
11. Beckert, B., Hähnle, R., Schmitt, P.H. (eds.): Verification of Object-Oriented Software. The KeY Approach – Foreword by K. Rustan M. Leino, No. 4334 in LNCS, Springer (2007)
12. Beckert, B., Kirsten, M., Klamroth, J., Ulbrich, M.: Modular verification of JML contracts using bounded model checking. In: Margaria, T., Steffen, B. (eds.) ISoLA 2020. LNCS, vol. 12476, pp. 60–80. Springer, Cham (2020). https://doi.org/10.1007/978-3-030-61362-4_4
13. Beckert, B., Klebanov, V., Weiß, B.: Dynamic logic for Java. In: Deductive Software Verification – The KeY Book. LNCS, vol. 10001, pp. 49–106. Springer, Cham (2016). https://doi.org/10.1007/978-3-319-49812-6_3
14. Beckert, B., Sanders, P., Ulbrich, M.: Formally verifying an efficient sorter. In: Finkbeiner, B., Kovács, L. (eds.) Tools and Algorithms for the Construction and Analysis of Systems - 30th International Conference on TACAS, Luxembourg City, Luxembourg, LNCS, Springer (2024). https://doi.org/10.1007/978-3-031-57246-3_15
15. Beckert, B., Schiffl, J., Schmitt, P.H., Ulbrich, M.: Proving JDK's dual pivot quicksort correct. In: Paskevich, A., Wies, T. (eds.) VSTTE 2017. LNCS, vol. 10712, pp. 35–48. Springer, Cham (2017). https://doi.org/10.1007/978-3-319-72308-2_3
16. de Boer, M., de Gouw, S., Klamroth, J., Jung, C., Ulbrich, M., Weigl, A.: Formal specification and verification of JDK's identity hash map implementation. In: ter Beek, M.H., Monahan, R. (eds.) Integrated Formal Methods, pp. 45–62. No. 13274 in LNCS, Springer, Cham (2022). https://doi.org/10.1007/978-3-031-07727-2_4
17. Börger, E., Stärk, R.: Abstract state machines: a method for high-level system design and analysis. Springer (2003)
18. Boyer, R.S., Elspas, B., Levitt, K.N.: SELECT–a formal system for testing and debugging programs by symbolic execution. ACM SIGPLAN Notices **10**(6), 234–245 (1975)
19. Bubel, R., Hähnle, R., Weiß, B.: Abstract interpretation of symbolic execution with explicit state updates. In: de Boer, F.S., Bonsangue, M.M., Madelaine, E. (eds.) FMCO 2008. LNCS, vol. 5751, pp. 247–277. Springer, Heidelberg (2009). https://doi.org/10.1007/978-3-642-04167-9_13
20. Burstall, R.M.: Program proving as hand simulation with a little induction. In: Information Processing '74, pp. 308–312. Elsevier/North-Holland (1974)
21. Clarkson, M.R., Schneider, F.B.: Hyperproperties. J. Comput. Secur. **18**(6), 1157–1210 (2010). https://doi.org/10.3233/JCS-2009-0393
22. Cleaveland, R., Mitsch, S., Platzer, A.: Formally verified next-generation airborne collision avoidance games in ACAS X. ACM Trans. Embed. Comput. Syst. **22**(1), 1–30 (2023). https://doi.org/10.1145/3544970

23. Cok, D.R.: OpenJML: JML for Java 7 by extending OpenJDK. In: Bobaru, M., Havelund, K., Holzmann, G.J., Joshi, R. (eds.) NFM 2011. LNCS, vol. 6617, pp. 472–479. Springer, Heidelberg (2011). https://doi.org/10.1007/978-3-642-20398-5_35

24. Cousot, P., Cousot, R.: Abstract interpretation: a unified lattice model for static analysis of programs by construction or approximation of fixpoints. In: Fourth ACM Symposium on Principles of Programming Language, Los Angeles, pp. 238–252. ACM Press, New York, January 1977

25. Darvas, A., Hähnle, R., Sands, D.: A theorem proving approach to analysis of secure information flow. In: Gorrieri, R. (ed.) Workshop on Issues in the Theory of Security, WITS. IFIP WG 1.7, ACM SIGPLAN and GI FoMSESS (2003)

26. Darvas, Á., Hähnle, R., Sands, D.: A theorem proving approach to analysis of secure information flow. In: Hutter, D., Ullmann, M. (eds.) SPC 2005. LNCS, vol. 3450, pp. 193–209. Springer, Heidelberg (2005). https://doi.org/10.1007/978-3-540-32004-3_20

27. De Gouw, S., De Boer, F.S., Bubel, R., Hähnle, R., Rot, J., Steinhöfel, D.: Verifying OpenJDK's sort method for generic collections. J. Autom. Reason. **62**(6) (2019)

28. De Gouw, S., Rot, J., De Boer, F.S., Bubel, R., Hähnle, R.: OpenJDK's java.utils.collection.sort() is broken: the good, the bad and the worst case. In: Kroening, D., Pasareanu, C. (eds.) Proc. 27th International Conference on Computer Aided Verification (CAV), San Francisco, pp. 273–289. No. 9206 in LNCS, Springer, July 2015

29. Dijkstra, E.W.: Guarded commands, non-determinacy and a calculus for the derivation of programs. In: Shooman, M.L., Yeh, R.T. (eds.) Proceedings of the International Conference on Reliable Software 1975, Los Angeles, California, USA, 21–23 April 1975, p. 2. ACM (1975). https://doi.org/10.1145/800027.808417

30. Dijkstra, E.W.: A Discipline of Programming. Prentice-Hall (1976)

31. Ernst, G., Schellhorn G. a nd Haneberg, D., J., P., Reif, W.: Verification of a virtual filesystem switch. In: Proceedings, Verified Software: Theories, Tools, Experiments, LNCS 8164, Springer (2014)

32. Ernst, G., Pfähler, J., Schellhorn, G., Haneberg, D., Reif, W.: KIV: overview and VerifyThis competition. Int. J. Softw. Tools Technol. Transfer **17**(6), 677–694 (2014). https://doi.org/10.1007/s10009-014-0308-3

33. Fischer, M.J., Ladner, R.E.: Propositional dynamic logic of regular programs. J. Comput. Syst. Sci. **18**(2), 194–211 (1979). https://doi.org/10.1016/0022-0000(79)90046-1

34. Fulton, N., Mitsch, S., Quesel, J.-D., Völp, M., Platzer, A.: KeYmaera X: an axiomatic tactical theorem prover for hybrid systems. In: Felty, A.P., Middeldorp, A. (eds.) CADE 2015. LNCS (LNAI), vol. 9195, pp. 527–538. Springer, Cham (2015). https://doi.org/10.1007/978-3-319-21401-6_36

35. Godlin, B., Strichman, O.: Inference rules for proving the equivalence of recursive procedures. In: Manna, Z., Peled, D.A. (eds.) Time for Verification. LNCS, vol. 6200, pp. 167–184. Springer, Heidelberg (2010). https://doi.org/10.1007/978-3-642-13754-9_8

36. Goldblatt, R.: Axiomatising the Logic of Computer Programming. LNCS, vol. 130. Springer, Heidelberg (1982). https://doi.org/10.1007/BFb0022481

37. Hähnle, R.: Dijkstra's legacy on program verification. In: Apt, K.R., Hoare, T. (eds.) Edsger Wybe Dijkstra: His Life, Work, and Legacy, pp. 105–140. ACM/-Morgan & Claypool (2022). https://doi.org/10.1145/3544585.3544593

38. Hähnle, R., Heisel, M., Reif, W., Stephan, W.: An interactive verification system based on dynamic logic. In: Siekmann, J.H. (ed.) CADE 1986. LNCS, vol. 230, pp. 306–315. Springer, Heidelberg (1986). https://doi.org/10.1007/3-540-16780-3_99
39. Hähnle, R., Menzel, W., Schmitt, P.H.: Integrierter Deduktiver Software-Entwurf. Künstl. Intell. **12**(4), 40–41 (1998)
40. Haneberg, D., Moebius, N., Reif, W., Schellhorn, G., Stenzel, K.: Mondex: engineering a provable secure electronic purse. Int. J. of Softw. Inf. **5**(1), 159–184 (2011)
41. Harel, D. (ed.): First-Order Dynamic Logic. LNCS, vol. 68. Springer, Heidelberg (1979). https://doi.org/10.1007/3-540-09237-4
42. Harel, D., Meyer, A.R., Pratt, V.R.: Computability and completeness in logics of programs (preliminary report). In: Hopcroft, J.E., Friedman, E.P., Harrison, M.A. (eds.) Proceedings of 9th Annual ACM Symposium on Theory of Computing, Boulder, Colorado, USA. pp. 261–268. ACM (1977). https://doi.org/10.1145/800105.803416
43. Harel, D., Pratt, V.R.: Nondeterminism in logics of programs. In: Aho, A.V., Zilles, S.N., Szymanski, T.G. (eds.) Conference on Record of the Fifth Annual ACM Symposium on Principles of Programming Languages, Tucson, AZ, USA. pp. 203–213. ACM Press, New York, NY (1978). https://doi.org/10.1145/512760.512782
44. Heisel, M., Reif, W., Stephan, W.: Program verification by symbolic execution and induction. In: Morik, K. (ed.) Proceedings of 11th German Workshop on Artifical Intelligence Informatik Fachberichte, vol. 152. Springer (1987)
45. Hentschel, M., Bubel, R., Hähnle, R.: The symbolic execution debugger (SED): a platform for interactive symbolic execution, debugging, verification and more. STTT **21**(5), 485–513 (2018)
46. Hiep, H.A., Maathuis, O., Bian, J., Boer, F.S.D., van Eekelen, M.C.J.D., Gouw, S.D.: Verifying openJDK's LinkedList using KeY. In: Biere, A., Parker, D. (eds.) Tools and Algorithms for the Construction and Analysis of Systems, 26th International Conference on TACAS, Dublin, Ireland, Part II, pp. 217–234. No. 12079 in LNCS, Springer, Cham (2020)
47. Hoare, C.A.R.: An axiomatic basis for computer programming. Commun. ACM **12**(10), 576–580, 583 (1969)
48. ISO: IEC 13568: 2002: Information technology–Z formal specification notation–Syntax, type system and semantics. Standard, International Organization for Standardization, Geneva, CH (2002)
49. Jónsson, B., Tarski, A.: Boolean algebra with operators I and II. Am. J. Math. 73: 891–939 and 74: 129–62 (1951+1952)
50. King, J.C.: A new approach to program testing. In: Shooman, M.L., Yeh, R.T. (eds.) Proceedings of the International Conference on Reliable Software 1975, Los Angeles, California, USA, 21–23 April 1975, pp. 228–233. ACM (1975). https://doi.org/10.1145/800027.808444,
51. King, J.C.: Symbolic execution and program testing. Commun. ACM **19**(7), 385–394 (1976)
52. Kozen, D., Tiuryn, J.: Logics of programs. In: van Leeuwen, J. (ed.) Formal Models and Semantics, pp. 789–840. Handbook of Theoretical Computer Science, Elsevier, Amsterdam (1990). https://doi.org/10.1016/B978-0-444-88074-1.50019-6
53. Kripke, S.: Semantical considerations on modal logic. Acta Philosophica Fennica **16**, 83–94 (1963)
54. Leavens, G.T., et al.: JML reference manual (May 2013), draft revision 2344
55. Lewis, C.I.: Implication and the algebra of logic. Mind **21**(84), 522–531 (1912). http://www.jstor.org/stable/2249157

56. Lin, Q., Mitsch, S., Platzer, A., Dolan, J.M.: Safe and resilient practical waypoint-following for autonomous vehicles. IEEE Control Syst. Lett. **6**, 1574–1579 (2022). https://doi.org/10.1109/LCSYS.2021.3125717
57. Litvintchouk, S.D., Pratt, V.R.: A proof-checker for dynamic logic. In: Reddy, R. (ed.) Proceedings of 5th International Joint Conference on Artificial Intelligence. Cambridge, MA, USA, pp. 552–558. William Kaufmann, San Mateo CA (1977). http://ijcai.org/Proceedings/77-1/Papers/098.pdf
58. Manna, Z., Pnueli, A.: Axiomatic approach to total correctness of programs. Acta Informatica **3**, 243–263 (1974). https://doi.org/10.1007/BF00288637
59. Mitsch, S., Ghorbal, K., Vogelbacher, D., Platzer, A.: Formal verification of obstacle avoidance and navigation of ground robots. I. J. Robotics Res. **36**(12), 1312–1340 (2017). https://doi.org/10.1177/0278364917733549
60. O'Hearn, P.W.: Incorrectness logic. Proc. ACM Program. Lang. **4**(POPL), 10:1–10:32 (2020). https://doi.org/10.1145/3371078
61. Platzer, A.: Differential dynamic logic for hybrid systems. J. Autom. Reas. **41**(2), 143–189 (2008). https://doi.org/10.1007/s10817-008-9103-8
62. Platzer, A.: Differential game logic. ACM Trans. Comput. Log. **17**(1), 1:1–1:51 (2015). https://doi.org/10.1145/2817824
63. Platzer, A.: A complete uniform substitution calculus for differential dynamic logic. J. Autom. Reason. **59**(2), 219–265 (2016). https://doi.org/10.1007/s10817-016-9385-1
64. Platzer, A.: Logical foundations of cyber-physical systems. Springer (2018). https://doi.org/10.1007/978-3-319-63588-0
65. Pnueli, A.: The temporal logic of programs. In: 18th Annual Symposium on Foundations of Computer Science (1977). https://doi.org/10.1109/SFCS.1977.32
66. Pratt, V.R.: Semantical considerations on Floyd-Hoare logic. In: 17th Annual Symposium on Foundations of Computer Science (1976). https://doi.org/10.1109/SFCS.1976.27
67. Reif, W.: Vollständigkeit einer modifizierten Goldblatt-Logik und Approximation der Omegaregel durch Induktion. Master's thesis, Fakultät für Informatik, Universität Karlsruhe (1984)
68. Reif, W.: Korrektheit von Spezifikationen und generischen Moduln. Ph.D. thesis, Karlsruhe Institute of Technology, Germany (1991). https://d-nb.info/931877865
69. Reif, W.: Correctness of generic modules. In: Nerode, A., Taitslin, M.A. (eds.) Logical Foundations of Computer Science, Second International Symposium Tver, Russia, LNCS, vol. 620, pp. 406–417. Springer (1992). https://doi.org/10.1007/BFB0023857
70. Reif, W.: The KIV-approach to software verification. In: Broy, M., Jähnichen, S. (eds.) KORSO: Methods, Languages, and Tools for the Construction of Correct Software, LNCS, vol. 1009, pp. 339–370. Springer (1995). https://doi.org/10.1007/BFB0015452
71. Schellhorn, G., Ahrendt, W.: The WAM case study: verifying compiler correctness for prolog with KIV. In: Bibel, W., Schmitt, P. (eds.) Automated Deduction: A Basis for Applications. Kluwer Academic Publishers (1998)
72. Schellhorn, G., Ernst, G., Pfähler, J., Haneberg, D., Reif, W.: Development of a verified flash file system. In: Ameur, Y.A., Schewe, K. (eds.) Abstract State Machines, Alloy, B, TLA, VDM, and Z: 4th International Conference on ABZ, Toulouse, France. LNCS, vol. 8477, pp. 9–24. Springer (2014). https://doi.org/10.1007/978-3-662-43652-3

73. van Eijck, J., Stokhof, M.: The gamut of dynamic logic. In: Gabbay, D.M., Woods, J. (eds.) Volume 7: Logic and the Modalities in the Twentieth Century, pp. 499–600. Handbook of the History of Logic, Elsevier, Amsterdam (2006). https://doi. org/10.1016/S1874-5857(06)80033-6

74. Vardi, M.Y.: The taming of converse: reasoning about two-way computations. In: Parikh, R. (ed.) Logic of Programs 1985. LNCS, vol. 193, pp. 413–424. Springer, Heidelberg (1985). https://doi.org/10.1007/3-540-15648-8_31

A Largely Automated Verification
of GHC's Natural Mergesort

Christoph Walther$^{(\boxtimes)}$ ⓘ

Fachbereich Informatik, Technische Universität Darmstadt, Darmstadt, Germany
Christoph.Walther@tu-darmstadt.de

Abstract. We report on a machine supported verification of a sorting algorithm from Haskell's GHC library based on Natural Mergesort. Our work was motivated by previous endeavors with the Dafny system and with Isabelle/HOL, where a skillful elimination of mutual recursion is required to enable verification. We replicate the Isabelle/HOL solution with the VeriFun system and also give a direct proof for verifying correctness and stability of GHC-Mergesort. Based on our experiences when working on the proofs, we suggest a simpler implementation of Natural Mergesort. We conclude by comparing the degree of machine support of the considered systems when creating the proofs.

Keywords: Program Verification · Automated Reasoning · Theorem Proving by Induction · Natural Mergesort · VeriFun

1 Introduction

Formal verification of sorting algorithms dates back to the late 1960s and early 1970s when researchers began to apply mathematical logic and formal methods to the correctness of software, and research continues with the development of proof tools to the present [1, 3, 4, 6]. In this paper, we report on computer assisted proofs for verifying a version of Natural Mergesort found in the GHC library of Haskell. Our work was motivated by previous endeavors with the interactive proof assistant Isabelle/HOL [5] and the Dafny system [2].

The formal proofs presented here are performed with the √eriFun system [9].[1] This system was designed and developed as an easy to learn and easy to use tool for teaching automated reasoning, semantics, verification and similar subjects and has been used in beginner courses about formal methods as well as in practical courses about program verification for about 15 years [11].

The system's object language consists of principles for defining polymorphic data types, procedures operating on them, and for statements (called "lemmas") about the data types and procedures. The language uses Unicode and offers in-, out-, pre- and postfix notation for function symbols so that readability is increased by use of the familiar mathematical notation. Figure 1 displays some

---

[1] Short for "A Verifier for Functional Programs".

ⓒ The Author(s), under exclusive license to Springer Nature Switzerland AG 2025
G. Ernst et al. (Eds.): Wolfgang Reif Festschrift, LNCS 15765, pp. 83–103, 2025.
https://doi.org/10.1007/978-3-031-92196-4_5

```
structure bool            <= true, false
structure ℕ               <= 0, ⁺(⁻ : ℕ)
structure list[@l]        <= ø, [infixr,10] :: (hd : @l, tl : list[@l])
structure rec[@N, @D]  <= [outfix] [ : ](key : @N, data : @D)


function merge(k, l : list[rec[ℕ, @D]]) : list[rec[ℕ,@D]] <=
if k = ø
  then l
  else if l = ø
         then k
         else if key(hd(k)) > key(hd(l))
                 then hd(l) :: merge(k, tl(l))
                 else hd(k) :: merge(tl(k), l)
              end_if
       end_if
end_if
lemma n ⊵ k ∧ n ⊵ l → n ⊵ k <> l <= ∀ n : ℕ, k, l : list[rec[ℕ, @D]]
if{n ⊵ k, if{n ⊵ l, n ⊵ k <> l, true}, true}
```

Fig. 1. Data types, procedures and lemmas in √eriFun

examples: Data type bool and data type ℕ for natural numbers built with the constructors 0 and ⁺(...) for the successor function are the only predefined data types in the system. ⁻(...) is the selector of ⁺(...) thus representing the predecessor function. Data type list is user defined and required for the case studies of this paper. Identifiers preceded by @ denote type variables, and therefore polymorphic lists are defined here. Lists are built with the constructors :: and ø for the empty list. The functions hd and tl (for *head* and *tail*) are the selectors of :: yielding the leftmost list element and the list with the leftmost element removed respectively if applied to a non-empty list.[2] Subsequently we will use lists of *records* defined by the polymorphic data type rec of Fig. 1.

Procedures are defined by *if*- and *case*-conditionals, functional composition and recursion. E.g., procedure merge of Fig. 1 merges two lists of records k and l to an ordered list whenever k and l are ordered. Procedure calls are evaluated eagerly, i.e. call-by-value. Predicates are defined by procedures with result type bool. Procedure function [infix] > (x, y : ℕ) : bool <= ... for deciding the greater-than relation is the only predefined procedure in the system. As > denotes a well-founded relation, termination and induction proofs may be based on >.

Lemmas are defined with conditionals *if* : *bool* × *bool* × *bool* → *bool* as the main connective, but negation ¬ and *case*-conditionals may be used as well. Only universal quantification is allowed for the variables of a lemma. Figure 1 displays a lemma about list concatenation <> and the upper bound ⊵ of (the keys in) a record list which is used in subsequent proofs. The string in the

---

[2] Selectors applied to constructors they do not belong to as e.g. ⁻(0), $hd(ø)$ and $tl(ø)$ remain unevaluated and denote an unknown value of appropriate type.

headline (between "lemma" and "<=") is just an identifier assigning a name to the lemma for reference and must not be confused with the lemma statement which is given as a boolean term in the lemma body. Some basic lemmas about equality (e.g. transitivity) and > (e.g. asymmetry) are predefined in the system. These lemmas are frequently used in almost every case study so that work is eased by having them always available instead of importing them from some library. For easing readability, we will subsequently write lemmas in the usual mathematical notation.

Lemmas are proved with the *HPL*-calculus (abbreviating *Hypotheses, Programs* and *Lemmas*) [11]. The most relevant proof rules of this calculus are *Induction, Use Lemma, Apply Equation, Unfold Procedure, Case Analysis* and *Simplification*. Formulas are given as sequents of form $\langle H, IH \vdash goal \rangle$, where H is a finite set of literals, i.e. negated or unnegated predicate calls and equations called the *hypotheses* of the sequent, *IH* is a finite set of *induction hypotheses* given as possibly quantified boolean terms and *goal* is a boolean term, called the *goalterm* of the sequent. Induction hypotheses are treated like verified lemmas being available only for the sequent they belong to. A sequent $S = \langle H, IH \vdash goal \rangle$ represents the formula $\phi_S := \forall \dots \left[\bigwedge_{h \in H} h \wedge \bigwedge_{ih \in IH} (\forall \dots ih) \right] \rightarrow goal$ with universally quantified non-induction variables in the induction hypotheses, cf. Fig. 9.

The *Induction* rule creates the base and step sequents for the initial sequent $\langle \{\}, \{\} \vdash goal \rangle$ of a lemma with body *goal* from an induction axiom. By choosing *Simplification*, the system's first-order theorem prover, called the *Symbolic Evaluator*, is started for rewriting a sequent's goalterm by using the hypotheses and induction hypotheses of the sequent, the definitions of the data types and procedures and the lemmas already verified. It is guided by heuristics, e.g. for deciding whether to use a procedure definition and for speeding up proof search by filtering out useless lemmas. Equality reasoning is implemented by conditional term rewriting with *ACI*-matching, where the orientation of equations is heuristically determined. The Symbolic Evaluator is a completely automated tool over which the user has no control, thus leaving the *HPL*-proof rules as the only means to guide the system to a proof.

Also the *HPL*-calculus is controlled by heuristics. By applying the *Verify* command to a lemma, the system starts to compute a proof tree by choosing appropriate *HPL*-proof rules heuristically. If a proof attempt gets stuck, the user must step in by applying a proof rule to some leaf of the proof tree (sometimes after pruning some unwanted branch of the tree), and the system then takes over control again. Also it may happen that a further lemma must be formulated by the user before the proof under consideration can be completed.

Upon the definition of a procedure, ✓eriFun's automated termination analysis (based on the method of *Argument-Bounded Functions* [10]) is invoked for generating termination hypotheses which are sufficient for the procedure's termination and are proved like lemmas. If the system fails to compute termination hypotheses, the user must provide an appropriate termination function. Afterwards induction axioms are computed from the terminating procedures' recursion structure to be on stock for future use.

```
function merge_pairs(K : list[list[rec[ℕ, @D]]]) : list[list[rec[ℕ,@D]]] <=
if K = ø
then K
 else if tl(K) = ø then K else merge(hd(K), hd(tl(K))) :: merge_pairs(tl(tl(K))) end_if
end_if

function merge_all(K : list[list[rec[ℕ, @D]]]) : list[rec[ℕ,@D]] <=
if K = ø
then ø
 else if tl(K) = ø then hd(K) else merge_all(merge_pairs(K)) end_if
end_if
```

Fig. 2. Procedures for merging lists of lists (from [8])

Usability of the system is supported by a high degree of automation. ✔eriFun
is implemented in JAVA and installers for running the system under *Windows*,
Unix/Linux and *Mac* are available from the web [9].

2   Natural Mergesort

We illustrate the principle of Natural Mergesort being implemented by *GHC-Mergesort* and compare it with Classical Mergesort before we discuss the proofs
of correctness and stability. In our formalization, the lists to be sorted consist of
records r of form $[n, d]$ as formally defined in Fig. 1, where $n \in \mathbb{N}$ is the *key* of r
and d denotes some data of arbitrary type stored by r. *Correctness* of a sorting
algorithm means that an ordered permutation of an input list is computed, where
a record list k is *ordered* iff $key(a) \leq key(b)$ for all records a preceding record
b in k. The *stability* requirement is satisfied iff all records with identical key
appear in the sorted list in the same order as in the initial one.

Classical Mergesort as well as Natural Mergesort are instances of the *Divide-and-Conquer* paradigm. When sorting a list k with 2 or more elements by Classical Mergesort, the list is recursively halved in the divide pass until singleton
lists are obtained. The recursive calls implicitly create a binary tree where the
nodes consist of lists such that the root node is given by k, each inner node
has two children obtained by halving their parent node and all leaves consist
of singleton lists. Then the tree is traversed bottom up in the conquer pass by
merging all siblings at the leaves of the tree with procedure merge from Fig. 1
and replacement of the common parent by the merged children. In this way the
replaced parents become leaves of the modified tree which in turn are merged
by merge with their sisters. The traversal ends when the root node is replaced
by the merge of its children then holding an ordered permutation of list k.

Natural Mergesort differs from the classical version in that the divide pass
is organized differently. Instead of computing singleton lists to start with the
merging, list k is searched for the longest prefix which is ordered with respect
to either $\leq$ or $>$, where such a prefix is called a *le-sequence* or a *gt-sequence*
respectively. When a sequence of either type is found, it is kept in a list K of

lists. Then a new search is started with the remains of list k, another sequence is inserted into K and so on until k is completely separated into *le-* and *gt* -sequences. As it is demanded that *gt*-sequences are stored in K in reverse order, K consists of ordered lists, and these lists provide the starting point for merging in the conquer pass. This is performed by the Haskell implementation with procedure merge_all of Fig. 2 which calls procedure merge_pairs for a pairwise merging of the lists in K. Having successively merged all lists in K, an ordered permutation of list k is eventually obtained with the final merge.

For instance, record list $k = \langle [2, A], [2, B], [3, C], [1, D], [5, E], [4, F], [2, G], [5, H] \rangle$ is separated into the list of sequences $K = \langle \langle [2, A], [2, B], [3, C] \rangle, \langle [1, D], [5, E] \rangle, \langle [2, G], [4, F] \rangle, \langle [5, H] \rangle \rangle$, and pairwise merging yields the ordered permutation $\langle [1, D], [2, A], [2, B], [2, G], [3, C], [4, F], [5, E], [5, H] \rangle$ of k. As the records of k with identical key appear in the sorted list in the same order as in the initial one, the stability requirement is satisfied as well.

For a list k with $n \geq 2$ elements, the computation of K in the divide pass requires $n - 1$ comparisons and the time for the merges of the lists in K in the conquer pass is proportional to $n \cdot \lceil log_2(|K|) \rceil$. The best case is given if k is a sequence as then $|K| = 1$ and the conquer pass is void. $|K| = \lceil n/2 \rceil$ in the worst case, e.g. for lists like $\langle [1, A], [2, B], [1, C], [2, D], \ldots \rangle$, so that the effort for the conquer pass then is proportional to $n \cdot (\lceil log_2(n) \rceil - 1)$. Classical Mergesort needs $n - 1$ halving steps in the divide pass and an effort for merging proportional to $n \cdot \lceil log_2(n) \rceil$ in the conquer pass. Hence Natural Mergesort always outperforms Classical Mergesort, saving nothing upon divide and n steps at least upon conquer.

3   A Replication of The Isabelle/HOL Proof

A Haskell implementation of Natural Mergesort from the GHC library is considered in [8], and proofs for correctness and stability of this algorithm obtained with the interactive proof assistant Isabelle/HOL are presented. To this effect, the Haskell code is translated into Isabelle notation as displayed in Fig. 3.[3] Procedure sequences computes the list K of sequences from a list k as illustrated in Sect. 2. If k has two elements at least, i.e. $k = a \# b \# xs$, the keys of the first two elements a and b of k are compared to decide whether to start with the search for a *gt-* or a *le*-sequence.[4] This search is implemented by the procedures desc and asc (for *descending* and *ascending*) which are called by sequences.

---

[3] Haskell's lazy evaluation is irrelevant for verification here, because the Haskell code for Natural Mergesort also terminates with eager evaluation. Consequently the computed result is independent of the used evaluation regime.

[4] # is used in Isabelle/HOL for the list constructor.

```
sequences key (a # b # xs) =
  (if gt key a b then desc key b [a] xs else asc key b (op # a) xs)
sequences key [] = [[]]
sequences key [v] = [[v]]

desc key a as (b # bs) =
  (if gt key a b then desc key b (a # as) bs else (a # as) # sequences key (b # bs))
desc key a as [] = (a # as) # sequences key []

asc key a as (b # bs) =
  (if ¬gt key a b then asc key b (λ x. as (a # x)) bs else as [a] # sequences key (b # bs))
asc key a as [] = as [a] # sequences key []
```

Fig. 3. Formalization of sequences, desc and asc in Isabelle/HOL (from [8])

Procedure desc tries to extend a gt-sequence as by exploring list $b \# bs$, where a is a strict lower bound of as.[5] If a is greater than b, the gt-sequence as is extended by a and a further extension is tried by a recursive call to desc, now with b as a strict lower bound of $a \# as$. Otherwise the gt-sequence $a \# as$ cannot be extended further and therefore sequences is called to decide how to continue. The requirement of storing gt-sequences in reverse order is satisfied in desc by inserting a in front of as.

Similarly procedure asc tries to extend a le-sequence with (not necessarily strict) upper bound a by exploring list $b \# bs$. However, different to procedure desc parameter as is a function (parameter) here. The motivation for doing so is to extend a le-sequence with minimal effort. A naive implementation would extend a le-sequence as with upper bound a by placing a at the end of as for keeping the le-sequence property. However, the costs for computing a le-sequence with length n in this way are proportional to n^2, whereas the given implementation cares for costs proportional to n. If a is greater than b, then le-sequence as [a] cannot be extended further and therefore sequences is called to decide how to continue. Otherwise the le-sequence is extended by a and a further extension is tried by a recursive call to asc, now with b as an upper bound of the extended le-sequence. Having computed the list of all sequences, these sequences are merged with procedure merge_all of Fig. 2 so that an implementation of *GHC-Mergesort* is obtained in our notation as

function sort(k:list[rec[ℕ, @D]]):list[rec[ℕ,@D]] $<$ = merge_all(sequences(k)) .

But instead to prove correctness and stability directly for procedure sort, a modified version of the sorting algorithm is used for verification: Two further procedures take_chain and drop_chain are introduced in [8] such that a call of $take\_chain(a, gt, k)$ computes a gt-sequence as a prefix of list k where a is

---

[5] We call some record a a lower bound of record list k and say that a is greater than record b if $key(a)$ is a lower bound of the list of keys in k or $key(a)$ is greater than $key(b)$ respectively.

```
function sequences(k : list[rec[ℕ, @D]]) : list[list[rec[ℕ,@D]]] <=
if k = ø
 then k :: ø
 else if tl(k) = ø
       then k :: ø
       else let a := hd(k), b := hd(tl(k)) in
             if key(a) > key(b)
              then (rev( take_chain(b, gt, tl(tl(k)))) <> b :: a :: ø)
                     :: sequences(drop_chain(b, gt, tl(tl(k))))
              else (a :: b :: take_chain(b, le, tl(tl(k))))
                    :: sequences(drop_chain(b, le, tl(tl(k))))
             end_if
            end_let
      end_if
end_if
```

Fig. 4. Modified version of procedure sequences

a strict upper bound of this sequence, and similarly $take_chain(a, le, k)$ computes a le-sequence of k with a as a lower bound of this sequence. By calling $drop_chain(a, op, k)$ with $op \in \{gt, le\}$ the remains of list k after eliminating the sequence computed by $take_chain(a, op, k)$ are obtained. Now two lemmas given in our notation as

$$\begin{aligned}&\forall k, l{:}list[rec[\mathbb{N}, @D]], b{:}rec[\mathbb{N}, @D] \quad desc(b, l, k) = \\ &\quad (rev(take_chain(b, gt, k)) <> b :: l) :: sequences(drop_chain(b, gt, k))\end{aligned} \tag{1}$$

$$\begin{aligned}&\forall k, l{:}list[rec[\mathbb{N}, @D]], b{:}rec[\mathbb{N}, @D] \quad asc(b, l, k) = \\ &\quad (rev(l) <> b :: take_chain(b, le, k)) :: sequences(drop_chain(b, le, k))\end{aligned} \tag{2}$$

are proven in [8] (where list reversal is implemented by some procedure rev) which justify the replacement of procedure call $desc(b, a :: \emptyset, tl(tl(k)))$ in sequences by

$$\begin{aligned}&(rev(take_chain(b, gt, tl(tl(k)))) <> b :: a :: \emptyset) \\ &:: sequences(drop_chain(b, gt, tl(tl(k))))\end{aligned}$$

and the replacement of procedure call $asc(b, a :: \emptyset, tl(tl(k)))$ by

$$\begin{aligned}&(a :: b :: take_chain(b, le, tl(tl(k)))) \\ &:: sequences(drop_chain(b, le, tl(tl(k))))\end{aligned}$$

yielding the modified version of sequences displayed in Fig. 4.

This indirect approach for proving properties of sort is motivated in [8] by the impracticality of the induction axiom computed by Isabelle/HOL from the original definition of sequences. By modifying sequences, one gets rid of the mutual recursion between sequences, desc and asc so that the induction axiom thus obtained eases verification considerably [8]. As it is proved with lemmas (1) and

(2) that the original procedure and its modified version compute the same function, the modified version of sequences can be used for proving that procedure sort is correct and stable.

However, it seems strange at first glance that the proof technical problems of Isabelle/HOL to cope with the mutual recursion between sequences, desc and asc given in Fig. 3 when trying to prove properties of sort do not show up when proving (1) and (2). The answer is that the definition of sequences is only needed in the proofs of the induction bases (i.e. when k is the empty or a singleton list) as all other calls of $sequences(drop\_chain(...))$ in the induction steps can be replaced by the induction hypotheses.

As ✓eriFun's syntax does not allow mutual recursion, we cannot prove lemmas (1) and (2) directly. Instead we use an incompletely defined procedure

$$\text{function } \mathsf{seq}(\mathsf{k}: \mathsf{list}[\mathsf{rec}[\mathbb{N}, @\mathsf{D}]]): \mathsf{list}[\mathsf{list}[\mathsf{rec}[\mathbb{N}, @\mathsf{D}]]] <=$$
$$\text{if } \mathsf{k} = \emptyset \text{ then } \mathsf{k} :: \emptyset \text{ else if } \mathsf{tl}(\mathsf{k}) = \emptyset \text{ then } \mathsf{k} :: \emptyset \text{ end\_if end\_if}$$

for replacing sequences by seq in the lemmas (1) and (2) as well as in procedures desc and asc. The system then automatically proves both lemmas thus verifying that both versions of sequences compute the same function.[6] Consequently we also may use sort defined with sequences from Fig. 4 for proving correctness and stability of sort defined with the original version of sequences. Our definition directly corresponds to the modification from [8], except that the function parameters in sequences and asc must be replaced appropriately because our system does not support higher-order functions. But our implementation of asc cares for computation of le-sequences in linear time also here, because list reversal can be performed with linear effort.

Sorting To prove the sorting property

$$\boxed{\forall k : list[rec[\mathbb{N}, @D]]\ ordered(sort(k))} \tag{3}$$

for sort, the lemmas

$$\forall k, l : list[rec[\mathbb{N}, @D]]\ ordered(k) \wedge ordered(l) \rightarrow ordered(merge(k, l)) \tag{4}$$

$$\forall K : list[list[rec[\mathbb{N}, @D]]]\ ORDERED(K) \rightarrow ORDERED(merge\_pairs(K)) \tag{5}$$

$$\forall K : list[list[rec[\mathbb{N}, @D]]]\ ORDERED(K) \rightarrow ordered(merge\_all(K)) \tag{6}$$

$$\forall k : list[rec[\mathbb{N}, @D]], a : rec[\mathbb{N}, @D]\ key(a) \trianglerighteq take\_chain(a, gt, k) \tag{7}$$

$$\forall k : list[rec[\mathbb{N}, @D]], a : rec[\mathbb{N}, @D]\ ordered(rev(take\_chain(a, gt, k))) \tag{8}$$

$$\forall k : list[rec[\mathbb{N}, @D]], a : rec[\mathbb{N}, @D]\ ordered(take\_chain(a, le, k)) \tag{9}$$

---

[6] Procedure seq is underspecified as no result is stipulated if $k \neq \emptyset$ and $tl(k) \neq \emptyset$ hold for the input list k. Consequently each true statement about seq (not referring to definedness)—as (1) and (2) in the present case—holds for all procedures P : list[rec[$\mathbb{N}$, @D]] $\rightarrow$ list[list[rec[$\mathbb{N}$, @D]]] satisfying $P(\emptyset)=\emptyset :: \emptyset$ and $P(r :: \emptyset) = (r :: \emptyset) :: \emptyset$. Therefore (1) and (2) hold for sequences in particular

$$\forall k\mathpunct{:}list[rec[\mathbb{N}, @D]] \ ORDERED(sequences(k)) \qquad (10)$$

are required, where $ORDERED(K)$ decides whether a list K of lists consists of ordered lists only. For proving (3), the lemmas (6) and (10) are needed. The proof of (6) requires lemma (5) which is proved in turn with lemma (4). The proof of (10) uses the lemmas (7), (8) and (9). All lemmas and the main statement (3) are automatically proven by ✔eriFun, where induction is needed except that first-order reasoning is enough for (3).

Permutation. The permutation property of sort is proven in [8] by showing that $sort(k)$ equals k if both lists are considered as multisets. A similar proof is obtained by proving that each record r appears exactly as often in list $sort(k)$ as in list k, which is expressed by lemma

$$\boxed{\forall k\mathpunct{:}list[rec[\mathbb{N}, @D]], r\mathpunct{:}rec[\mathbb{N}, @D] \ r\#sort(k) = r\#k \ .} \qquad (11)$$

For proving this statement, the lemmas

$$\forall k\mathpunct{:}list[rec[\mathbb{N}, @D]], r, a\mathpunct{:}rec[\mathbb{N}, @D], op\mathpunct{:}cmp$$
$$r\#take\_chain(a, op, k) + r\#drop\_chain(a, op, k) = r\#k \qquad (12)$$

$$\forall k, l\mathpunct{:}list[rec[\mathbb{N}, @D]], r\mathpunct{:}rec[\mathbb{N}, @D] \ r\#merge(k, l) = r\#k + r\#l \qquad (13)$$

$$\forall K\mathpunct{:}list[list[rec[\mathbb{N}, @D]]], r\mathpunct{:}rec[\mathbb{N}, @D] \ r\#\#merge\_pairs(K) = r\#\#K \qquad (14)$$

$$\forall K\mathpunct{:}list[list[rec[\mathbb{N}, @D]]], r\mathpunct{:}rec[\mathbb{N}, @D] \ r\#merge\_all(K) = r\#\#K \qquad (15)$$

$$\forall k\mathpunct{:}list[rec[\mathbb{N}, @D]], r\mathpunct{:}rec[\mathbb{N}, @D] \ r\#\#sequences(k) = r\#k \qquad (16)$$

are required, where $r\#\#K$ counts the number of all occurrences of a record r in the lists of a list K of lists. For proving (11), the lemmas (15) and (16) are needed. Lemma (16) requires lemma (12), and (15) is proven with lemma (14) which in turn uses lemma (13). Apart from lemma (15), ✔eriFun's heuristic orients all equations in the above lemmas left-to-right thus enabling the automatic induction proofs of all lemmas, except for the main statement (11). The proof of (11) succeeds by first-order reasoning after the system is instructed interactively to apply the equation of lemma (15).

Stability Like for the use of Isabelle/HOL, the proof of stability necessitates more work in the amount of lemmas to be spotted as well as in the number of required user interactions. Stability of sort is expressed by lemma

$$\boxed{\forall k\mathpunct{:}list[rec[\mathbb{N}, @D]], n\mathpunct{:}\mathbb{N} \ get(n, sort(k)) = get(n, k)} \qquad (17)$$

where $get(n, k)$ eliminates all records with a key different from n from list k thus guaranteeing that the elements in list $get(n, k)$ appear in the same order as in k. The stability proof demands the lemmas

$$\forall k, l\mathpunct{:}list[rec[\mathbb{N}, @D]], n\mathpunct{:}\mathbb{N} \ get(n, k <> l) = get(n, k) <> get(n, l) \qquad (18)$$

$$\forall k{:}list[rec[\mathbb{N}, @D]], n{:}\mathbb{N}, a{:}rec[\mathbb{N}, @D], op{:}cmp$$
$$get(n, take\_chain(a, op, k)) <> get(n, drop\_chain(a, op, k)) = get(n, k) \tag{19}$$

$$\forall k{:}list[rec[\mathbb{N}, @D]], n{:}\mathbb{N}, a{:}rec[\mathbb{N}, @D]$$
$$n \geq key(a) \rightarrow get(n, drop\_chain(a, gt, k)) = get(n, k) \tag{20}$$

$$\forall k, l{:}list[rec[\mathbb{N}, @D]], n{:}\mathbb{N}, a{:}rec[\mathbb{N}, @D]$$
$$n \geq key(a) \rightarrow get(n, take\_chain(a, gt, k) <> l) = get(n, l) \tag{21}$$

$$\forall k, l{:}list[rec[\mathbb{N}, @D]], n{:}\mathbb{N}, a{:}rec[\mathbb{N}, @D]$$
$$get(n, rev(take\_chain(a, gt, k)) <> l) = get(n, take\_chain(a, gt, k) <> l) \tag{22}$$

$$\forall k{:}list[rec[\mathbb{N}, @D]], n{:}\mathbb{N}$$
$$k \neq \emptyset \wedge n < key(hd(k)) \wedge ordered(k) \rightarrow get(n, k) = \emptyset \tag{23}$$

$$\forall k, l{:}list[rec[\mathbb{N}, @D]], n{:}\mathbb{N}$$
$$ordered(k) \rightarrow get(n, merge(k, l)) = get(n, k) <> get(n, l) \tag{24}$$

$$\forall K{:}list[list[rec[\mathbb{N}, @D]]], n{:}\mathbb{N}$$
$$ORDERED(K) \rightarrow GET(n, merge\_pairs(K)) = GET(n, K) \tag{25}$$

$$\forall K{:}list[list[rec[\mathbb{N}, @D]]], n{:}\mathbb{N}$$
$$ORDERED(K) \rightarrow get(n, merge\_all(K)) = GET(n, K) \tag{26}$$

$$\forall k{:}list[rec[\mathbb{N}, @D]], a, b{:}rec[\mathbb{N}, @D], n{:}\mathbb{N}$$
$$n \neq key(a) \wedge n \neq key(b) \rightarrow get(n, k <> a :: b :: \emptyset) = get(n, k) \tag{27}$$

$$\forall k{:}list[rec[\mathbb{N}, @D]], n{:}\mathbb{N} \ GET(n, sequences(k)) = get(n, k) \tag{28}$$

where $GET(n, K)$ eliminates all records with a key different from n in the lists of a list K of lists. Figure 5 displays the dependencies between the lemmas when proving the stability statement (17).

Except for (18) and (26), ✓eriFun's heuristic orients all equations in the above lemmas left-to-right thus significantly supporting the proofs also here. All proofs are by induction except that the main statement (17) is proved by first-order reasoning only. The system has to be instructed interactively to apply the equation of lemma (18) when proving (22), to use an induction hypothesis and lemma (23) in the proof of (24), and to use (26) for the proof of (17). The remaining lemmas are proved automatically.

4   A Direct Verification of GHC-Mergesort

Correctness and stability of *GHC-Mergesort* can also be verified directly with ✓eriFun, i.e. without the detour via the procedures take_chain and drop_chain. We have to eliminate mutual recursion also here because our system's object language does not allow mutual recursive definitions.

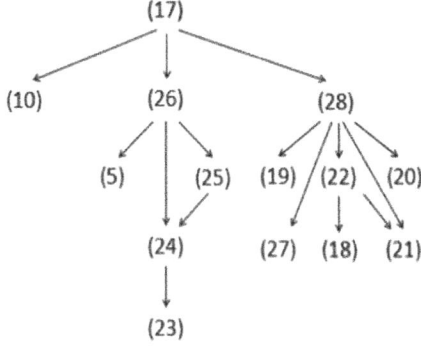

Fig. 5. Lemma dependencies when proving stability of sort.

Mutual recursion can be eliminated by joining the code of two mutual recursively defined procedures f and g into one procedure $f \oplus g$ with an additional flag parameter stipulating whether f or g is meant upon a call of $f \oplus g$. However, we do not get rid of the proof technical obstacles coming with mutual recursion like in the approach of Sect. 3. Joining procedures is only a workaround for implementing mutual recursive procedures in a programming language not allowing mutual recursive definitions, however without easing verification. This means that the mutual dependencies still are present and therefore the proof technical problems coming with mutual recursion persist. In particular, for verifying some property ϕ_f of procedure f, a corresponding property ϕ_g has to be formulated for g so that $\phi_f \wedge \phi_g$ can be proved by induction. This is required because by the mutual recursion the induction hypothesis stemming from ϕ_g is needed to prove the induction conclusion for ϕ_f (and vice versa). This problem still exists if f and g are merged into one procedure as can be observed from the lemmas (31), (32) and (34) formulated subsequently.

Any number of procedures can be joined into a common one, of course, and we do so with the procedures sequences, desc and asc of Fig. 3 yielding procedure call displayed in Fig. 6 as the translation of the Isabelle/HOL definitions.

User support is needed for proving termination of procedure call because $desc(\ldots, k)$ and $asc(\ldots, k)$ call $sequences(k)$ so that an automatic termination proof fails as parameter k remains unchanged. Therefore we have to provide the system with a pair

$$\lambda k{:}list[rec[\mathbb{N}, @D]]).\,|\,k\,|\,,\ \lambda f{:}flag.\ case\ f\ of\ seq:0, other:1\,end\_case$$

of termination functions, where $|\ldots|$ computes the length of a list. This instructs ✓eriFun to generate termination hypotheses for each call of seq, desc and asc

```
function call(f: flag, a: rec[N, @D], l, k: list[rec[N, @D]]): list[list[rec[N,@D]]] <=
case f of
  seq : if k = ∅
          then k :: ∅
          else if tl(k) = ∅
                  then k :: ∅
                  else let a' := hd(k), b := hd(tl(k)) in
                          if key(a') > key(b)
                            then call(desc, b, a' :: ∅, tl(tl(k))) else call(asc, b, a' :: ∅, tl(tl(k)))
                          end_if
                        end_let
                end_if
        end_if,
  desc : if k = ∅
          then (a :: l) :: ∅
          else let b := hd(k) in
                  if key(a) > key(b)
                    then call(desc, b, a :: l, tl(k))  else (a :: l) :: call(seq, b, ∅, k)
                  end_if
                end_let
        end_if,
  asc : if k = ∅
          then rev(a :: l) :: ∅
          else let b := hd(k) in
                  if key(a) > key(b)
                    then rev(a :: l) :: call(seq, b, ∅, k) else call(asc, b, a :: l, tl(k))
                  end_if
                end_let
        end_if
end_case
```

Fig. 6. Procedure call implementing procedures sequences, desc and asc.

based on the lexicographic combination of the termination functions which the system then automatically proves.[7]

Sorting. For verifying the sorting property (3) for procedure

$$
\text{function sort}(k: \text{list}[rec[N, @D]]): \text{list}[rec[N, @D]]) <=
$$
$$
\text{merge\_all}(\text{call}(seq, hd(k), ∅, k))
$$

we command √eriFun to compute a proof. However, the system immediately gets stuck with proof obligation $ordered(merge\_all(call(seq, hd(k), ∅, k)))$ which is proven with lemma (6) if

$$\forall k: list[rec[N, @D]] \ ORDERED(call(seq, hd(k), ∅, k)) \tag{29}$$

---

[7] We have replaced the needless recursive calls of *sequences*.(∅) in the base cases of desc and asc of Fig. 3 (being introduced by the translation presented in [8]) with ∅ in procedure call. This modification has no effects on any of the proofs. The formal parameters a and l of procedure call are not considered if $f = seq$, but are required by the signature of procedures desc and asc.

can be verified. When trying to prove this subgoal by induction, the system gets stuck in the step case, this time with proof obligation

$$k \neq \text{\o} \wedge tl(k) \neq \text{\o} \rightarrow$$
$$(key(hd(k)) > key(hd(tl(k))))$$
$$\rightarrow ORDERED(call(desc, hd(tl(k)), hd(k) :: \text{\o}, tl(tl(k)))))$$
$$\wedge (key(hd(k)) \leq key(hd(tl(k))))$$
$$\rightarrow ORDERED(call(asc, hd(tl(k)), hd(k) :: \text{\o}, tl(tl(k)))))$$

which is straightforwardly generalized to

$$\forall k : list[rec[\mathbb{N}, @D]], a, b : rec[\mathbb{N}, @D]$$
$$(key(a) > key(b) \rightarrow ORDERED(call(desc, b, a :: \text{\o}, k))) \tag{30}$$
$$\wedge (key(a) \leq key(b) \rightarrow ORDERED(call(asc, b, a :: \text{\o}, k))) \ .$$

But an induction proof of (30) will not succeed as only $a :: \text{\o}$ appears in the induction hypotheses which does not match $b :: a :: \text{\o}$ coming with the recursive calls $call(desc, hd(k), b :: a :: \text{\o}, tl(k))$ and $call(asc, hd(k), b :: a :: \text{\o}, tl(k))$ in the evaluated induction conclusion.

This is a well-known problem in formal verification: Parameter l of procedures desc and asc is called an *accumulator* because it stores intermediate values and holds the final result when the computation has finished. Methods for generalization necessitated by the presence of accumulators correspond to the invention of loop invariants when verifying imperative programs. The solution is to replace a term like $a :: \text{\o}$ yielding an unusable induction hypothesis by a fresh variable satisfying certain properties which are *(i)* as specific as necessary to preserve truth of the generalization, but are *(ii)* as general as possible so that an induction proof succeeds. Spotting properties satisfying *(i)* and *(ii)* frequently require a "eureka moment" thus challenging the creativity of the human theorem prover.

Based on the analysis of the previous sections, a solution is not that difficult to find in the present case: We have to demand that parameter l of desc holds an *ordered* list with *strict lower bound* $key(b)$ when computing a *gt*-sequence, and that the reversal of l with *upper bound* $key(b)$ is *ordered* upon the search of a *le*-sequence by asc. We have to claim in addition that (29) holds as this provides the induction hypothesis needed for proving the induction conclusions for desc and asc. And, of course, (29) is the statement required for proving the sorting property, whereas the statements about desc and asc are only needed to make the induction work. Thus we formulate the accumulator generalization

$$\forall k, l : list[rec[\mathbb{N}, @D]], b : rec[\mathbb{N}, @D]$$
$$ORDERED(call(seq, b, l, k)) \wedge (l \neq \text{\o} \rightarrow$$
$$(key(hd(l)) > key(b) \wedge ordered(l) \rightarrow ORDERED(call(desc, b, l, k)))$$
$$\wedge (key(hd(l)) \leq key(b) \wedge ordered(rev(l)) \rightarrow ORDERED(call(asc, b, l, k))))$$
$$\tag{31}$$

of (29) and (30) which ✔eriFun then automatically proves by induction corresponding to the recursion structure of procedure call, cf. Fig. 9.[8] Given

---

[8] The (strict) lower and the upper bound of l need not be stated explicitly in (31) because $key(b) < key(hd(l)) \wedge ordered(l)$ entails $key(b) \lhd l$ and $key(b) \geq key(hd(l)) \wedge ordered(rev(l))$ entails $key(b) \unrhd rev(l)$.

$$if\{k = \emptyset,$$
$$\quad true,$$
$$\quad if\{tl(k) = \emptyset,$$
$$\qquad true,$$
$$\quad if\{key(hd(k)) > key(hd(tl(k))),$$
$$\qquad if\{n = key(hd(k)),$$
$$\qquad\quad GET(n, call(desc, hd(tl(k)), hd(k) :: \emptyset, tl(tl(k)))) = hd(k) :: get(n, tl(tl(k))),$$
$$\qquad\quad if\{n = key(hd(tl(k))),$$
$$\qquad\qquad GET(n, call(desc, hd(tl(k)), hd(k) :: \emptyset, tl(tl(k))))$$
$$\qquad\qquad\quad = hd(tl(k)) :: get(n, tl(tl(k))),$$
$$\qquad\qquad GET(n, call(desc, hd(tl(k)), hd(k) :: \emptyset, tl(tl(k)))) = get(n, tl(tl(k)))\}\},$$
$$\qquad if\{n = key(hd(k)),$$
$$\qquad\quad if\{n = key(hd(tl(k))),$$
$$\qquad\qquad GET(n, call(asc, hd(tl(k)), hd(k) :: \emptyset, tl(tl(k))))$$
$$\qquad\qquad\quad = hd(k) :: hd(tl(k)) :: get(n, tl(tl(k))),$$
$$\qquad\qquad GET(n, call(asc, hd(tl(k)), hd(k) :: \emptyset, tl(tl(k))))$$
$$\qquad\qquad\quad = hd(k) :: get(n, tl(tl(k)))\},$$
$$\qquad\quad if\{n = key(hd(tl(k))),$$
$$\qquad\qquad GET(n, call(asc, hd(tl(k)), hd(k) :: \emptyset, tl(tl(k))))$$
$$\qquad\qquad\quad = hd(tl(k)) :: get(n, tl(tl(k))),$$
$$\qquad\qquad GET(n, call(asc, hd(tl(k)), hd(k) :: \emptyset, tl(tl(k)))) = get(n, tl(tl(k)))\}\}\}$$

Fig. 7. Proof obligation obtained when trying to prove lemma (33)

(31) and lemma (6), the main statement (3) has an automatic (first-order) proof too.

Permutation. For proving the permutation property (11) for sort, we formulate the lemma

$$\forall k, l{:}list[rec[\mathbb{N}, @D]], a, r{:}rec[\mathbb{N}, @D], f{:}flag$$
$$f = seq \rightarrow r\#\#call(f, a, l, k) = r\#k \tag{32}$$
$$\wedge\, f \neq seq \rightarrow r\#\#call(f, a, l, k) = r\#a :: (l <> k) \,.$$

The system orients both equations in the lemma left-to-right so that an automated proof by induction upon the recursion structure of procedure call is obtained. The proof of (11) succeeds with the use of (32) after instructing the system to use lemma (15).

Stability. When attempting to prove statement (17), the proof gets stuck with proof obligation $get(n, merge\_all(call(seq, hd(k), \emptyset, k))) = get(n, k)$ which we can prove with lemma (26) if

$$\forall k{:}list[rec[\mathbb{N}, @D]], n{:}\mathbb{N}\ GET(n, call(seq, hd(k), \emptyset, k)) = get(n, k) \tag{33}$$

can be verified. When trying to prove this subgoal by induction, the system gets stuck in the step case, this time with the proof obligation displayed in Fig. 7 (presented in the system's notation for saving space).

It is obvious that accumulator generalization is required also here, however a bit more complex than in the sorting case. We formulate

$$
\begin{aligned}
&\forall k, l{:}list[rec[\mathbb{N}, @D]], b{:}rec[\mathbb{N}, @D], n{:}\mathbb{N} \\
&GET(n, call(seq, b, l, k)) = get(n, k) \wedge (l \neq \o \rightarrow \\
&\quad (key(hd(l)) > key(b) \wedge ordered(l) \\
&\quad\quad \rightarrow GET(n, call(desc, b, l, k)) = get(n, b :: l) <> get(n, k)) \\
&\wedge (key(hd(l)) \leq key(b) \wedge ordered(rev(l)) \\
&\quad\quad \rightarrow GET(n, call(asc, b, l, k)) = get(n, rev(b :: l)) <> get(n, k)))
\end{aligned}
\tag{34}
$$

and command ✓eriFun to compute a proof. The system uses induction corresponding to the recursion structure of procedure call and succeeds after being instructed to use an induction hypothesis in three of the step cases and lemma (23) in one of them. When computing the proofs, ✓eriFun also uses lemma

$$
\begin{aligned}
&\forall k, l, m{:}list[rec[\mathbb{N}, @D]], n{:}\mathbb{N} \\
&\quad get(n, k <> l) <> m = get(n, k) <> get(n, l) <> m
\end{aligned}
\tag{35}
$$

which is oriented left-to-right and proved automatically by structural induction. The proof of the stability statement (17) succeeds with the lemmas (31) and (34) by first-order reasoning, where we command the system to use lemma (26).

5   A Simplification of GHC-Mergesort

Intricate programming may hamper verification. In the present case, two simple modifications of the original Haskell code ease reasoning, however without changing the algorithm idea and its underlying mathematics.

First we eliminate an unhandy use of recursion in the original definition: We provide procedure sequences with a second formal parameter $l{:}list[rec[\mathbb{N}, @D]]$ initially set to the singleton list containing the first element of a non-empty record list k to be sorted. We also eliminate the formal parameter a of desc and asc and replace all references to a in the procedure bodies by references to the first element of record list l. These changes result in procedure call* of Fig. 8, and lemma

$$
\begin{aligned}
&\forall k, l{:}list[rec[\mathbb{N}, @D]], a{:}rec[\mathbb{N}, @D], f{:}flag \\
&\quad f = seq \rightarrow call(f, a, \o, a :: k) = call^*(f, a :: \o, k) \\
&\wedge f \neq seq \rightarrow call(f, a, l, k) = call^*(f, a :: l, k)
\end{aligned}
\tag{36}
$$

(which the system easily proves by structural induction upon k) states the relationship between procedure call and its simplification call*. As no sequence can be computed by call* for an initially empty record list k, procedure

```
function sort*(k : list[rec[ℕ, @D]]) : list[rec[ℕ, @D]]) <=
  if k = ø then ø else merge_all(call*(seq, hd(k) :: ø, tl(k))) end_if
```

sorts the empty list directly instead of merging a list of lists containing the empty list only. Since procedure call* only uses structural list recursion, the termination

```
function call*(f: flag, l, k: list[rec[N, @D]]) : list[list[rec[N, @D]]] <=
if k = ∅
then case f of asc : rev(l) :: ∅, other : l :: ∅ end_case
else if l ≠ ∅
      then let a := hd(l); b := hd(k) in
           if key(a) > key(b)
              then case f of
                     asc : rev(l) :: call*(seq, b :: ∅, tl(k)), other : call*(desc, b :: l, tl(k))
                     end_case
                else case f of
                     desc : l :: call*(seq, b :: ∅, tl(k)), other : call*(asc, b :: l, tl(k))
                     end_case
              end_if
           end_let
      end_if
end_if
```

Fig. 8. Simplified implementation of sequences, desc and asc by procedure call*.

proof is trivial (and automatic of course). Therefore the induction axiom computed by ✔eriFun for a statement $\forall f{:}flag, k{:}list[rec[N, @D]], a{:}rec[N, @D]\ \phi(f, a, k)$ from the recursion structure of procedure call* is structural list induction only, hence much simpler than the induction axiom computed from the recursion structure of procedure call, cf. Fig. 9. Using the induction axiom computed from call*, ✔eriFun proves correctness and stability of sort* as illustrated in Sect. 4. Theorem proving is eased because less user interactions are required for the step case when proving the stability statement (17) as compared to the direct proof of Sect. 4 where the call-induction must be used.

The implementations of sequences, desc and asc by our procedures call and call* differ from the original procedures of Fig. 3 as we do not have function parameters in our definitions. However, these function parameters are irrelevant when justifying termination and consequently for the corresponding induction axioms computed from call and call* in turn, so that these induction axioms would be also successful in presence of function arguments.

A second simplification of the original Haskell code is obtained by replacing all calls of procedure merge_all with calls of procedure

```
function merge_lists(K : list[list[rec[N, @D]]]) : list[rec[N,@D]] <=
if K = ∅ then ∅ else merge(hd(K), merge_lists(tl(K))) end_if
```

in the procedure definitions and in all lemmas referring to merge_all. This eases work as procedure merge_pairs becomes needless and consequently the lemmas (5), (14) and (25) need not to be spotted by the user. For the proofs of Sect. 3 even more can be saved because procedure ## is not needed here. Therefore lemmas (15) and (16) become needless as well, and lemma (11) now has an automatic proof.

$\forall k{:}list[rec[\mathbb{N}, @D]], a{:}rec[\mathbb{N}, @D] \quad k = \emptyset \to \phi(seq, a, k)$
$\wedge\ \forall k{:}list[rec[\mathbb{N}, @D]], a{:}rec[\mathbb{N}, @D] \quad k \neq \emptyset \wedge tl(k) = \emptyset \to \phi(seq, a, k)$
$\wedge\ \forall k{:}list[rec[\mathbb{N}, @D]], a{:}rec[\mathbb{N}, @D] \quad k \neq \emptyset \wedge tl(k) \neq \emptyset$
$\qquad \wedge\ \forall a'{:}rec[\mathbb{N}, @D]\ \phi(desc, a', tl(tl(k))) \wedge \forall a'{:}rec[\mathbb{N}, @D]\ \phi(asc, a', tl(tl(k)))$
$\qquad \to \phi(seq, a, k)$
$\wedge\ \forall k{:}list[rec[\mathbb{N}, @D]], a{:}rec[\mathbb{N}, @D] \quad k = \emptyset \to \phi(desc, a, k)$
$\wedge\ \forall k{:}list[rec[\mathbb{N}, @D]], a{:}rec[\mathbb{N}, @D] \quad k \neq \emptyset \wedge tl(k) = \emptyset \to \phi(desc, a, k)$
$\wedge\ \forall k{:}list[rec[\mathbb{N}, @D]], a{:}rec[\mathbb{N}, @D] \quad k \neq \emptyset \wedge tl(k) \neq \emptyset$
$\qquad \wedge\ \forall a'{:}rec[\mathbb{N}, @D]\ \phi(desc, a', tl(tl(k))) \wedge \forall a'{:}rec[\mathbb{N}, @D]\ \phi(asc, a', tl(tl(k)))$
$\qquad \wedge\ \forall a'{:}rec[\mathbb{N}, @D]\ \phi(desc, a', tl(k)) \to \phi(desc, a, k)$
$\wedge\ \forall k{:}list[rec[\mathbb{N}, @D]], a{:}rec[\mathbb{N}, @D] \quad k = \emptyset \to \phi(asc, a, k)$
$\wedge\ \forall k{:}list[rec[\mathbb{N}, @D]], a{:}rec[\mathbb{N}, @D] \quad k \neq \emptyset \wedge tl(k) = \emptyset \to \phi(asc, a, k)$
$\wedge\ \forall k{:}list[rec[\mathbb{N}, @D]], a{:}rec[\mathbb{N}, @D] \quad k \neq \emptyset \wedge tl(k) \neq \emptyset$
$\qquad \wedge\ \forall a'{:}rec[\mathbb{N}, @D]\ \phi(desc, a', tl(tl(k))) \wedge \forall a'{:}rec[\mathbb{N}, @D]\ \phi(asc, a', tl(tl(k)))$
$\qquad \wedge\ \forall a'{:}rec[\mathbb{N}, @D]\ \phi(asc, a', tl(k)) \to \phi(asc, a, k)$
$\to \forall f{:}flag, k{:}list[rec[\mathbb{N}, @D]], a{:}rec[\mathbb{N}, @D]\ \phi(f, a, k)$

$\forall f{:}flag, k{:}list[rec[\mathbb{N}, @D]], a{:}rec[\mathbb{N}, @D] \quad k = \emptyset \to \phi(f, a, k)$
$\wedge\ \forall f{:}flag, k{:}list[rec[\mathbb{N}, @D]], a{:}rec[\mathbb{N}, @D] \quad k \neq \emptyset$
$\qquad \wedge\ \forall f'{:}flag, a'{:}rec[\mathbb{N}, @D]\ \phi(f', a', tl(k)) \to \phi(f, a, k)$
$\to \forall f{:}flag, k{:}list[rec[\mathbb{N}, @D]], a{:}rec[\mathbb{N}, @D]\ \phi(f, a, k)$

Fig. 9. Induction axioms obtained from procedures call and call*

6   Proof Statistics

Table 1 displays the effort required for proving correctness and stability of the 3 versions of *GHC-Mergesort* by ✓eriFun as well as some statistics of the proofs obtained with the Dafny system and with Isabelle/HOL.[9] Column *Proc.* displays the number of user defined procedures (with the number of user provided termination functions in parentheses), *Lem.* is the number of user defined lemmas (with the number of those requiring user support in parentheses), and *Rules* counts the total number of *HPL*-proof rule applications, separated into user invoked (*User*) and system initiated (*Sys.*) ones. Column *%* gives the ratio between *Sys.* and *Rules* thus measuring the quality of the system's heuristic for choosing the right *HPL*-rule. The number of first-order proof steps and runtime (in seconds) of the Symbolic Evaluator are displayed under *Steps* and *Sec.*[10]

Row *Simplified (VeriFun)* displays the values for the simplified implementation of Sect. 5, row *Direct (VeriFun)* gives the proof statistics of the direct proof from Sect. 4 and row *Modified (VeriFun)* displays the values obtained with ✓eriFun upon the replication of the Isabelle/HOL proof as illustrated in Sect. 3.

---

[9] Apart from the presented lemmas, lemmas like in Fig. 1 and basic lemmas like associativity of append <>, associativity and commutativity of addition + etc. have been imported from a proof library and are counted in the proof statistics as well.

[10] Time refers to running ✓eriFun 3.5 under *Windows 7 Enterprise* with an INTEL Core i7-2640M 2.80 GHz CPU using JAVA 1.8.0_311.

Table 1. Proof statistics for *GHC-Mergesort*

	Proc.	*Lem.*	*Rules*	*User*	*Sys.*	*%*	*Steps*	*Sec.*
Simplified (VeriFun)	13 (-)	23 (6)	110	8	102	92,7	4380	10
Direct (VeriFun)	15 (1)	27 (5)	180	9	171	95,0	8081	41
Direct (Dafny)	17 (3)	31 (11)	–	50	–	–	–	25
Modified (VeriFun)	20 (-)	39 (4)	173	5	168	97,1	3823	5
Modified (Isa/HOL)	27 (2)	55 (48)	–	198	–	–	–	–

As the statistics reveal, the simplified version requires least user support in the definition of procedures and in the formulation of required lemmas. The modified version is most costly as further procedures are defined which requires to spot additional lemmas, but the need to suggest proof steps is lower than for the other solutions. The effort for the direct approach is somewhat in the middle as more lemmas are required than in the simplified version but less than in the modified one. Although the direct approach necessitates the formulation of a termination function and 4 more proof rule suggestions than needed for the replicated proof, this proof is more elementary as no additional procedures have to be defined and the invented lemmas are directly related to the original definition.

Except for procedure call, √eriFun proved termination of all procedures automatically, where merge of Fig. 1 , merge_pairs and merge_all of Fig. 2 and sequences of Fig. 4 are the only procedures not defined by structural recursion. Each required induction axiom had been computed by the system, and the right induction axiom as well as the variables to induct upon had been chosen automatically. *Use Lemma* and *Apply Equation* were the only *HPL*-proof rules needed when user interaction was required.

An assertional proof for verification of the original sorting algorithm with the *Dafny* system [2] is presented in [7], where an imperative version of sequences is used. The system's object language is a mixture of imperative, object-oriented and functional programming. Program statements are given (mainly) by contracts of pre/postconditions for the procedures as well as by lemmas. Verification is supported by including assertions, invariants and hints for proving termination into the program code. Detailed proof steps (called *calculations*) may be inserted into the code as well. Dafny must be supported for proving termination similar as explained in Sect. 4. Additionally, the system needs user guidance for proving termination of merge_pairs and merge_all. Instead of proving the permutation property directly, a stronger result is proven by showing that stability entails multiset equality. The required generalizations of lemmas (31) and (34) are stipulated in form of *requires* and *ensures* assertions for the procedures.

Table 2. Proof statistics for a collection of sorting algorithms

	Proc.	Lem.	Rules	User	Sys.	%	Steps	Sec.
Insertion Sort	4 (-)	4 (-)	13	-	13	100,0	320	<1
Bubble Sort	6 (-)	4 (-)	26	-	26	100,0	1497	5
Selection Sort	5 (-)	10 (-)	49	-	49	100,0	1145	8
Minimum Sort	5 (-)	6 (-)	26	-	26	100,0	1105	6
Quicksort	9 (-)	16 (-)	63	-	63	100,0	775	<1
Mergesort	7 (-)	6 (-)	36	-	36	100,0	990	1
Heapsort	16 (-)	29 (8)	174	16	158	90,8	15495	58
Simplified (VeriFun)*	11 (-)	17 (3)	77	3	74	96,1	1885	2
Direct (VeriFun)*	13 (1)	19 (1)	108	2	106	98,1	4538	10
Modified (VeriFun)*	18 (-)	26 (1)	107	1	106	99,1	2278	2

Different to the use of our procedure *GET*, lists of lists are flattened for proving stability which necessitates some additional lemmas and user support for proving them. Verification with Dafny required more than 5 times user assistance as needed for the ✓eriFun proofs, cf. rows *Direct (VeriFun)* and *Direct (Dafny)* in Table 1.[11]

Row *Modified (Isa/HOL)* gives the statistics of the verification published in [8]. We count 48 lemmas in the Isabelle/HOL proof (plus 7 lemmas which are used but defined elsewhere), none of which had been proven automatically. In fact, all proofs are hand crafted (except for proofs of subgoals which had been handed over to some first-order reasoner) resulting in a large *User* count. However, 36 lemmas required only little user support, mostly by stipulating an induction axiom, the variables to induct upon and then calling some first-order theorem prover for verification of the base and step cases. The remaining 12 lemmas needed some more work for the user to design a proof. Two lemmas had to be formulated to guide Isabelle/HOL for proving termination of sequences from Fig. 4 and for procedure merge_all respectively. User support was given in addition by annotating most of the lemmas with hints how to be used in subsequent proofs. Automatic support is weak as the presented solution requires more procedure and lemma definitions and the need for the user to guide the

---

[11] The numbers are obtained from the tables in [7], where we count all functions, predicates, methods and function methods for the value of *Proc.*, all lemmas as well as all *ensure* contracts stipulated for the methods for the value of *Lem.*, and all lemma calls, asserts and calculation commands for the value of *User*. Time is given in [7] without reference to machine details.

system to a proof is increased by a factor of almost *40* as compared to the ✓eriFun proofs.[12]

We also compare verification of the three versions of *GHC-Mergesort* with proofs of the ordering and permutation property for a collection of classical sorting algorithms obtained with ✓eriFun, see the upper part of Table 2. Termination is proved automatically for all procedures in the collection. No user support is required as well when proving the lemmas of the collection, except for *Heapsort*: Whereas the balanced trees are implicitly obtained with the recursive calls of the sorting procedures in *Quicksort* and *Mergesort*, creation and modification of the heap is explicitly programmed in *Heapsort*. This necessitates far more procedures and lemmas than needed for the other sorting algorithms, and consequently the amount of user support and system resources is much higher. The lower part of Fig. 2 show the proof statistics for the three versions of *GHC-Mergesort* without stability (indicated by *) to enable a comparison with the proofs from the collection. As the statistics reveal, the three versions of *GHC-Mergesort** require significantly less user and machine effort than necessitated for *Heapsort* but much more than for the other sorting algorithms in the collection. The statistics also show that much of the user and machine effort for *GHC-Mergesort* is necessitated by verifying the stability property.

The files with the proofs computed by ✓eriFun for the sorting algorithms from Tables 1 and 2 are available from [9].

Acknowledgments. We are grateful to Christian Sternagel who patiently answered several questions about his proofs and the use of Isabelle/HOL. We also thank Rustan M. Leino for explaining certain features of Dafny and for some clarifications of his proofs.

References

1. Beckert, B., Sanders, P., Ulbrich, M., Wiesler, J., Witt, S.: Formally verifying an efficient sorter. In: Finkbeiner, B., Kovács, L. (eds.) TACAS 2024. LNCS, vol. 14570, pp. 268–287. Springer, Cham (2024). https://doi.org/10.1007/978-3-031-57246-3_15
2. Dafny. https://dafny.org/

---

[12] The numbers are obtained from the proof script found at https://www.isa-afp. org/browser_info/current/AFP/Efficient-Mergesort/Efficient_Sort.html. The number given under *Proc.* refers to functions which are algorithmically defined in the proof script plus those which are used in the script but defined elsewhere. The number in parentheses is given under the assumption that no user support is required for proving termination of the elsewhere defined procedures. The number given under *Lem.* refers to lemmata which are defined in the proof script plus 7 lemmas which are used in the script but defined elsewhere. The number in parentheses is given under the assumption that no user support is required for proving the elsewhere defined lemmas. The number given under *User* refers to commands for proving the lemmas given in the proof script plus the number of hints how a lemma should be used.

3. de Gouw, S., de Boer, F.S., Bubel, R., Hähnle, R., Rot, J., Steinhöfel, D.: Verifying open JDK's sort method for generic collections. J. Autom. Reason. **62**(1), 93–126 (2019). https://doi.org/10.1007/s10817-017-9426-4

4. Dramnesc, I., Jebelean, T., Stratulat, S.: Certification of sorting algorithms using THEOREMA and COQ. In: Watt, S.M., Ida, T. (eds.) SCSS 2024. LNCS, vol. 14991, pp. 38–56. Springer, Cham (2024). https://doi.org/10.1007/978-3-031-69042-6_3

5. Isabelle/HOL. https://isabelle.in.tum.de

6. Lammich, P.: Efficient verified implementation of introsort and pdqsort. In: Peltier, N., Sofronie-Stokkermans, V. (eds.) IJCAR 2020. LNCS (LNAI), vol. 12167, pp. 307–323. Springer, Cham (2020). https://doi.org/10.1007/978-3-030-51054-1_18

7. Leino, K.R.M., Lucio, P.: An assertional proof of the stability and correctness of natural mergesort. ACM Trans. Comput. Log. **17**(1), 6:1–6:22 (2015). https://doi.org/10.1145/2814571

8. Sternagel, C.: Proof Pearl - a mechanized proof of GHC's mergesort. J. Autom. Reason. **51**(4), 357–370 (2013). https://doi.org/10.1007/s10817-012-9260-7

9. VeriFun. http://www.verifun.de

10. Walther, C.: On proving the termination of algorithms by machine. Artif. Intell. **71**(1), 101–157 (1994). https://doi.org/10.1016/0004-3702(94)90063-9

11. Walther, C., Schweitzer, S.: Verification in the classroom. J. Autom. Reason. **32**(1), 35–73 (2004). https://doi.org/10.1023/B:JARS.0000021872.64036.41

Software Engineering and Society Engineering for Self-organising Multi-agent Systems

Jeremy Pitt$^{(\boxtimes)}$(iD), Matthew Scott(iD), Mikayel Suvaryan, Ana Dimoska,
Asimina Mertzani(iD), and Ciske Smit(iD)

Department of Electrical and Electronic Engineering, Imperial College, London, UK
j.pitt@imperial.ac.uk

Abstract. To address the need for deliberation, decentralisation and democratisation in the next generation of cyber-physical systems, we converge the paradigms of self-organisation and multi-agent systems. This demands education in two dimensions for student engineers learning to design, implement and operationalise such systems. The first dimension is education in software engineering for a new style of system: dynamic, interactive, unpredictable, prone to error, self re-configurable, etc. The second dimension is education in 'society engineering', including elements of political science, psychology and economics. This covers an eclectic range of subjects including governance styles and structures; institutions and other social arrangements; deliberative assemblies and social choice; social influence and social networking; and collective action, civic engagement, and socially-constructed conceptual resources. This article reports on the design and delivery of an advanced, and inter-disciplinary, software engineering course in *Self-Organising Multi-Agent Systems*. This offers graduate computer scientists and electronic engineers a theoretical and practical grounding in both educational dimensions.

1 Introduction

The Digital Transformation is producing a demand for large-scale cyber-physical systems, in which uncertainty and dynamism preclude traditional network provisioning at design-time, while complexity of decision-making demands intelligence, but its speed and frequency exclude operator intervention at run-time. Instead, autonomous computational processes have to collaborate in order to satisfice both individual and collective goals, especially in the face of unexpected errors, competing values, conflict, complexity and scale.

This puts a new application emphasis on deliberative self-governance and reflective self-improvement, in order to resolve repeated social problems involving collective action and even cooperative survival, with imperfect information, internal competition, no centralised authority and no obvious terminating condition. A theoretically well-grounded and practically tractable approach to these problems can be found in the intersection of two sub-fields of Artificial Intelligence (AI): self-organisation and multi-agent systems.

G. Ernst et al. (Eds.): Wolfgang Reif Festschrift, LNCS 15765, pp. 104–131, 2025.
https://doi.org/10.1007/978-3-031-92196-4_6

The challenge then is to provide students with the theoretical knowledge, technical skills and practical experience to design, implement and experiment with a self-organising multi-agent system, laying the foundations for specifying, implementing and operationalising complex cyber-physical systems. Meeting this challenge, though, requires education in two dimensions: firstly, education in software engineering for a new style of system: dynamic, interactive, unpredictable, prone to error, self re-configurable, etc.; and secondly, education in 'society engineering', including elements of political science (e.g. governance styles and structures, institutions and other social arrangements, voting, deliberation, etc.), psychology (e.g. social influence, social networking, etc.) and economics (e.g. collective action, socially-constructed conceptual resources, etc.).

This article describes how, through a taught course and a (relatively) large-scale group project, often much larger than anything encountered in their education to date, students learn about these critical concepts in both software engineering and 'society engineering' for self-organising multi-agent systems. Moreover, through a process of self-reflection, students also gain a better understanding of themselves, the benefits of civic participation, the societal implications of AI, and an insight into the issues of operationalisation of future socio-technical systems.

Supporting the taught course, this article also proposes an architecture for a modular, customisable, multi-agent framework called the *base platform*. This platform serves to reduce the 'barrier to entry' that aspiring multi-agent system developers may face when engineering such systems. The use of an existing framework helps both in the preliminary planning phase, and in the subsequent programming phase. During planning, this framework reduces the number of prerequisite decisions, such as the choice of programming language and design paradigm. Furthermore, in the programming phase, a framework helps with structuring any boilerplate code, and provides a set of modules and components that can be used and extended to produce more complex systems.

2   Background and Motivation

The motivation for a taught course that emphasises both software and society engineering is based on a series of observations. The first observation is that, in the process of the "Digital Transformation", there is a significant challenge to engineer ever more complex cyber-physical systems to support and enhance the full spectrum of human activities and capabilities. Secondly, the system components will be autonomous and heterogeneous, but they will need to cooperate to achieve collective goals, but may have conflicting individual goals. Thirdly, system operation will be too fast, frequent and entangled for either human operator intervention, or some other centralised authority directing actions, so the components will have to make decisions for themselves, at run time, and exhibit some kind of computational intelligence in doing so. Finally, the components will need to make these decisions, and voluntarily agree to regulate their behaviour, with respect to a set of conventional rules or *social arrangements*: however, being conventional rather than physical these arrangements are mutable, mutually-agreed, and potentially breakable.

In the context of these four observations, we propose to converge and study *self-organising multi-agent systems* (SOMAS) in order to address problems of *self-governance* that inevitably arise when a group of autonomous individuals have to 'live' and 'work' together, whether in a community of people or, as proposed here, in an *artificial society*. The requirement for computational intelligence encapsulating and maintaining its own state and being 'responsible' for its own decisions and actions is the motivation for *autonomous agents*. Their need to communicate and coordinate is the motivation for *multi-agent systems*. And the requirement for determining, for and by themselves, the selection, modification and enforcement of the social arrangements is the motivation for *self-organisation*. Within this paradigm, this course needs to focus on issues for overcoming specific problems of *self-governance*.

For example, in *ad hoc* or sensor networks, each node in the network has a finite set of resources (battery power, bandwidth, CPU time, buffer memory, etc.). The goal of each node (or rather, the agent at each node) is to get its packets transmitted across the network; however each node is also dependent on the other nodes using their resources to achieve its goal. The 'optimal' strategy would then appear to be free riding: a node should transmit its own packets and dump everyone else's. But if every node adopts this strategy, then no packets at all get transmitted. This is a classic problem of *collective action* for managing a *common-pool resource*: there is an incentive to maximise individual self-interest in the short-term which can deplete (destroy) both the resource and relational cooperation in the long term, which is in nobody's interest or benefit.

Self-governance is one approach to solving such a problem, which involves defining a set of rules i.e., social arrangements (also called an *institution*). The agents mutually agree to comply voluntarily for the benefit of everybody. However, several issues need to be considered in evaluating the 'quality' of the social arrangements, for example:

- Are the social arrangements *sustainable*, that is, are the arrangements necessary and sufficient to maintain the common-pool resource and maintain the continued participation of the agents (even as new agents enter the system)?
- Are the social arrangements *fair*, that is, do the arrangements enable the benefits of the common-pool resources distributed to each participating agent to its own satisfaction, given that there are many different possible definitions of 'fairness', and different agents may have different definitions? and
- Are the social arrangements *legitimate*, that is, do the arrangements have the meaningful consent of the participating agents, in the sense that they were consulted about, and understood the consequences of, agreeing to them?

In answering these questions (from a 'Western' perspective), it might be intuitively supposed that 'democratic' self-governance is the appropriate approach, but this actually raises are a number of further issues concerning the application of the rules, including:

- knowledge management: in an open system in which each agent only has partial knowledge of the overall system, how is that knowledge made available for socially-productive purposes, enabling a diverse collective of individuals to make 'correct' decisions and act upon those decisions effectively;
- majority will vs. epistemic competence: there is a question of how to balance majority preference and expert judgement, so that from a community of individuals expertise can emerge, without being suppressed by the 'ignorance of the crowd';
- tolerance of dissent: which mechanisms are in place that enable critics to expose inconsistencies between core values and current practices, without being excluded or punished for criticism;
- constitutional choice and the 'dilemma of the rules': on the one hand, social arrangements need to be sufficiently unrestricted to allow 'freedom of (collective) action'; but on the other hand, they need to be sufficiently restricted to resist the 'iron law of oligarchy', i.e. (the entropic tendency for human self-governance to be reduced to some form of 'tyranny');
- the 'art' of compromise, and leveraging compromise: on the one hand, consensus formation is the ideal, but is potentially impractical; on the other hand, plurality (majority vote) is efficient and effective but exposes the risk of majoritarian tyranny (a completely different form of political regime).

In the SOMAS course, the primary focus is the transposition of theories for addressing these problems, as developed in the social sciences (philosophy, psychology and political science), into the specification of data structures and algorithms for representation, reasoning and learning, for solving similar problems as they might arise in artificial societies.

Therefore, the educational rationale for the inter-disciplinary SOMAS course is to introduce students to the theories, concepts and technologies in the fields of Multi-Agent Systems and Self-Organising Systems; but also to learn about the intersection of these fields with the social sciences. Given the increasing prevalence of these new cyber-physical systems in the Digital Society, which are at the forefront of research, innovation, and application, the knowledge and understanding of this three-fold intersection will contribute significantly to developing well-educated, well-informed and well-trained graduate students with deep competence in critical, timely and relevant skills. Moreover, with a better understanding of the automation of self-governance in artificial societies, students may also get a better understanding of how to organise social arrangements in their personal, professional and political lives.

As mentioned in Sect. 1, the primary motivation for supplying a supporting platform is to reduce the 'barrier to entry' that students may face in engineering a MAS for the first time. This is achieved both by making decisions for them in the planning phase, and offering pre-built software components for use in the programming phase. A similar approach is seen in front-end web development where users can structure front-end applications with the paradigm provided by the *React* toolkit, and customise these applications with pre-built components from its compatible frameworks such as *Material UI* (MUI). Furthermore,

Python's *Robot Operating System* (ROS) library also shows a similar methodology. Both frameworks use component-based modularity for helping structure builds: this platform offers pre-built components and modules that users can compose, similarly to how ROS supplies nodes and topics to allow independent components to communicate. These frameworks also share support for encapsulation and reusability: this platform offers a class-based composition structure, similar to the way that ROS provides packages and nodes.

3   Structure and Process

The Self-Organising Multi-Agent Systems (SOMAS) course, as taught at Imperial College London, consists of two components: organised lectures and a self-organised practical. This section looks at each component in turn, and then summarises the learning objectives.

3.1   Organised Lectures

There are ten two-hour lectures, divided essentially into three parts. The first part provides foundations, covering multi-agent systems and self-organisation. The second part addresses strategic interaction, covering game theory, social choice theory, Ostrom's theory of self-governing institutions, [18], and dispute resolution. The third part addresses social interaction, covering justice, social construction, knowledge management and self-governance. The detailed lecture content is as follows.

L1 & L2: SOMAS Foundations. This pair of lectures presents motivating examples of collective action and cooperative survival, from actual examples such as *ad hoc* networks and SmartGrids to experimental testbeds, and discusses the foundations of rule-based self-organisation in norm-governed multi-agent systems. It surveys some key concepts like institutions, institutionalised power, and social construction, which will be used as fundamental building blocks for defining algorithms for self-organisation, and also defines a methodology for formalising theories of social science in computational logic.

L3 & L4: Game Theory and Social Choice Theory. This pair of lectures presents two foundational theories of strategic interaction: game theory and social choice theory. Game theory can be used to analyse any situation in which two or more entities (*players*, i.e., agents) each have to select one action from a set of possible actions; best known perhaps through the Prisoner's Dilemma, which has two players and a set of two possible actions (cooperate or defect). Social choice theory can be used to analyse any situation in which two or more entities (*voters*) have to convert a set of individually expressed preferences over a set of candidates into a collective choice. It presents a number of paradoxes in collective preference selection, from Condorcet's Jury Theorem to Arrow's Impossibility Theorem via Simpson's and Anscombe's paradoxes.

L5: Alternative Dispute Resolution. This lecture focuses on two particular features of open systems, namely the existence of conflict and the expectation of errors, and considers a range of techniques to deal with them, including negotiation, mediation, arbitration and litigation, including a protocol, and institutionalised formalisation of the principles, for trial by jury.

L6: Institutions and Sustainability: This lecture shows how Elinor Ostrom's Nobel Prize-winning economic theory of self-governing institutions for sustainable common-pool resource management can be formalised as an executable specification. A group of autonomous agents can use this specification in a self-governing electronic institution, to ensure that a common-pool resource can be sustained. In particular, it demonstrates how Ostrom's institutional rules can be complemented by institutionalised power as a central concept in the algorithmic specification of protocols for selecting, modifying and enforcing rules.

L7: Resource Allocation and Distributive Justice. This lecture shows how Nicholas Rescher's philosophical theory of distributive justice, which proposes that there are seven canonical ways for allocating rewards, some of which are relevant in a specific context and are called legitimate claims, can be formalised as an executable specification. A group of autonomous agents can use this specification in a self-governing electronic institution, to resolve the problem of 'fairness' in resource allocation. Critically, the agents can introspect on the outcome of the allocation and vote to change the priorities assigned to the agreed legitimate claims.

L8: Artificial Social Construction. This lecture effectively extends the notion of institutionalised power to other forms of socially-constructed conceptual resource and socially-informed decision-making. Therefore, formal models for representing and reasoning about trust, forgiveness, other forms of social capital (such as 'favours'), and values are developed.

L9: Knowledge Management and Interactional Justice. This lecture shows how political scientist Josiah Ober's insights from classical Athenian democracy, in particular its knowledge management processes which enabled Athens to out-perform other city states, can be used to formalise a process of interactional justice. Based on a formal analysis of social networks and social networking, an algorithm is given for mapping each agent's individual subjective perception of its treatment by an institution into a collective objective evaluation, as an input to possible rule re-organisation or reformation.

L10: Self-Governance and Legitimacy. This lecture shows how key principles of Josiah Ober's theory of Basic Democracy, in particular equal participation in self-governance activities and a preference for the avoidance of tyranny, can be used in a multi-agent system to avoid the 'entropic tendency' to backslide from democracy to some form of autocracy. The analytic simulation inspires the specification of eight principles of Democracy-by-Design for the development of cyber-physical systems using mutable rules for self-organisation.

3.2   Self-organised Practical

There are two key principles underpinning the self-organised practical exercise. Firstly, that the students should apply the principles studied in the lectures to the design and implementation of a self-organising multi-agent system to address a Model Assignment (see the next section). Secondly, that the students should, in effect, externalise the same principles in the self-organisation of themselves in the design and implementation of said system.

This translates to the following features:

- This is a full class exercise: everyone taking the course is involved in the same project. The project is defined by the Model Assignment;
- The class is, however, divided into teams: the teams are self-formed and self-assigned. This is to introduce voluntary association, as a feature of the Model Assignment, although there is prior knowledge available to Undergraduates, that generally is not available to M.Sc. students;
- The Model Assignment is deliberately under-specified: this demands that as well as intra-team working, the students also have to engage in inter-team working, in order to define the common elements of the project, e.g. agent communication language, rules of the "game(s)", environment, platform services, APIs, etc.;
- The team members have to decide amongst and by themselves who is assigned to which tasks, whether this is collective tasks, or team-specific tasks involving the design and implementation of an agent strategy for the game.
- In the first years of running the course, the students also had to decide for themselves which programming language to use. While this always provoked a 'lively' debate, it proved more cohesive, less distracting, and less exclusive, to pre-empt that debate and 'even the playing field' by providing a re-configurable and extensible Base Platform (see Sect. 5).

Although a wide range of assessment methods are possible, the assessment methods applied in practice has involved a presentation, demonstration and a written report. There are three components to the marking:

- Instructor assessment: evaluation of each team's achievement and contribution according to their project management, technical competence and material quality. Each member of the group is awarded this mark;
- Peer assessment (absolute): each team ranks all the other teams according to their subjective judgement of collective contribution. A Borda point score is calculated according to the rankings, which are converted into a normal distribution of marks. Each member is awarded this mark as well: a provisional mark is given by the sum of the instructor assessment mark and the peer assessment mark;
- Peer assessment (relative): each team member ranks each other team member according to their subjective judgement of individual contribution. This ranking is used to modify the provisional mark to produce the final mark.

Note that in the relative peer assessment, accommodation is made for both compensation and exceptional contribution. For the former, somebody doing more "in the normal scheme of things" to compensate for somebody else doing less moderates the provisional mark up (or down, respectively). Somebody being identified for "going the extra mile" for unique insight moderates their mark up[1].

---

[1] At the time of writing, no-one has been so disruptive that their "exceptional contribution" has caused them to be marked down.

3.3   Learning Objectives

The specific learning objectives from the course are that, on expected participation and successful completion of this module, students should be able to:

- Analyse situations of coordination and strategic interaction in distributed systems of autonomous components (agents);
- Account for basic concepts of computational agency, self-organisation, and the social construction of conceptual resources;
- Design and operationalise systems which are regulated by mutually-agreed and mutable conventional rules;
- Specify and implement algorithms for strategic decision-making related to action selection, preference selection, dispute resolution, opinion formation, and collective action; and
- Create self-organised solutions to "social" problems in multi-agent systems, such as sustainability, fairness, knowledge management and legitimate self-governance.

These learning objectives are aligned with five of the six major categories of Bloom's Taxonomy (knowledge, comprehension, application, analysis and synthesis). The sixth, evaluation, is implicit in the reflection essay, and is explicit in the experimental design that is a key feature of the Model Assignments, as presented in the next section.

4   Model Assignments

This section presents some Model Assignments, including the *Archipelago* scenario and the *Megabike* scenario, which exhibit particular features of scalability and sustainability (survivability). These assignments have all been used by different cohorts of students to produce a self-organising multi-agent system that uses social concepts covered in the lectures to address the inter-dependent collective action situations demanded by the assignment.

Four variants have already been proposed (see below). Instructors are encouraged to propose their own variants, with a setting just to add colour and framing: the basic requirements are that the variant should have autonomous agents exercising independent reasoning over their choice of action in a certain environment with an economy of scarcity, which, in the absence of an external or central authority enforcing choice, exhibits some conflict between collective and individual goals; for example collective utility is maximised if everyone does X; an individual's utility is maximised if everyone else does X (so it derives all the benefits of collective action), and it alone does something else (solely beneficial to itself). Therefore, particularly desirable features of a variant include: knowledge aggregation (communication over a social network), deliberation and decision-making, knowledge alignment (collective action), an economy of scarcity, institutions (sets of mutually-agreed, mutable, conventional rules), and conceptual resources (e.g. trust, reputation, etc.).

Depending on several factors (time constraints, numbers, programming experience), instructors can either use the model platform provided (see links in the Appendix) and ask student teams to concentrate on agent interaction and strategies; or, they can ask student teams also to specify and implement such a platform between them. This can be done either by having a specialised *infra* team, or by making a cross-team initiative. Either way it raises interesting meta-level issues of collective action, such as deliberation and defining standards, contributing resources to a collective effort, collaboration between notionally competing groups, and so on. However, instructors do have to make judicious use of Ostrom's seventh principle of self-governing institutions, "the right to self-organise is respected by external authorities", and know when to, and when not to, intervene. There is no hard-and-fast rule for this, as it is highly situation dependent and can be influenced by many variable factors, such as the dynamics of the cohort, the open-endedness of the assignment and specific constraints on submission of examinable materials, but never on the instructors themselves.

4.1   The Archipelago

In a Cooperative Survival 'game', the fundamental dilemma is that none of the 'players' survive unless all of them do. In practice this means that without a critical mass, who can gather sufficient resources to sustain or protect every member of the 'mass' from potentially catastrophic environmental hazards, the entropic tendency to elimination cannot be resisted.

In the *Archipelago* Scenario, it was envisioned that each player was an island[2] (see Fig. 1). Moreover, the cooperative survival dilemma was constructed from three other, inter-dependent sub-games of operational choice. The first sub-game was, a resource-extraction stag-hunt game, where each island has to decide to forage for resources individually (low risk, low reward) or together with others (higher risk, high reward); moreover, excess of one activity could deplete the resource altogether. The second sub-game was a resource provision and appropriation game, where islands provision to and appropriate from a common-pool resource (i.e. to even out risk and reward from the resource extraction sub-game). The third sub-game was a collective risk dilemma (CRD). In this CRD, the islands must invest sufficient resources to mitigate the impact of an environmental disaster (storm, earthquake, volcano, etc.) when the timing (deadline D), location and magnitude of the disaster, and the threshold (T) amount to mitigate its impact, may both be unknown.

In [26], the islands were represented by agents in multi-agent systems. To address each operational choice game, the agents used three meta-level, political choice games in the context of mutually-agreed institutions (cf. [18]), one based on parliamentary deliberation, one on trading, and one on forecasting. The results showed that if sufficient resources were invested in all three institutions as a basis for cooperation, then the probability of collective survival was increased.

---

[2] The reference to "No man is an island" from Donne's *For Whom The Bell Tolls* was intentional.

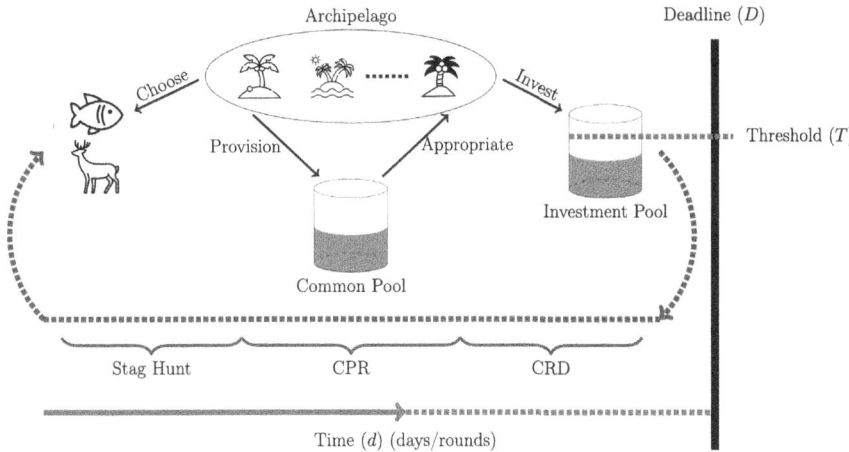

Fig. 1. The *Archipelago* Scenario: an iterated cooperative survival 'game' in which the agents have to establish institutions to deal with three inter-dependent decision arenas (sub-games): resource extraction (foraging with potential depletion), resource distribution (common-pool provision and appropriation), and resource investment (collective risk dilemma).

If the bespoke multi-agent simulator built to animate this scenario was a sort of Rube Goldberg machine – although it was actually a complicated system to perform a complicated task – at least it worked fairly well. The results showed that if sufficient resources were invested in all three institutions as a basis for cooperation, then the probability of collective survival was increased. However, this did not mean that the tasks to design, implement, simulate, calibrate, reproduce, validate and communicate these results were any simpler.

For example, with so many parameters, the experimenter is obliged to find a subset of the independent variables which produce 'stable' systems, such that a 'meaningful' experiment to test the relationship between the remaining independent variables and the dependent ones can be designed and executed. In addition, as soon as a number is chosen, e.g., three for the games, three for the mechanisms, there is an automatic (but not especially helpful) question whether or not this number is a necessary and sufficient condition for the results. Moreover, the implementation of a monolithic system with switches and command-line flags to test all the combinations of games and mechanisms was, with hindsight, another example of *feature-creep* in product design.

4.2   The Platform

For this paper, we consider the social dilemma presented in the 2019 film 'The Platform' (*El Hoyo*). This film envisions a tower consisting of N floors with a pair of prisoners on each floor (Fig. 2).

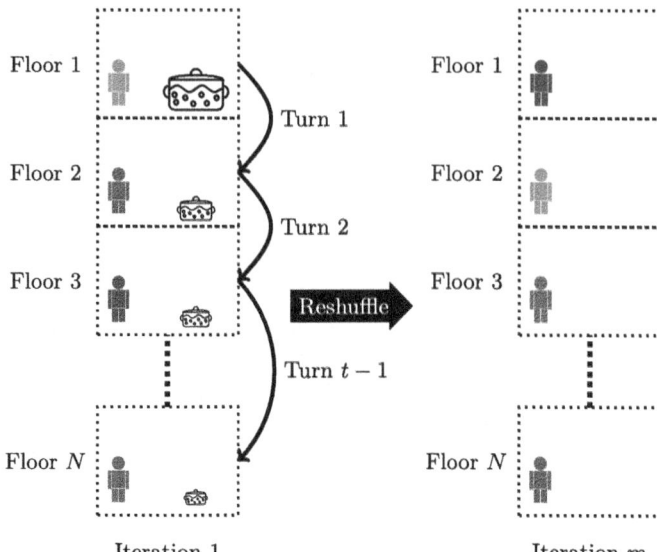

Fig. 2. *The Platform* Scenario. Resources (food) descends from one level to the next, with no restraint on resource appropriation. At intervals, the agents are re-assigned to different levels. The central dilemma is: what social arrangements should the agents choose if they did not know to which level they would be assigned.

A platform laden with food descends through a central shaft in the tower, stopping at consecutive floors. The prisoners are allowed to eat as much as they want while the platform has stopped on their floor, but cannot save food "for later". Obviously it is advantageous to be on a low-numbered floor to have first access to the food on the platform; however there is a 'reshuffle' after D days, with all the agents are randomly re-assigned to new floors, and with no knowledge of which floor they will be re-assigned.

It has been shown that by taking an approach inspired by moral philosophy there are solutions to the *social contract design problem* [7]. This means that, for any non-cooperative game, it is theoretically feasible to define a social contract which produces a modified game that optimises for a moral imperative. Ostrom's work, as previously mentioned [18], provides empirical evidence that it is practically possible for groups of people to resolve collective action situations through the social construction of self-governing institutions. Effectively this is identifying the institutions, understood as a set of rules, as the social contract and sustainability of the common-pool resource as the moral imperative.

The question addressed in this paper is under what conditions is it practically possible for groups of *agents* to resolve a collective action situation, specifically that posed by *The Platform* scenario. In this scenario, we presume that the motivation for creating a social contract comes from an abstraction of the psychological concept of *social motives* [17,23], which Folmer describes as "the psychological processes that drive people's thinking, feeling and behavior in interactions with other people." Social motives are further identified as a potential source of conflict, with Folmer also claiming that "the actions that are dictated by one individual's motives are incompatible with, or even harmful to, the interests of others." In other words, the social contract must not only solve the social dilemma, but must also resolve any residual tension between potentially conflicting social motives.

Although, without loss of generality, we make some modifications to the scenario from the film – for example, we assume one prisoner per floor rather than a pair (although that is only required for dramatic effect), no movement between floors, and communication allowed between adjacent floors – we are assuming strict constraints of no prior knowledge, no pre-existing social network and no external authority, with the additional complications of a dynamic population and periodic floor re-assignment. The challenge is then to determine whether, despite the combination of limited communication, social motives and the propensity for social construction, the agents can 'find' a social contract which is a solution to the current formulation of the game and perpetuates across subsequent re-formulations.

4.3   Escape the Dark Pit(t)

For the purposes of this paper, we consider a social dilemma where a group of players start at the bottom of a pit, each level of which contains an enemy to be fought. The group must battle and defeat the enemy before they can ascend to the next level. However, any deaths incurred in the battle reduce the group's ability to defeat increasingly strong enemies. With each enemy defeated, players get access to a stash of *loot*, containing weapons, shields and potions, which can be divided amongst the group. Weapons are used to attack the enemy, shields are used to defend against the enemy, and potions are used to regenerate health.

Furthermore, this game is designed to be played in an economy of scarcity, meaning that the allocation of loot cannot fully satisfy all of the players' individual desires, leading to biased decision formations, reinforced by increased individual utility [10] [6]. This condition sets the stage for Ostrom institutions to be formalised for solving a common-pool resource (CPR) management situation [20] within a norm-governed society. These societies take into account the permissions and obligations of its members, as well as the possibility of a deviation from the expected action [1], creating a framework for sanctions, forgiveness to inspire reconciliation [32] and defiance to incite change [5].

These social norms can be formalised by social contracts, which specify the conditions under which these norms must be obeyed. It has been shown that it is always theoretically possible to design an optimal social contract for the moral imperative [7], although designing this contract is often not a trivial task [25].

As well as defining the conditions by which the social contract must be obeyed, the contract also defines the punishment for not doing so. The breaking of a contract often merits a *sanction* [19], which comes as a detriment to the disobedient actor involved. Such sanctions can vary drastically in severity, such as with their duration, so must be carefully constructed, since "unfair sanctions" [8] can have detrimental impacts on human co-operation. To prevent this, designing effective sanctions has seen a computational approach [3] [15]. In this scenario where sanctions entail exclusion, a negative feedback loop is formed, where sanctioning a defector becomes detrimental to the collective. It is important to prevent free-riders from appropriating the shared resource yet refusing to fight (the risk-averse approach), however over-exclusion will leave them more susceptible to damage, thereby hindering the possibility of co-operative survival.

4.4   Megabike

The *Archipelago* scenario effectively provided two parameters for scalability, i.e. the number of agents (islands) and the number of institutions. We are interested in situations with more parameters, indirect relationships between independent and dependent variables, and the emergence of properties at scale. Examples include the effect of scale on monitoring, and when mutual observation is replaced by appointed agencies, or when time pressure effectively prohibits deliberation as a phase in 'democratic' decision-making.

Accordingly, the *Megabike* scenario, inspired by multi-rider bikes as illustrated in Fig. 3, has been proposed.

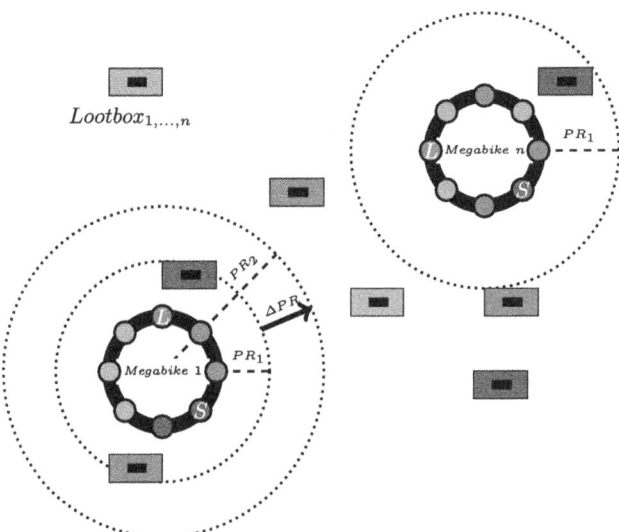

Fig. 3. The *Megabike* Scenario: the agents need to agree 'articles of association' which specify perception radius, pedalling power and steering direction, in order to detect and collect lootboxes to replenish their energy, avoid an existential threat, and prevent – literally – free-riding.

A *megabike* then has n riders, i.e., agents, each of whom has a set of pedals, brakes, and steering controls (see Fig. 3). Each agent has limited energy, which is depleted over time, and faster by pedalling, braking or steering. Agents can replenish their energy by accessing a lootbox; but each agent needs a particular colour of lootbox energy.

Each agent has a goal of maximising its energy intake and ensuring its own survival. The obvious strategy of (literally) free-riding is therefore not advisable, since non-cooperation causes co-riders to avoid moving towards their lootboxes, and besides, if every agent free-rides then either its energy will exhaust naturally or its *megabike* will be destroyed by an *Existential-Threat* that patrols the micro-world in pursuit of the slowest *megabike*.

Moreover, a *Megabike* scenario proceeds in a series of iterations. In each iteration, the agents must firstly form a voluntary association to take control of a *megabike*, and secondly agree the 'articles of association', including the selection of rules for: modifying rules (meta-rules), appointment to roles with institutionalised power [11], such as the designated Leader (L in Fig. 3) deciding the next target lootbox, a designated steerer (S) responsible for navigation, and a designated agent enacting rules for inclusion and exclusion, etc. In other words, the agents mutually agree an institution with the features identified by [18] under the constraints of Ober's thought experiment Demopolis [16].

In the final phase of an iteration, the scenario proceeds in a series of rounds. In each round the agents apply the articles of association rules to collectively agree on the target lootbox, direction of travel and the speed of the bike (e.g. to avoid the *Existential-Threat*, or if they appear to be in competition with another *megabike* for their preferred lootbox).

Thus, the *Megabike* scenario also comprises of a number of inter-dependent games, including a linear public goods game, a resource allocation game, a cooperative survival game, and a constitutional choice game. It also features opportunities for supervised and social learning that can affect scalability.

5 Supporting Platform

This section is concerned with the specification and implementation of a multi-agent simulator called the *base platform*. To satisfy the wide range of functional requirements of the Model Assignments while addressing non-functional requirements of scalability issues, we take a generic approach. In contrast to the plug-in architecture of PreSage-2 [12], we propose a modular, architectural 'framework' for self-organising, multi-agent systems, that we call the *base platform*. This platform abstracts some of the key components common to multi-agent simulation, such that increasingly complex simulators, e.g. for the *Megabike* scenario, can be built on top of it.

5.1 Overview

The implementation of this framework is based on three components. The first component is the *base server* (see Sect. 5.2), which is an extensible *server* architecture used to regulate the synchronous state of the *base agents*, and advance the game loop. The second component is the *base agent,* described in Sect. 5.3, which is an extensible (multi-)agent building block that gives a simple implementation of the core functions that an agent performs in a given scenario. For example, agents may advance the physical game-state through interactions with the *base server,* or update their own internal state through sending a *base message* to other agents. Therefore, finally, we provide an abstract, extensible *base message* component that serves as a common language for multi-agent communication (Sect. 5.4). This building block can be extended to more complex, message types to communicate more information between agents. Figure 4 gives a high-level class diagram for both how the components of the base platform work together, and how they can be composed to produce a more complex, application-specific *extended platform,* i.e. a platform customised and configured for different Model Assignments.

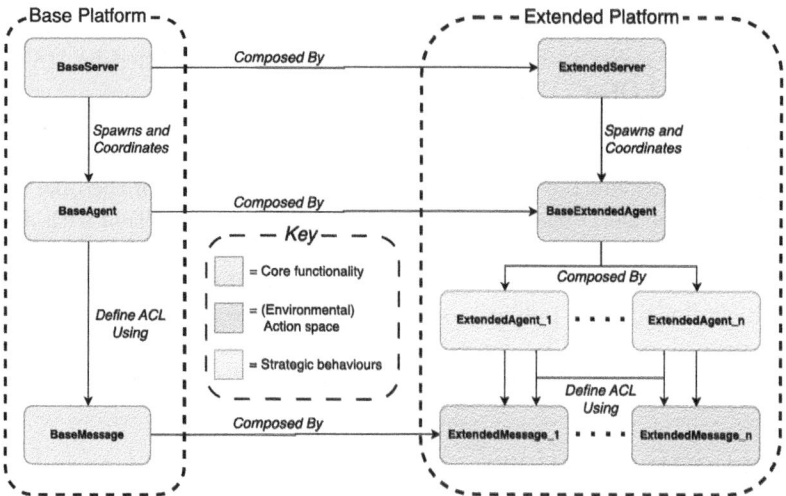

Fig. 4. Full class diagram and recommended composition structure for the *base platform. Agent Common Language* is abbreviated to *ACL,* for brevity

For this implementation, we consider a difference between internal, 'mental' state and external 'physical' state. The 'mental' state is the internal, subjective reality that an agent builds of its surroundings, and is built from communication between agents, and interactions with its environment. The 'physical' state, conversely, is the objective reality that is moderated between agents, and defines the current state of the environment.

When formalising the composition pattern recommended for the *base platform,* we only show how the relevant interface methods are implemented by an appropriate data structure, for brevity (i.e., not the parameters that would be required for implementing these functions or maintaining state).

5.2   Base Server

In order to synchronise the physical state of a simulator (the 'environment') between agents, a server is used to process an agent's actions, enact them on the environment to update it, and advance the current state of the game and each iteration of the game loop. This server can be interpreted as a 'coordinator agent' [13,30], or a conventional server that may be used in distributed systems, say [31].

This server must be engineered to simplify the design process and reduce the 'decision budget' of the students undertaking the course. As such, we propose an abstract, extensible architecture for a 'base server' component, that allows for easy extension, so long as the method for defining more complex components is well defined and clear. With this, the *base platform* can handle the low-level 'flow' of the system (game-state regulation, point in game loop, etc.), while allowing for the students to build upon the core functionality with functionality that is specific to the scenario they wish to produce. Ultimately, we can reduce the initial 'barrier to entry' confronting the students by enforcing and supplying a pre-made architecture that already gives methods for injecting agents and supporting agent-to-agent communication.

The *base platform* supplies both a *base server interface*, which abstracts the core functionality for a server, and a *base server*, which supplies a (base) implementation of the interface to allow for extension/composition, giving the students core functionality and a data structure for building more complex systems. Formally, the server interface *IServer*, which is implicitly implemented by the **BaseServer**, is defined in Fig. 5.

In this implementation, we grant the server access to/control over the agents in the system, as well as the advancement of the game-state. Agents are stored in a hashmap, allowing for $O(1)$ insertion, removal and retrieval operations, through *AddAgent*, *RemoveAgent*, and *GetAgentMap*, respectively. By using a logical set of data structures, and building an abstract framework that lends itself well to inheritance, this architectural component also becomes *experimentally scalable* [27]. That is, independent variables can be scaled without impacting performance, since the time complexity of agent indexing is unaffected by the number of agents.

In Fig. 5, we recommend a possible composition structure for extending the functionality of the base server. Taking **BaseServer** as the data structure that implements the methods in *IServer* (i.e., gives a base implementation of the interface for composition), we can extend the "core" functionality by supplying an **ExtendedServer**. This data structure composes/extends the base server object to define scenario-specific functionality, while still retaining the core features supplied in the *base platform* package.

Figure 5 also demonstrates the **BaseServer** structure implementing a second interface: **IGameRunnerUtils**. This interface provides the relevant, overridable methods for controlling the game loop on both a low frequency *iteration* level, and a high frequency *turn* level. The additional functions supplied in this interface are formalised in Table 1.

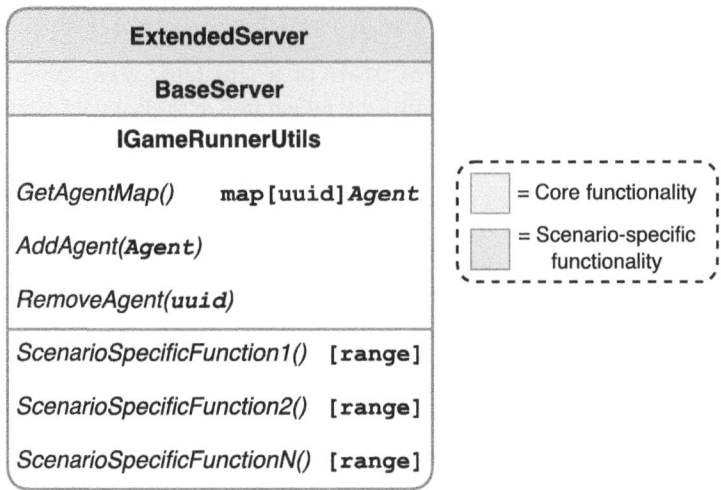

Fig. 5. Suggested composition structure for the *base server*

Table 1. *IGameRunner* interface: core server functionality provided by the base platform for advancing and controlling the game loop

Function	Range
RunStartOfIteration(*int*)	*void*
RunTurn(*int,int*)	*void*
RunEndOfIteration(*int*)	*void*
GetIterations()	*int*
GetTurns()	*int*
Start()	*void*

By decomposing the game loop into M iterations, where each iteration itself runs N turns, we give students the ability to parametrise the simulator at both a high-frequency and low-frequency level. *RunStartOfIteration* and *RunEndOfIteration* allow for students to define the low-frequency behaviours that occur once at the start and end of the iteration, respectively, whereas *RunTurn* defines the high-frequency behaviour. We also supply getters, for students to access the number of turns and iterations in the server, and a method for running the simulator, in *Start*.

5.3   Base Agent

Agents serve as the core interactive component in a multi-agent system. In this paradigm, we provide a distinction between *physical state* and *mental state*, described previously. As such, the server is implemented to maintain and control the physical, external state, whereas the agents are implemented to update their mental, internal state.

This parallels the *beliefs-desires-intentions* (BDI) agent model [9], where agents have a subjective perception of the external reality (their mental image of the physical state or *beliefs*). Agents also have a set of changes they would like to enact on the physical state (their *desires*), which are controlled through some coordinator component (the server, in this context). The resulting changes enacted on the physical state parallel their *intentions*.

To again simplify the initial design process, similarly to the *base server* we also supply a *base agent* component to abstract the core functionality and allow for the design of more complex agents through composition. The *IAgent* interface abstracts these functions, and is formalised in Fig. 6, where the **BaseAgent** data structure is shown to implement it. This interface not only encompasses the functionality for performing local agent state updates, but the functionality required for performing agent-to-agent communication. The latter is discussed in Sect. 5.4.

Given how abstract an agent implementation may be, the only commonalities we provide for maintaining agent state are 1.) a means of identifying the agent through *GetID*, and 2.) a means of arbitrarily updating an agent's state through *UpdateInternalState*, which is designed to be overridden according to the relevant scenario and called during the game loop by the server. We also provide two methods for interacting with the other agents in the system, *ViewAgentIDSet* returns an (immutable) hashset of the agent IDs, while *AccessAgentByID* returns a (read-only) agent instance for a given ID.

Figure 6 shows a suggested composition structure for implementing an agent with the *base platform*. As with the *base server*, we offer the **BaseAgent** data structure to implement the *IAgent* interface and provide the "core" functionality. This data structure can be composed using an **ExtendedAgent**, say, which defines all common functionality that would be required for the scenario-specific implementation.

Unlike with the *base server*, we supply an additional level of composition for the agent strategy to give greater expressiveness to the simulator. One can imagine that the scenario will give some commonality to the agents, in the form of some function that must be executed each round (all agents must move or rest, perhaps). The actual strategy that informs an agent of how to perform this action may be a level deeper, however. As such, students may wish to define further, agent-specific functions, which are implemented in the **StrategicAgent** data structure. For brevity, we detail the implementation of the messaging functionality later in Sect. 5.4, and represent them with the placeholder **IMessaging** in Fig. 6.

5.4   Base Message

For agents to communicate, we define an *agent communication language* (ACL). As with the previous components of the *base platform*, this language must comprise abstract building blocks, such that the language can be customised for the relevant scenario.

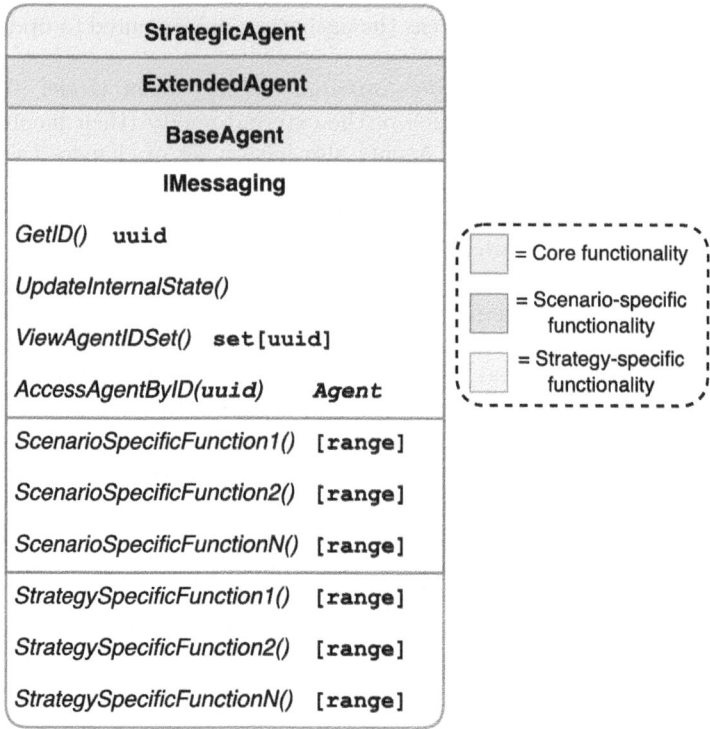

Fig. 6. Suggested composition structure for the *base agent*

As such, the final component of the *base platform* is the *base message* component, which characterises an abstract messaging component that can be sent and received by *base agents*, and coordinated by the *base server*. Once again, to address the needs for simplifying the students' design process, we provide an architecture that encourages composition to support more complex, scenario-specific data structures, whilst still being operational inside of the *base server* and shared with the *base agents* described above. The core functionality we supply for the *base message* class is formalised in Fig. 7 as the *IMessage* interface, and implicitly implemented by the **BaseMessage** data structure.

The *IMessage* interface provides two functions to abstract the data structure providing an ACL. The *base message* building block, which implements this interface, provides 1.) a means of accessing the sender of a message (through *GetSender*), and 2.) a means of handling a message (through *InvokeMessage-Handler*).

Given that each type of message may need to be handled independently, we utilise the *visitor design pattern* [21]. As such, each message type extending the base message should override the *InvokeMessageHandler* function, to call the appropriate handler function in the agent class. This explains the importance of the *IMessaging* shown in Fig. 6, and is formalised further in Fig. 8.

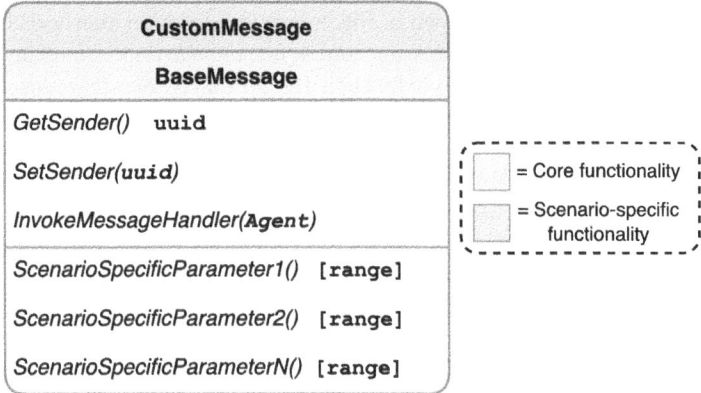

Fig. 7. Suggested composition structure for the *base message*

We envision a composition structure similar to that in Fig. 7, where each message contains scenario-specific 'getter' functions, and overrides the definition of *InvokeMessageHandler*. The resulting agent functions are shown in Fig. 8, where we imagine *N customMessages*, all of which have a relevant handler implemented by the **ExtendedAgent**.

Additionally, the *base agent* has the capacity to send a message, as well as handle it. As such, we also introduce the *IMessaging* interface (not to be confused with the *IMessage* interface implemented in Fig. 7), which provides the appropriate methods for sending messages over the system. Here, the server method is invoked to take an agent's message, and call the relevant handler on the recipient.

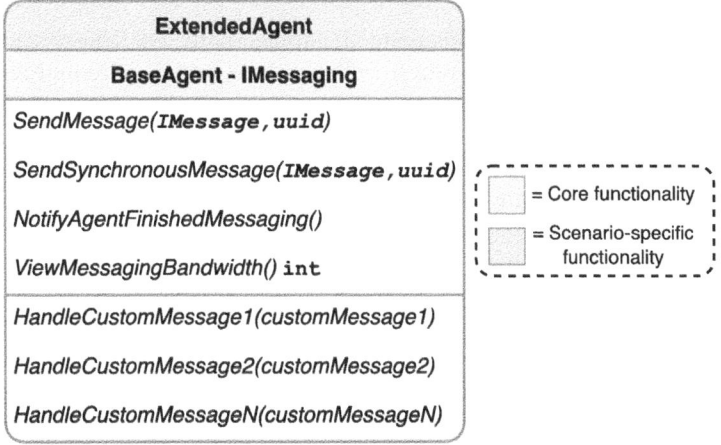

Fig. 8. Appropriate handler functions added to **ExtendedAgent** data structure

The *IMessaging* interface shown in Fig. 8 provides four key methods for agent-to-agent communication. *SendMessage* sends an *asynchronous* message of type *IMessage* to a recipient, identifiable by their ID. Similarly, *SendSynchronousMessage* sends an *IMessage* to a recipient but pauses the game loop (the 'main thread') to wait for the recipient to handle it, unlike with the asynchronous case, where the handler is threaded in parallel to the game loop. We supply this method to handle cases where communication *must* be handled as a separate 'phase', before the main thread advances. For example, in the *Megabike* scenario outlined in Sect. 4.4, there is a "voluntary association phase" where agents must negotiate bike membership. If users wished to have a phase running prior to this, where agents first *communicate* about membership, synchronous messaging enables this. Simply, this allows user to differentiate between synchronous messages that *must* be handled, thereby delaying the main thread, and asynchronous messages that *may* be handled, instead running in parallel to the main thread.

NotifyAgentFinishedMessaging is a 'quality of life' (QoL) method used to communicate to the main thread that the agent has finished its communication for the turn, and that the game loop can (and should) continue safely with its next turn. Finally, *ViewMessagingBandwidth* returns the maximum number of messages a single agent can simultaneously handle (controlled by its personal semaphore), to avoid oversaturating the simulator with thread-spawnage.

6   Self-assessment

In this section, we consider the strengths and weaknesses of the course, the model assignments, and the software platform.

6.1   Strengths

Openness: the project is an open-ended, intellectually challenging, curiosity-driven undertaking, giving free-rein to collaborative, imaginative and innovative solutions, and develops engineering competence in system architecture design, algorithm specification, and programming skills. By use of different assignments, instructors can ensure that there is very little code re-use between different cohorts from one year to another (and the students still have to work with each other).

Extensible: although care needs to be taken to ensure that the projects end goal and deliverables are realistic and achievable in the time available, the group project lends itself as an excellent starting point for further development in, say, an individual project.

Experience: students gain experience of writing their own algorithms, giving them an understanding of multi-agent systems from the lowest level of abstraction (knowledge encapsulated in autonomous decision-making process communicating with others) to higher (social entities with shared and congruent values, conventional rules and institutionalised speech acts) via some of the 'paradoxes' of self-organisation (e.g. Arrow's impossibility theorem about voting systems, Condorcet's paradox on voting preferences, the paradox of self-amendment (when a rule used to justify its own amendment, etc.).

Transferable skills: each academic year, the cohort is divided into groups (which they should self-organise themselves, as a first meta-application of self-organisation). These groups then need to work together as a team to develop their own agent strategy to solve the problem and work with other teams to develop standards, e.g. to define scenario "rules", the agent communication language, etc. This encourages development of teamwork, organisational, time-management and other inter-personal skills.

Inter-disciplinarity: the project requires reading and understanding ideas from multiple disciplines. The exposure to theories and thought-processes from other disciplines can be very formative for engineering students: not at all what they expected, taken out of "comfort zones", forced to interact with others, and so can be both eye-opening and thought-provoking.

Ownership: in the best spirit of self-determination, students are encouraged to take ownership of the Model Assignment. In the spirit of "those who are affected by the rules participate in their selection", modification and application of those rules (applying a particular feature of sustainable self-governance [18]), the students are encouraged to demonstrate creativity and exercise autonomy in the definition of the rules of the Model Assignment, provided the reason(s) for the changes is (are) in the interests of advancing self-organised solutions to strategic or social interaction.

Multiple paths to success: the best programmers don't necessarily get the highest marks. The element of self-assessment results in those who contribute the most getting the highest marks. While that could be in programming, it could equally be in some other aspect of the project, for example game design, agent algorithm specification, or even management.

"Ah-ha moment": in a moment of reflection (which is perhaps made obvious by the interleaving of lectures and laboratory work), the students realise that the concepts and mechanisms they are taught, and required to implement in their self-organising multi-agent system of computational agents, also apply to themselves as a self-organising system of human agents. This realisation (the "ah-ha moment") crystallises and grounds the importance and significance of otherwise potentially abstract theoretical concepts—even if those concepts are being practically implemented in an experimental setting, the model assignments are closer to games than serious games or 'real' applications.

Potential/Actual publication: there is significant person-power in a cohort (e.g. for 40 students working one day a week for five weeks, this is 200 person-days, or 3/4 of a person-year). Collectively, the group can make significant progress with that much available effort, although as per the law of requisite variety [24], the limiting factor of organisational performance is the finite processing power of the collective, and so appropriate self-regulation is required to realise the benefit of numbers. However, committed and properly self-organised students can prepare a research paper at least suitable for an AAMAS conference Workshop, given that they can address a fundamental problem in collective action, formalise it algorithmically, implement a simulator, and produce insightful experimental results. Publications produced from, inspired by, or based on the various Model assignments include: [4, 14, 25, 26, 28, 29]

6.2 Weaknesses

The project's potential weaknesses are rather a set of risks to a successful outcome, of which instructors need to be aware at the outset, and monitor and manage during the project. These risks are a kind of corollary of its open, self-organised, student-driven nature, and so more concerned with the administration of the project. However, in the interests of quality assurance – that the module as a 'product' meets the expectations of the students and the standards of the educational establishment – instructors do have to be both intuitive and experienced about when to guide and when to intervene directly as the "external authority" in order to meet these requirements.

These risks include: timing, boundaries, background, inclusivity, inspiration, personality, engagement and (unsurprisingly, given the collective action content of the project) free-riding.

Timing: presentation of theoretical material depends on being able to "front load" the technical material in the first half of an academic session so that the practical work can be done in the second half. Otherwise, if the lectures are distributed evenly, under time pressure students tend to choose practical coding over technical content, even though they need (but think they don't need) that technical content to inform their practical solutions.

Awareness of boundaries: is a product of under-specification and open-ended nature of the Model Assignments. Like the Tolstoy short story, *How much land does man need?*, the temptation to add "just one more feature" sometimes needs to be resisted if a complete working system is to be developed in the time available. Instructors should insist on a Milestone for an MVP (Minimum Viable Product) so that students have a working platform and can proceed from there.

Variable backgrounds: relates to the distribution of programming experience and skills within the group. Ideally the cohort will be of about the same standard: if not, care has to be taken to ensure that the highly skilled students do not dominate the others, and even better, that the more-skilled can impart their skills to the others. Use of a new programming language for developing and extending the given base platform is a good leveller in a mixed ability group.

Personality: refers to the variation in character types which might affect both intra-team and inter-team dynamics. Although the formal lectures specifically stress mechanisms for alternative dispute resolution, instructors should be aware of maverick or (unfortunately) disruptive behaviour.

Inclusivity: is another aspect of the background problem, and depends on the possibly asymmetric distribution of male, female and non-binary students in the cohort. Care has to be taken to ensure that those students, who may be in a minority in the cohort, are properly represented, are able to express their views, and are not talked over or talked down by more vocal individuals.

Inspiration: instructors should be aware that effective Model Assignments are so straightforward to identify and specify. We have taken inspiration from many diverse sources (including board games and films), and it does benefit from several design sessions to ensure that the scenario demonstrates the features that demand self-organisation of social arrangements in an 'interesting' way, but that it also properly scoped and feasible (for example, the Archipelago Scenario was particularly challenging, especially given that cohort had to develop their own platform as well).

Engagement: the practical element requires that students get involved in deliberation and decision-making from the outset. Otherwise, if late-comers subsequently try to join in, their colleagues may find it easier to do tasks for themselves than spend time bringing others "up to speed". There is an element of self-motivation that students need to bring to the project themselves.

The ultimate end-point of lack of engagement is intentional free-riding, i.e. deriving the benefits of collective action without incurring the costs of participation. It can, though, be very difficult for the instructors to detect free-riding, especially if the free-riding on this course is reciprocally compensated elsewhere and the students don't object. Students themselves can be made aware of the problem in lectures, and mechanisms for dealing with it (dispute resolution), but the ultimate sanction resides in the assessment mechanism, where a sufficiently significant component of the final mark is determined by the students themselves. Of course, if they don't want to punish, they won't [2].

6.3   Student Feedback

Ultimately, the most significant test is whether or not the intended beneficiaries – the students themselves – meet the learning objectives.

In an attempt to self-evaluate this, there is a fourth component of (unmarked) assessment, which is that the students are asked to submit a 500-word reflective essay on their experience. These essays have not been systematically analysed, but they are generally positive. However, these two statements do seem to encapsulate the spirit of the entire enterprise, and the inevitable frustrations—and benefits—of students working with ill-defined boundaries and less control, and having to engage with each other, since they really have to self-organise:

Anonymous 1: Envision yourself as a train driver tasked with not only driving but also laying the tracks as you go. This already daunting task is compounded by the responsibility of deciding the track's path, introducing a new problem where you have to appease one or two shouting: "Trains are noisy!" or "Train stations are too far!" If you are like most people you think that this task is impossible for one person to deal with, if you did SOMAS you would wish it was a one-person job so you could bang it out in solitude as you descend into madness. Although my depiction of

SOMAS leans towards the grim, it had its share of bright spots. Given another chance, I'd probably choose it again.

Anonymous 2: The atmosphere in the classroom shifted palpably after the final lecture, leaving us to navigate the uncharted waters of self-organization. It was suggested that anyone interested in creating the infrastructure should come to the front of the classroom. This was when reality set in, and there were twenty or thirty teammates shouting at each other about how they interpreted the norms given to us in the form of Jeremy's Mega-bike proposal. It was a scene of chaos turning into order as we swiftly established a central governance, spearheaded by team captains. As the collective risk increased, in the form of a deadline, a hierarchy naturally formed, and the cohort generally took on one of two decisions: either take initiative or be led by others. If we continued just for a few more weeks, inevitably we would have started to see the iron law of oligarchy take place, and then, just maybe, we could have finally enforced a decision for Audi, Awdi, or Owdi.

Mostly, though, most students come to the realisation that all the theory that they have been taught in the lectures is not just to be implemented in their simulator, but applied to themselves in their practice. Indeed, one student once asked "are we [the students] just lab rats in an elaborate experiment in self-organisation that you're running"?[3]

7   Summary and Conclusions

The international workshop series *Engineering Societies in the Agents World* (ESAW) ran for ten years from 1999 to 2008. It is arguable that the issues that the workshop series was intended to address – social coordination between autonomous agents in multi-agent systems using computational intelligence to solve 'wicked' problems – is as least as relevant at the time of writing (2024) as it was when the workshop series was initiated. We would argue that now, perhaps, the issues are even more pressing, not just because of the increasing scale and complexity of the applications being addressed, but also because of the inevitable inclusion of mixed-initiative socio-technical systems, in which the autonomous agents are both computational and human, and the social coordination is between both computational (artificial) and natural (human) intelligences.

For these reasons, we have developed a course in Self-Organising Multi-Agent Systems for undergraduate and (taught) post-graduate students, that addresses two educational dimensions: software engineering and, as per ESAW, *society engineering*, based on the codification of deep social knowledge from the social sciences. This course is supported by a textbook [22], and a software platform

---

[3] To which the answer was: "No, because there is chance that scientists can become fond of lab rats". :).

that is extensible and customisable for different Model Assignments, available open source (see Appendix). The platform is written in Go (`https://go.dev/`), so in the *Megabike* scenario, there is Go where the bikes are.

However, this is not an end-point, and a second volume of algorithms codifying aspects from theories of social influence, psycholinguistics, political deliberation and innovation is planned. This extended material would further emphasise the inter-disciplinary requirements of the technical challenge: the need to have an understanding of how some mechanisms of philosophy, psychology, economics, politics and sociology operate in human societies, and an understanding of how to formalise those mechanisms in tractable algorithm. This is essential not just for multi-agent cyber-physical systems with interacting computational intelligence, but will be equally essential for multi-agent socio-technical systems with interacting computational and human intelligences. This endeavour is, of course, highly complementary to the research and educational programmes of Professor Wolfgang Reif, reflecting his interests in and emphasis on complex systems, self-organisation and software engineering.

Acknowledgments. The first author would particularly like to thank Professor Wolfgang Reif and many researchers in his team for much convivial collaboration down the years. All the authors are very grateful to the editors for the invitation and opportunity to contribute to this volume, and are especially grateful for the excellent feedback from the two anonymous reviewers that has substantially improved this paper. Images in Fig. 1 are courtesy of flaticon.com.

To the 200+ students (including three of the co-authors) who have benefited from (suffered from?) the SOMAS course since 2019 – thank you, we couldn't have done it without you. And whatever might have been said about lab rats earlier, we are actually quite fond of you. Well, some of you :)

Appendix

List of project source code:

- *The Archipelago*: https://github.com/MattSScott/SOMAS2020
- *The Platform*: https://github.com/SOMAS2021/SOMAS2021
- *Escape the Dark Pit(t)*: https://github.com/antonypap/SOMAS2022
- *Megabike*: https://github.com/MattSScott/MegaBike
- *Base Platform*: https://github.com/MattSScott/basePlatformSOMAS

References

1. Artikis, A., Sergot, M., Pitt, J.: Specifying norm-governed computational societies. ACM Trans. Comput. Logic **10**(1), 1–42 (2009)
2. Axelrod, R.: An evolutionary approach to norms. Am. Polit. Sci. Rev. **4**(80), 1095–1111 (2009)
3. Balke, T., De Vos, M., Padget, J.: I-abm: combining institutional frameworks and agent-based modelling for the design of enforcement policies. Artif. Intell. Law **21**, 371–398 (2013)

4. Blackledge, B., et al.: Incentivising participation with exclusionary sanctions (Full). In: Fornara, N., Cheriyan, J., Mertzani, A. (eds.) Coordination, Organizations, Institutions, Norms, and Ethics for Governance of Multi-Agent Systems XVI, COINE 2023, LNCS, vol. 14002, pp. 37–54. Springer, Cham (2023). https://doi.org/10.1007/978-3-031-49133-7_3

5. Kurka, D.B., Pitt, J., Ober, J.: Knowledge management for self-organised resource allocation. ACM Trans. Auton. Adapt. Syst. **14**(1), 1:1–1:41 (2019)

6. Cialdini, R.: Influence: The Psychology of Persuasion. William Morrow & Company, New York, NY (1984)

7. Davoust, A., Rovatsos, M.: Social contracts for non-cooperative games. In: Third AAAI/ACM Conference on Artificial Intelligence, Ethics, and Society, pp. 43–49 (2020)

8. Fehr, E., Rockenbach, B.: Detrimental effects of sanctions on human altruism. Nature **422**(6928), 137–140 (2003)

9. Georgeff, M., Pell, B., Pollack, M., Tambe, M. and Wooldridge, M.: The belief-desire-intention model of agency. In: Intelligent Agents V: Agents Theories, Architectures, and Languages (ATAL): 5th International Workshop, pp. 1–10. Springer (1999)

10. Gigerenzer, G.: How to make cognitive illusions disappear: Beyond "heuristics and biases". Eur. Rev. Soc. Psychol. **2**(1), 83–115 (1991)

11. Jones, A., Sergot, M.: A formal characterisation of institutionalised power. J. IGPL **4**(3), 427–443 (1996)

12. Macbeth, S., Busquets, D., Pitt, J.: System modeling: principled operationalization of social systems using PreSage-2. In: Gianni, D., D'Ambrogio, A., Tolk, A. (eds.) Modeling and Simulation-Based Systems Engineering Handbook, pp. 43–66. CRC Press, Boca Raton, FL (2014)

13. Maturana, F., Norrie, D.: Multi-agent mediator architecture for distributed manufacturing. J. Intell. Manuf. **7**, 257–270 (1996)

14. Mertzani, A., Pitt, J., Sarkadi, S., Sas, M., Scott, M., Smit, C.: Cohesion and the explanation of constitutional choice in self-governing systems. In: A Life Workshop Agent-Based Modelling oh Human Behaviour (ABMHuB) (2024)

15. Nardin, L., Balke-Visser, T., Ajmeri, N., Kalia, A.K., Sichman, J.S., Singh, M.P.: Classifying sanctions and designing a conceptual sanctioning process model for socio-technical systems. Knowl. Eng. Rev. **31**(2), 142–166 (2016)

16. Ober, J.: Demopolis: Democracy before Liberalism in Theory and Practice. Cambridge University Press, Cambridge, UK (2017)

17. Oppenheimer, O.: The origin of social motives. Educ. Theory **4**(2), 95–104 (1954)

18. Ostrom, E.: Governing the Commons: the Evolution of Institutions for Collective Action. Cambridge University Press, Cambridge, UK (1990)

19. Ostrom, E.: Common-pool resources and institutions: toward a revised theory. Handb. Agric. Econ. **2**, 1315–1339 (2002)

20. Ostrom, E.: The challenge of common-pool resources. Environ. Sci. Policy Sustain. Dev. **50**(4), 8–21 (2008)

21. Palsberg, J., Jay, C.B.: The essence of the visitor pattern. In: Proceedings. 22nd Annual International Computer Software and Applications Conference (Compsac'98) (Cat. No. 98CB 36241), pp. 9–15 (1998)

22. Pitt, J.: Self-Organising Multi-Agent Systems. World Scientific Press, Singapore (2021)

23. Folmer, C.R.: Social motives. In: The SAGE Encyclopedia of Theory in Psychology, pp. 886–890. SAGE (2016)

24. Ross Ashby, W.: Requisite variety and its implications for the control of complex systems. Cybernetica **1**(2), 83–89 (1958)
25. Scott, M., Dubied, M., Pitt, J.: Social motives and social contracts in cooperative survival games. In: International Workshop on Coordination, Organizations, Institutions, Norms, and Ethics for Governance of Multi-Agent Systems XV (COINE), pp. 148–166. Springer (2022)
26. Scott, M., Pitt, J.: Interdependent self-organizing mechanisms for cooperative survival. Artif. Life **29**(2), 198–234 (2023)
27. Scott, M., Pitt, J.: Meta-level scalability issues in analysing the scalability of self-organisation. In: 2024 IEEE International Conference on Autonomic Computing and Self-Organizing Systems Companion (ACSOS-C), pp. 73–78. IEEE (2024)
28. Scott, M., Sas, M., Pitt, J.: An information-theoretic analysis of leadership in self-organised collective action. In: IEEE International Conference on Autonomic Computing and Self-Organizing Systems ACSOS, pp. 101–110. IEEE (2024)
29. Scott, M., Mertzani, A., Smit, C., Sarkadi, S., Pitt, J.: Social deliberation vs. social contracts in self-governing voluntary organisations. *CoRR*, abs/2403.16329, (2024)
30. Tabares, V., Duque, N., Ovalle, D.A.: Multi-agent system for expert evaluation of learning objects from repository. In: Bajo, J., et al. (eds.) PAAMS 2015. CCIS, vol. 524, pp. 320–330. Springer, Cham (2015). https://doi.org/10.1007/978-3-319-19033-4_27
31. Tesauro, G., et al.: A multi-agent systems approach to autonomic computing. In: Proceedings of the Third International Joint Conference on Autonomous Agents and Multiagent Systems, vol. 1, pp. 464–471 (2004)
32. Vasalou, A., Hopfensitz, A., Pitt, J.: In praise of forgiveness: ways for repairing trust breakdowns in one-off online interactions. Int. J. Hum Comput Stud. **66**(6), 466–480 (2008)

Refactoring of LCMsim: A Lightweight Julia Package for Mould Filling Simulations in Liquid Composite Moulding

Christof Obertscheider[1], Leonard Heber[2], Carola Lenzen[2], and Ewald Fauster[3(✉)]

[1] Aerospace Engineering Department, University of Applied Sciences Wiener Neustadt, Wiener Neustadt, Austria
christof.obertscheider@fhwn.ac.at
[2] Institute for Software and Systems Engineering, University of Augsburg, Augsburg, Germany
[3] Processing of Composites Group, Department Polymer Engineering and Science, Montanuniversität Leoben, Leoben, Austria
ewald.fauster@unileoben.ac.at

Abstract. The Julia package LCMsim was developed to verify if the equations for compressible fluid flow through a thin curved cavity can be solved on a shell mesh. Although directly accessible as open source, the code was hard to understand, maintain and extend for outsiders. In this paper the general design of LCMsim after a refactoring step is discussed and general design principles for a numerical tool solving the equations of hydrodynamics are presented and implemented.

LCMsim is designed to run mold filling simulations for Liquid Composite Molding, a manufacturing process for fiber reinforced polymer composites where dry fibers are placed into a mold which is subsequently closed or sealed with a flexible bag by drawing a vacuum. Afterwards, the pressure difference between the injection gates and the vents pushes the resin through the porous preform.

The governing equations are specified as Initial-Boundary-Value problem with partial differential equations including temporal and spatial derivatives. Consequently, this paper can serve as a guidance for scientific computing applications.

The results section quantifies the decrease in computational time for a benchmark problem after refactoring (17 times faster). In addition, a methodology is addressed for modifying existing equations as well as for adding supplementary equations of phenomenological models.

Keywords: Julia · Software architecture · Computational Fluid Dynamics (CFD) · Shell mesh · Liquid composite moulding (LCM) · Filling simulation

© The Author(s), under exclusive license to Springer Nature Switzerland AG 2025
G. Ernst et al. (Eds.): Wolfgang Reif Festschrift, LNCS 15765, pp. 132–154, 2025.
https://doi.org/10.1007/978-3-031-92196-4_7

1   Introduction

Resin Transfer Molding (RTM) and Vacuum-Assisted Resin Infusion (VARI) are two examples for Liquid Composite Molding (LCM) technologies [9,19]. In RTM, dry fiber preforms are placed into a matched mold and resin is then injected under pressure. In VARI, dry fiber preforms are placed into a single-sided mold which is subsequently sealed with a flexible bag by drawing vacuum. The vacuum also pulls the resin through the fiber preform.

During mold design, filling simulations can study different manufacturing concepts (i.e. placement of injection gates and vents) to guarantee complete filling of the part and avoid air entrapment where flow fronts converge. Incomplete filling and voids result in a decrease of the mechanical properties (stiffness and strength) of the manufactured part.

An additional degree of freedom for the injection strategy comes into play if multiple injection gates in RTM or multiple vents in VARI are present. Controlling the injection pressure at the individual inlets or controlling the pressure in the individual vents can influence the flow front propagation in order to ensure a uniform, complete and fast impregnation. A more regular spreading of the resin can be achieved by data-driven controller designs using various machine learning models. Training of the machine learning models in a real-world setup is not feasible. A large number of time-consuming and costly experiments was required. Therefore, a numerical simulation of the resin flow is used to provide a simulative environment resembling the filling process.

A software tool which can be used as a simulative environment must be validated with experiments. This is true for PAM-RTM [6] and RTMsim [11] which were already used for this purpose [17,18]. The Julia module RTMsim and its successor LCMsim [12] are both open source and at any point in simulated time allow for changing the injection pressure and obtaining the state of the simulation. This ability is crucial for online control and needs to be possible without terminating the executing process to achieve high efficiency and stability. Multiple parallel instances of the simulation must run on a multi-node computer cluster. Therefore, efficiency is another requirement for such a simulation environment.

The main goal of this paper is to show how a software tool for scientific computing (if the model is defined in terms of an initial-boundary-value (IBV) problem) can be designed in a Julia-like programming language. The main requirements are computational efficiency, easy expandability of the open source code and possible integration into other projects such as a graphical use interface (GUI) or machine learning infrastructure. All requirements for the refactoring of LCMsim and how they are implemented are shown in Sect. 2.2.

The numerical solution of the IBV problem describing the flow front propagation mainly follows standard procedures from Computational Fluid Dynamics (CFD) [7,20]. Section 2.1 presents the IBV problem and discusses special aspects, for example how the mesh handling for the Finite Area Method is different than for the Finite Volume Method.

Finally, Sect. 3 presents quantitative results and how the requirements are fulfilled.

2   Materials and Methods

2.1   Numerical Solution of the IBV Problem

Processes in continuum mechanics are described by partial differential equations with initial and boundary conditions [16]. Methods from continuum mechanics are used to describe processes in solid mechanics, fluid mechanics and heat transfer.

The filling of the cavity is modeled as compressible, two-phase (incompressible oil/resin[1] and compressible air), viscous, laminar, iso-thermal fluid flow through porous media.

From a mathematical point of view resin flow through a porous cavity is described by the following IBV problem. All physical quantities (fluid mixture mass density ρ, filling fraction c, superficial velocity $\boldsymbol{u}$ which describes the resin flow through the porous cavity, pressure p) and all parameters (porosity ε, permeability $\boldsymbol{K}$, dynamic viscosity μ) are functions of space $\boldsymbol{x}$ and time t where $\boldsymbol{x} \in \Omega \subset \mathbb{R}^3$ and Ω is the cavity. Since the flow through the porous cavity is described by the superficial velocity with zero component in thickness direction, the assumption of plug flow is valid and the cavity is described by its mid-surface with cavity thickness as a local property of the geometric model [15].

The governing equations for the default filling model in LCMsim are [15]:

$$\frac{\partial \varepsilon \rho}{\partial t} + \nabla \cdot (\rho \boldsymbol{u}) = 0 \tag{1}$$

$$\frac{\partial \rho \boldsymbol{u}}{\partial t} + \nabla \cdot (\rho \boldsymbol{u} \boldsymbol{u}) = -\nabla p + S_u \tag{2}$$

with the two-component velocity vector $\boldsymbol{u} = (u, v)$ where u and v are aligned with the cell reference directions (see the mesh handling sections for details) and with the fluid mixture mass density $\rho = \gamma \rho_{\text{resin}} + (1 - \gamma)\rho_{\text{air}}$. ρ_{resin} is the resin mass density at injection pressure and ρ_{air} is the air mass density at initial cavity pressure. The binary filling fraction γ is equal to $\gamma = 0$ (i.e. empty) if $\rho < \bar{\rho}$ and $\gamma = 1$ (i.e. completely filled) if $\rho \geq \bar{\rho}$, and the threshold value is $\bar{\rho} = (\rho_{\text{resin}} + \rho_{\text{air}})/2$. S_u is a source term which contains the pressure loss from flow through a porous medium according to Darcy's law, i.e. $S_u = -\mu \boldsymbol{K}^{-1} \boldsymbol{u}$ with dynamic viscosity μ and permeability tensor $\boldsymbol{K}$. The equation of state (EOS) models the pressure rise as the fluid mixture mass density increases. It is assumed that the pressure build up is slow at low levels of fluid mixture mass density (i.e. zero slope $\partial p/\partial \rho$). Depending on the chosen EOS model the slope at ρ_{resin} is changed (smaller slope allows for larger time steps and a very large slope can model the real compressibility of the resin). Figure 1 shows curves for two different EOS.

---

[1] Oil and resin are used synonymously here. Part manufacturing is performed with resin whereas characterization tests often are performed with oil as a test fluid.

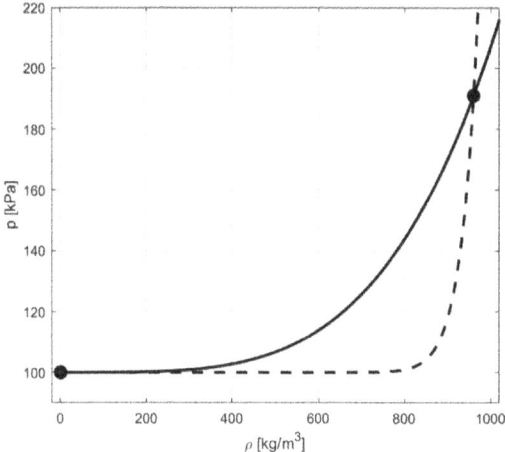

Fig. 1. The equation of state models the pressure build-up for the air-resin mixture. A polynomial fit $c_1 + c_2\rho^\alpha$ through the two mass density/pressure pairs for air at initial cavity pressure and resin at injection pressure is used. The solid curve is for air density $1.225\,\text{kg/m}^3$ at $100000\,\text{Pa}$ and resin density $960\,\text{kg/m}^3$ at $191000\,\text{Pa}$ which is used in the validation cases from [15]. The dashed EOS curve assumes real compressiblity. The solid EOS curve allows for larger time steps and some numerical diffusion.

Predominant numerical solution techniques for the IBV problem formulated by Eqs. (1) and (2) are the Finite-Element-Method (FEM), the Finite-Difference-Method (FDM) and the Finite-Volume-Method (FVM). The latter is used here. For all methods, the domain is divided into a set of discrete volumes V_i, called cells. This process is called meshing and is shown in Fig. 2. The cells (subfigure b) fill the whole domain Ω (subfigure a) without overlap. Since plug flow describes the filling of the thin-walled porous cavity, only one cell through the thickness is required (subfigure c). Thus, it is more convenient to extract a mid-surface model (subfigure d) first and then spilt this into cells to obtain a shell mesh (subfigure e) instead of a solid mesh.

The fluid flow at a certain time instance is described by values of the physical quantities in every cell. In order to calculate how these values change with the FVM, the fluid-flow equations are volume-integrated over each individual cell. If shell cells (with cavity thickness as cell property) are used instead of solid cells, the method is called Finite-Area-Method (FAM). All spatial derivatives are approximated by finite differences (algebraic relations between the physical quantities in the considered and neighbouring cells) and a set of ordinary differential equations including temporal derivative must be solved. Explicit methods calculate the values for the physical quantities at a later time instance from physical quantities at the current time. Implicit methods use physical quantities from the current and the later state.

The time step Δt in an explicit scheme is limited by the Courant-Friedrichs-Lewy (CFL) condition introduced in [4]. Implicit schemes allow for larger time

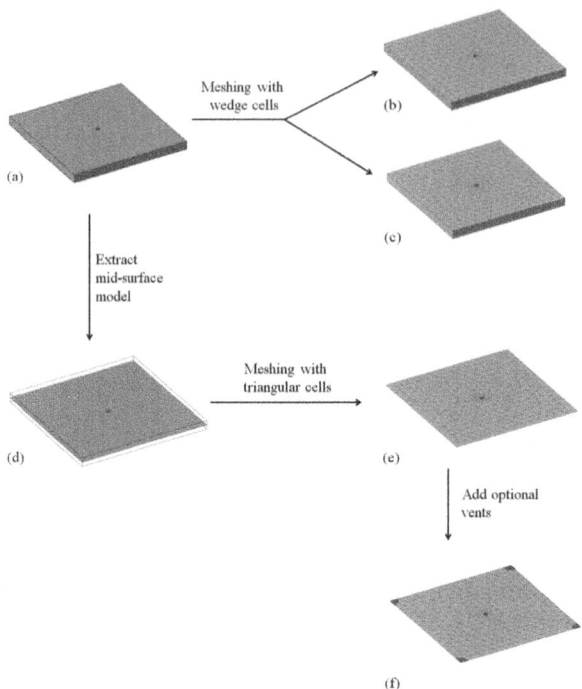

Fig. 2. Top: Solid domain with central inlet. Two different meshes (only one and four cells through the thickness) are shown. Bottom: Mid-surface domain with central inlet. A shell mesh is created. Furthermore, a setup with central inlet and optional vents is shown.

steps Δt at higher computational cost because a large linear system must be solved. An explicit scheme was chosen because it is much easier to add and modify partial differential equations.

A numerical scheme is the result of the FAM discretization. It describes how the values of a physical quantity in a considered cell i after a time increment Δt (indicated by time index $n + 1$) can be calculated from values of physical quantities in the considered and neighbouring cells. One can easily see similarities to the IBV problem in the implemented numerical scheme for cell V_i:

$$\rho_i^{n+1} = \left(\varepsilon_i \rho_i^n - \frac{\Delta t}{V_i} \sum_j \boldsymbol{n}_{i,j} \cdot \underbrace{\frac{1}{2} \left(\rho_i^n + \rho_j^n \right) \frac{1}{2} \left(\boldsymbol{u}_i^n + \boldsymbol{u}_i^n \right) A_{i,j}}_{(\rho \boldsymbol{u})_{i,j}^n} \right) / \varepsilon_i \tag{3}$$

$$u_i^{n+1} = \left(\rho_i^n u_i^n - \frac{\Delta t}{V_i} \sum_j \boldsymbol{n}_{i,j} \cdot (\rho \boldsymbol{u})_{i,j}^n u_{i,j}^n A_{i,j} - \left(\frac{\partial p}{\partial x} \right)_i \Delta t \right) / \left(\rho_i^{n+1} + (\mu_i / k_1) \, \Delta t \right) \tag{4}$$

$$v_i^{n+1} = \left(\rho_i^n v_i^n - \frac{\Delta t}{V_i} \sum_j \boldsymbol{n}_{i,j} \cdot (\rho \boldsymbol{u})_{i,j}^n v_{i,j}^n A_{i,j} - \left(\frac{\partial p}{\partial y} \right)_i \Delta t \right) / \left(\rho_i^{n+1} + (\mu_i / k_2) \, \Delta t \right) \tag{5}$$

where i is the index of the considered cell and $j \in S_i$ represents the index of a neighbouring cell and the combined index i, j represents the index of the cell interface between cell i and cell j. Only the Darcy source term is treated implicitly. Otherwise a much smaller time step than given by the CFL condition would be required. The iteration remains explicit because ρ_i^{n+1} is evaluated before u_i^{n+1} and v_i^{n+1}.

The physical quantities $u_{i,j}^n$ and $v_{i,j}^n$ at the cell interface are evaluated using first-order upwinding. For flow from cell i to cell j, the velocity components evaluated using first-order upwinding are $u_{i,j}^n = u_i^n$ and $v_{i,j}^n = v_i^n$. The j-sum runs over all neighbouring cells. When the neighbouring cell is a pressure inlet cell, the velocity in the factor without upwinding in the convective term is evaluated according to Darcy's law.

The solver section for an IBV problem consists of a `for`-loop over all interior and wall cells inside a `while`-loop for time evolution. After every time step, the time step size can be adjusted.

For every interior and wall cell, the following steps must be performed:

(1) The pressure and filling fraction gradients are evaluated.
(2) The time evolution for new mass density, x- and y-velocities with the discretized continuum, x- and y-momentum equations in the cell coordinate systems is performed.
(3) The pressure is evaluated according to the equation of state.
(4) The filling fraction is evaluated.

If boundary conditions in the pressure inlet and outlet cells are changed, a new `while`-loop has to be started.

All details of the described procedure are shown in Algorithm 1.

Parameters. The IBV problem is characterized by a set of parameters. The parameters can be split into process parameters (for example injection pressure p_a and initial cavity pressure p_{init}, mass density of air ρ_{air} at initial cavity pressure and mass density of resin ρ_{resin} at injection pressure, dynamic viscosity μ of the resin) and preform parameters for every cell group (for example: cavity thickness t at initial cavity pressure, permeability k_1 in the first reference direction, ratio k_2/k_1 of in-plane permeability for calculating the permeability k_2 in the second reference direction, porosity ε, directional vector which is projected on to cell to get the reference directions). All input parameters are summarized in Sect. 2.2.

Mesh Handling. As described in Sect. 2.1 and shown in Fig. 2e, the spatial domain is described by a mid-surface model which is meshed with triangles. The cell thickness is assigned as a cell property. A triangular cell represents a wedge cell where flow through the three lateral control surfaces is possible.

The mid-surface model can be curved and cells can have edges where more than two cells are connected to each other such as for handling T-junctions. This is illustrated in Fig. 3a.

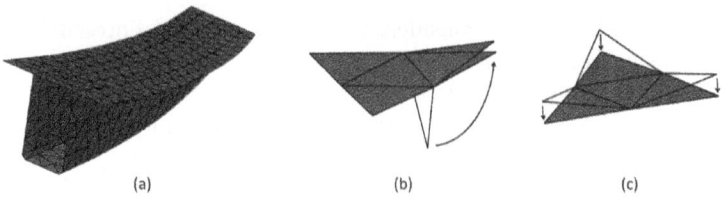

Fig. 3. Annulus filler [2] with T-sections and curved regions (a). Two examples for a locally flat mesh in (b) and (c).

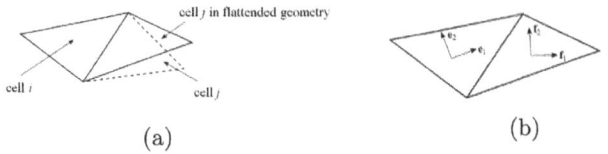

Fig. 4. Cells i and j after flattening (a) and local coordinate systems in cells i and j (b).

For discretizing the equations on the shell mesh of the part's mid-surface it is assumed that the geometry was locally flat there. The neighbouring cells are rotated about the common edges to lie in the plane of the considered cell. This is illustrated in Fig. 3b and 3c respectively. This local flattening of the geometry is an approximation. Pressure losses due to the change in flow direction are neglected which is a valid approximation for low-level flow velocity as in the considered application. The flow is considered two-dimensional and the velocity field is fully described by two in-plane components. Flow in thickness direction is neglected.

As one can see in the numerical scheme described by Equations (3) to (5), the values for the physical quantities in a considered cell i at the new time step $n+1$ after a time step Δt are calculated from the values of the physical quantities in the considered and neighbouring cells at the old time step n and by geometric properties, for example cell size V_i, interface area $A_{i,j}$ between cells i and j and also the distance vector $\mathbf{l}_{i,j}$ between neighbouring cell centers for the gradient evaluation.

Before calculating the flow velocity between cell V_i and neighbouring cell V_j at the cell boundary, the u- and v-coordinates of the neighbouring cells are rotated into the flattened geometry and transformed into the local coordinate system of cell i. Figure 4a shows cells i and j in the flattened setup and Fig. 4b shows the local coordinate systems in cells i and j.

Initialization. The cell values for the physical quantities are saved in arrays. Consequently, arrays for the physical quantities mass density, x- and y-velocity in the first and second direction of the local cell coordinate system, pressure and filling fraction are allocated for an old and a new time step. At the beginning of a simulation the arrays for the old time step are filled with the initial condition.

Handling of Boundary Conditions. In contrast to commercial CFD packages, inflow and outflow boundary conditions are not specified at cell boundaries. Additional cells for injection gates and vents must be added as indicated by the cells highlighted in red color in Subfigure e and f of Fig. 2. The pressure in these cells is prescribed and the governing equations are not solved for these cells.

If a cell face (actually a cell edge in the shell mesh) is not shared with another cell, wall boundary conditions are automatically assigned.

2.2   LCMsim Refactoring

The idea of a software tool for LCM filling simulations, which is easy-to-use for the mold engineer and gives maximum flexibility to the scientist, has led to the initial development of RTMsim and its successor LCMsim. During the attempt to use LCMsim for a Reinforcement Learning application [17], the software showed some architectural flaws, that hindered further development. To enable the usage of LCMsim in a broader range of applications and environments, the original source code was refactored to fit the newly identified needs.

In software engineering, the term refactoring describes the activity of restructuring an application's source code without changing its functionality. Refactoring aims to reduce symptoms of suboptimal software design by reverting imperfections that may have accumulated during development. As a cumulative effect, this usually results in better maintainability and extensibility of the code base as well as performance improvements [1].

This section describes why refactoring of LCMsim was needed, which goals were pursued and how the resulting software was structured. Furthermore, a short description of how to use the software module and the graphical user interface are provided. The refactored Julia module, called LCMsim v2, is available on GitHub alongside an extensive documentation [14].

Need for Refactoring. The original implementation of LCMsim showed some deficiencies which made the project hard to understand, maintain and extend, which in consequence impeded further development. Critical limitations are:

1. Former design decisions did not allow for a more general usage of the software, which will be listed below.
2. The overall structure of the software implementation showed little modularization, which caused these decisions to be spread throughout and to affect the whole monolithic architecture.

These limitations led to a refactoring of the whole project, because single problems could not be fixed separately. Some specific aspects that needed to be changed are:

1. One major problem was that the implementation only allowed the user to specify up to five sets of cells for parameterization, which is too low for some of the intended applications. This affected big parts of the preparation routine, since the five possible parameter sets were implemented statically.
2. The behaviour of the simulation in terms of writing to the file system and a graphical user interface was strongly entangled with the functionality, i.e. the numerical calculations.
3. The implementation of different physical models of calculation was distributed through the whole code and realized by conditional statements. Therefore, changing a model or adding a new one would have affected large parts of the code and this is prone to introducing errors.
4. During refactoring, some possibilities for performance optimization were found. The original software performed some disadvantageous loops where the number of iterations could be reduced or computationally expensive operations could be eliminated. The latter were identified by profiling. For example, it was found that the usage of the backslash operator for finding the least-square solution of the over-determined linear system Amat*xvec=bvec with a 3×2-matrix Amat is less efficient than a multiplication of a transposed matrix by a matrix Aplus=transpose(Amat)*Amat, a multiplication of a transposed matrix by a vector bvec_mod=transpose(Amat)*bvec and a few algebraic operations for explicitly inverting a 2×2 matrix Aplus and subsequent multiplication by a vector bvec_mod. Consequently, the code portion

```
xvec=Amat\bvec
dpdx=xvec[1]
dpdy=xvec[2]
```

was replaced by the mathematically equivalent code

```
Aplus=transpose(Amat)*Amat
a=Aplus[1,1]
b=Aplus[1,2]
c=Aplus[2,1]
d=Aplus[2,2]
bvec_mod=transpose(Amat)*bvec
inv = 1/(a * d - b * c)
dpdx = inv * d * bvec_mod[1] - inv * b * bvec_mod[2]
dpdy = -inv * c * bvec_mod[1] + inv * a * bvec_mod[2]
```

Requirements. Five high level requirements, including both functional and non-functional, were identified for the refactoring of LCMsim:

1. A critical non-functional requirement for numerical simulation is efficiency. During mold design, filling simulations can study different manufacturing concepts (i.e. placement of injection gates and vents) to guarantee complete filling. For the engineer it can be tiring to repeatedly wait for results. For machine learning applications it is mandatory to have a fast and efficient simulation to enable model training with realistic computational capabilities.
2. A needed functionality is the full access to the parameterization of each cell. This is important for:

(a) the free placement of injection gates and vents, i.e. specifying groups of cells as such,

(b) to turn these on/off, or to change the pressure level at gates and vents, and

(c) the temporal and spatial variation of the preform properties.

3. Allow for start and stop of the simulation at arbitrary time stamps.

4. The solver and the resulting data need to be efficiently accessible for different user groups. This includes the installation and the integration into other projects such as a machine learning infrastructure [17] or a graphical user interface.

5. A non-functional requirement regarding the source code is extensibility, as the goal is to create a software that fits the technical needs described above, while receiving updates from ongoing research in the fields of composite manufacturing and numerical methods. The simulation shall model different LCM technologies and support different numerical models but the source code should still be clearly readable and extendable by further models without having to change existing code (Open-Closed Principle) [8].

The Julia Programming Language [3] was chosen for this project because:

– Julia is a fast, high-performance language with just-in-time compilation.
– Julia is easy to install and available for various operating systems.
– Julia provides a simple and expressive syntax.
– LCMsim predecessors are available in Julia.

One arguable disadvantage of Julia is that it is not object-oriented and consequently does not allow for some usual and well proven software engineering practices. Due to the relatively small size of this project, the difficulties were maintainable and the advantages overruled.

Architecture. This section gives an overview of LCMsim v2's architecture. Code interna and data structures are explained when needed, but avoided if possible. The source code has been restructured by refactoring into four functional units:

1. Mesh preparation
2. Simulation setup
3. Flow problem solver
4. Collection of convenience and output functions

This modularization applies the Single-Responsibility-Principle, as single parts of the source code fulfill separated tasks and could be changed without affecting others [8]. The graphical user interface has been moved to an external package that accesses the public interface of the LCMsim v2 package.

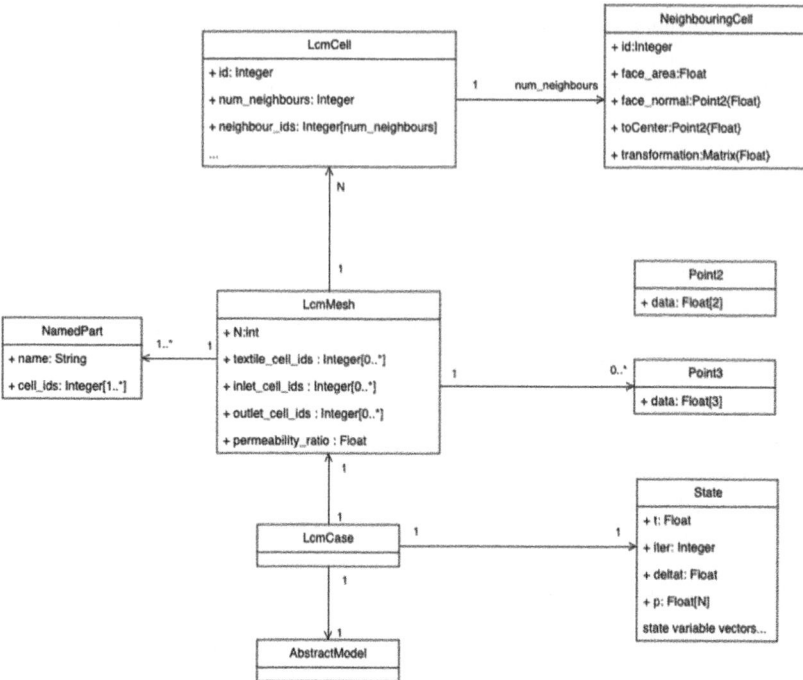

Fig. 5. The central data structures of LCMsim v2. Properties irrelevant for the overall understanding were omitted for readability. Julia structs are strictly seen different from classes, however a UML class diagram could be used here to model their relations.

Mesh Preparation: Reading different mesh formats is possible. Based on the file extension, the corresponding parsing routine is selected and the mesh is translated to a unified program representation.

The raw mesh data which just contains the positional, connectivity and grouping information of the mesh cells is enriched with user provided part information. Context and material parameters specified for sets are assigned to the cells.

Many geometric properties referring either to the individual cells or their connections are needed to solve the governing equations, see Sect. 2.1. These are calculated a priori to avoid their highly repeated computation. Opposing to the original implementation, the mesh preparation is done by a collection of subroutines which are called by a main routine. This makes this step easier to maintain and extend.

As Julia structs[2] are immutable[3] by default, the information is transferred to a structured representation of the mesh (i.e. LcmMesh) only at the very end of the mesh preparation. The data structures can be seen in Fig. 5. The main component of the LcmMesh is an array or, in Julia terms, a vector of LcmCells. Each LcmCell represents a triangular element and carries information about itself as well as information about its neighbours. The self-referring properties consist of the beforementioned material properties and geometric parameters (for example the cell's permeability or area). Those are left out in the LcmCell block for brevity. Additionally, each cell carries information about each of its neighbours, which are grouped in the NeighbouringCell struct. This allows to efficiently access these values when iterating over the cell vector to solve the governing equations. This can be seen as a trade-off between memory consumption and runtime.

Consider the following example. The pressure gradient must be evaluated at every time step for every cell. For this calculation the vectors between considered and neighbouring cell centers are required. If these vectors were not precalculated and accessible through the considered cell's struct, this would require multiple look-ups to retrieve the neighbouring cells and their centers. This pattern applies to many of the properties which are saved in LcmCell and NeighbouringCell.

As runtime efficiency was identified as a critical requirement, the decision easily fell towards higher memory consumption and reduced computational complexity. The same procedure was chosen for the index shortcut fields of LcmMesh (namely textile_cell_ids, inlet_cell_ids and outlet_cell_ids). These allow for efficient access to all cells of these types, without any extra checks and look-ups, but again at the cost of memory.

Simulation Setup: This step mainly consists of parsing the user-provided simulation parameter file and creating the initial state. The State struct describes the time varying variables of a simulation case. The vectorial fields can be matched to the cells of the mesh that this state has been created from.

A key concept of LCMsim v2 is the incorporation of multiple physical models. An argument to the simulation preparation routine indicates the choice.

The physical models are programmatically represented by structs, which carry model-specific physical parameters and whose type indicates the chosen physical model.

The model types are related to each other by inheritance. Similarly to the mesh creation, a struct of the chosen model's type is instantiated at the end of the preparation routine. The models, their relations and their properties are visualized in Fig. 6.

The initial state is then created using the newly created model's parameters.

---

[2] A struct in Julia is a composite data type that allows to store multiple values in a single object.

[3] Mutable structs are possible, but their usage is usually discouraged for efficiency reasons.

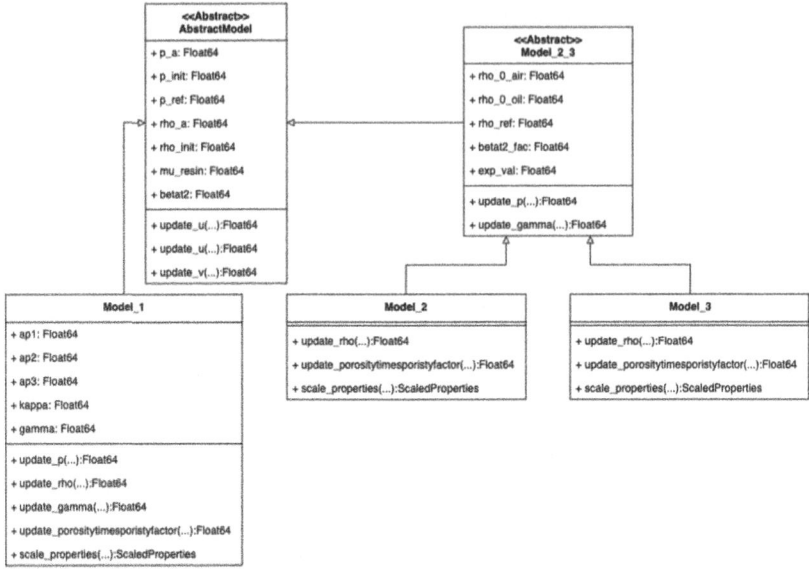

Fig. 6. Inheritance structure representing the physical models. Even though Julia not being object oriented, the struct types were modeled as classes where relevant functions mainly associated with a type were interpreted as methods of the respective class. Note `ScaledProperties`, which is a simple data structure that is created temporarily for each cell and used by the state update functions. It allows to apply a scaling to the parameters it contains, which can then retrieved by the update functions in a unified manner.

Flow Problem Solver: The core functionality of LCMsim v2 is contained in one function which takes a model instance, a mesh and a state as input. It solves the IBV problem for the chosen physical model up to a specified end time. It consists of a main time-evolution loop and the performed steps are typical for the numerical solution of an IBV problem.

An abstracted version of the main loop is depicted in Algorithm 1. It incorporates the different physical models in one clearly readable code snippet by using the programming pattern of template methods.

As introduced earlier, each physical model is represented by a subtype of `AbstractModel`. In the main loop, an update function for each state variable is called with an instance of the physical model. By supplying multiple implementations for each function signature, the developer can realize the different physical models with little effort. The implementation to use is chosen at runtime with respect to the given model type.

Algorithm 1. Abstract depiction of the solve function

Require: $mesh : LcmMesh, state_i : State, t_{next} : Float64, model : AbstractModel$

  $t_i \leftarrow state_i.t$

  $\Delta t \leftarrow state_i.\Delta t$

  $state_{i+1} \leftarrow state_i$

  while $t_i \leq t_{next}$ **do**

    for $cell \in mesh.textile\_cells$ **do**

      - Scale parameters

      $scaled\_properties \leftarrow scale\_properties(model, ...)$

      - Calculate the pressure gradient

      - Evaluate the numerical flux functions, i.e. sum term in Eqs. 3-5

      - Calculate new values for state variables according to 3-5:

      $state_{i+1}.rho[cell] \leftarrow update\_rho(model, scaled\_properties, \Delta t, ...)$

      ...

      - Overwrite old state of this cell

      $state_i.rho[cell] \leftarrow state_{i+1}.rho[cell]$

      ...

    end for

    - Calculate next time step

    $\Delta t \leftarrow adaptive\_timestep(model, \Delta t, ...)$

    $t_i \leftarrow t_i + \Delta t$

  end while

  - Return updated state

  $state_{i+1}.t \leftarrow t_i$

  $state_{i+1}.\Delta t \leftarrow \Delta t$ **return** $state_{i+1}$

In Algorithm 1, this mechanism is applied for example at `scale_properties()` and `update_rho()`, which behave differently according to the model type. For this to work, all implementations of one specific update formula need to share the exact same signature, i.e. the same argument and return types. They may only differ in the type of the model, which then determines which implementation is effectively used. Since it is not possible in Julia to force the implementations of certain functions, some discipline is required to make use of and maintain this structure. This is a shortcoming of Julia compared to most object-oriented languages where concepts such as abstract classes with abstract methods can be employed to force the implementation of these methods by specific subclasses already at compile time.

As some models share certain update formulas, it is possible to build a tree-like inheritance structure with `AbstractModel` as the root, where the models sharing these calculations are summarised under an abstract parent node. The Julia compiler chooses the implementation for the most specific type if multiple versions of a function are available. Thus the developer has to provide all implementations for a state variable update that are as specific as necessary but as general as possible according to the inheritance tree. In order to avoid confusion only one implementation should be provided per branch of the inheritance tree.

The currently implemented models can be seen in Fig. 6. This structure is compliant to the Open-Closed-Principle in a way that new numerical models can be added without modifications to the main simulation loop. Another benefit is a quick overview of the similarities and differences of the numerical models.

The class diagram in Fig. 6 lists the specific implementations of model functions. Note that every model of `Model1` - `Model3` has the same set of functions with implementations given either in the type itself or in a super type. Sometimes this leads to trivial implementations which we accept in favor of the simplicity and readability of the code.

For example, `scale_properties()` is only used in `Model3` and for `Model1` and `Model2` the function does not actually scale the values.

Convenience Functions: The end user is not meant to access directly the solve function described before. Therefore, different convenience functions are provided which serve as public interface to prepare and retrieve data and to invoke the solver. The mesh, the part description and the simulation parameters always need to be provided as files respectively their paths. The results of the preparation routine can be directly saved to a given path or retrieved as runtime objects. Depending on the application, it can be more handy to operate only on saved data or to keep everything in the memory. Both variants can be used to start or continue solving.

The datatype used for this high level communication is `LcmCase`, as can be seen in Fig. 5.

When saving results to the disk, two data formats are available: JLD2 and HDF. JLD2 is a Julia package which allows serialization of Julia objects. When using this option, a `LcmCase` instance is saved with the latest state being included. The other variant is to save data as a HDF file. HDF is data format used to efficiently store and load hierarchically organized data. When using this option, the resulting file contains sufficient information to recreate the mesh and the simulation parameters. It is also possible to save a series of states, which is for example useful to create animations of the flow front propagation. A complete documentation of the file structure of the HDF output can be found at the LCMsim v2 repository [14]. Both formats can be used to load and continue previously saved simulations.

Running a Simulation. This section shortly summarizes the necessary steps to perform a simulation with LCMsim v2.

LCMsim v2 does not include mesh generation as meshing is considered an external task[4]. A three-node triangular shell mesh must be imported. Depending on the mesh format, cell sets which are assigned with different preform properties have to be specified differently. Currently, three different mesh formats are implemented and the selection is based on the file extension:

---

[4] Mesh generation is possible with commercial software packages (for example Altair HyperWorks, Ansys or Abaqus) and free software packages (for example Salome-Meca, PrePoMax, Gmsh or Netgen).

Table 1. Additional input for the parts defined by the sets in the mesh file.

Column name	Description	Value range
name	Name of the part. Will be used later on to match action to inlets/outlets	Freely choosable (reasonable)
type	Type of this part	{base, inlet, outlet, patch}
part_id	The physical part ID assigned to the part in the mesh file.	Determined by mesh file.
thickness	Cavity thickness.	> 0.
permeability	Permeability in first principal direction.	> 0.
permeability_noise	Standard deviation applied to permeability values with a normal distribution.	0-0.5 (no definitive cap but unrealistic values break simulation).
porosity	Porosity.	$\in (0., 1.)$
porosity_noise	Standard deviation applied to porosity values with a normal distribution.	0-0.5 (no definitive cap but unrealistic values break simulation).
porosity_1	Porosity at pressure p_1.	$\in (0., 1.)$
p_1	Pressure corresponding to porosity p_1	> 0.
alpha	Factor for permeability in second principal direction.	Real
refdir_1	x-,	Real
refdir_2	y-,	Real
refdir_3	z-component of reference direction which will be projected onto the cells to give the first principal direction.	Real

- .dat: Nastran format with nodes, elements and element sets.
- .bdf: Nastran format with nodes, elements and property ID assigned to elements.
- .inp: Abaqus format with nodes, elements and element sets.

The parts defined by the sets in the mesh file need additional input parameters for the simulation. These are provided via a .csv-file with the content shown in Table 1. For example, the following snippet describes the properties for the validation case 1 from [13]:

```
name,type,part_id,thickness,porosity,porosity_noise,permeability,
    permeability_noise,alpha,refdir1,refdir2,refdir3,
    porosity_1,p_1
base,base,1,3e-3,0.7,0.0,3e-10,0.0,1,1,0,0,0.7,0.1e5
central_inlet,inlet,2,3e-3,0.7,0.0,3e-10,0.0,1,1,0,0,0.7,0.1e5
```

The physical parameters for the simulation need to be provided by a .csv-file with the content shown in Table 2. For example, the following snippet describes the parameters for the validation case 1 from [13]:

```
p_ref,rho_ref,gamma,mu_resin,p_a,p_init,rho_0_air,rho_0_oil
1.01325e5,1.225,1.4,0.06,1.35e5,1.0e5,1.2,960
```

With these three files prepared, the solver can be started using one of the functions from the public interface.

GUI. The graphical user interface is provided via a separate repository [10]. The GUI solely depends on functions from LCMsim v2's public interface, such that the functional internals of both modules are fully decoupled.

The GUI was developed for Windows operating systems and is launched via a batch file which starts Julia with a configuration file as argument, i.e. julia -L lcmsim_config.jl. The configuration file specifies:

- if LCMsim is started in batch mode where files for the part properties, simulations parameters and mesh are used or if parameters are filled in the mask,
- which model type is started (2: model from [15], 4: model 2 with degree-of-cure equation),
- which mesh type is used (1: .dat, 2: .inp, 3: .bdf),
- the path to the LCMsim_v2.jl repository,
- the path to the GUI repository.

The following configuration file is used to start the GUI shown in Fig. 7:

```
i_batch=1
i_model=2
i_mesh=1
mypath=joinpath(pwd())
repositorypath="D:\\work\\github\\LCMsim_v2.jl"
guipath="D:\\work\\github\\LCMsim_GUI\\gui_and_cases\\gui"
include(joinpath(guipath,"lcmsim_v2_gui_gtk4.jl"))
```

Clicking the button Run with input files calls the create_and_solve function from the public interface.

Table 2. Physical parameters for the simulation.

Column name	Description
p_ref	Reference pressure (only for model type 1)
rho_ref	Reference mass density (only for model type 1)
gamma	Adiabatic index (only for model type 1)
mu_resin	Dynamic viscosity of the oil/resin
p_a	Pressure at the inlets
p_init	Initial cavity pressure
rho_0_air	Density of air at p_init
rho_0_oil	density of oil/resin at p_a

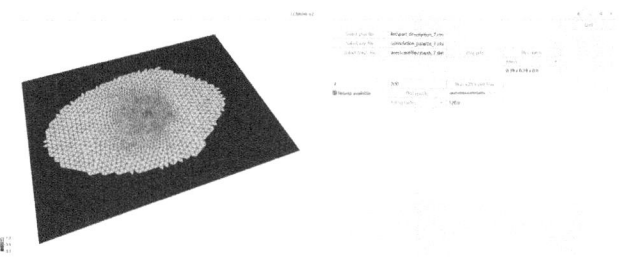

Fig. 7. GUI for starting a flow simulation run with files for part properties, simulation parameters and mesh.

3   Results and Discussion

The improvements from the refactoring are demonstrated by a comparison of the computational time for test case 1 of [15] (radial flow experiment in a permeameter with central inlet and two preform regions with different in-plane permeability) and by adding an additional conservation law for the degree of cure.

In Sect. 3.1, a first comparison verifies that the simulation results have not changed due to the refactoring step. In Sect. 3.2, a second study shows which code parts have to be adapted in order to integrate the effect of viscosity on mold filling when involving fast curing thermosets.

3.1   Performance Comparison

The relevant geometric parameters are as follows (test case 1 from [15]): Flat plate cavity ($390 \times 290 \times 3.14 \, \text{mm}^3$) with central injection port (13 mm diameter). The main preform covers the leftmost preform portion (245 mm wide) and the manipulated preform with lower permeability and porosity covers the rightmost section (145 mm wide). Initial cavity pressure is ambient pressure 100000 Pa and injection pressure is 191000 Pa. The fluid properties are mass density $960 \, \text{kg/m}^3$ and dynamic viscosity 0.071 Pas. Different levels of flow front propagation velocity are expected in the different preform zones.

The flow experiment is simulated with both tools, i.e. LCMsim and LCMsim v2, respectively, on a hardware which is state-of-the-art for mechanical engineering simulations (Processor Intel Core i7-12800H, 64 GB RAM). The simulation for the whole 160 s filling time takes 490 s with LCMsim v2, showing an improvement by a factor of approximately 17 over LCMsim. A similar factor is achieved for other cases since the improvement comes from the time per iteration (i.e. for one time step) and the adaptive time stepping was not changed.

The filling after 120 s simulated with LCMsim and LCMsim v2 is shown in Fig. 8 and shows negligible differences. The difference is measured by comparing the filled areas. The two filing contour plots from Fig. 8 are converted into binary pictures of equal size where the black area covers the filled area. Then, the numbers of black mesh cells are used to calculate a relative error in the filled area. The filled area is nearly identical ($< 0.5\,\%$ difference). This difference in the filled areas is illustrated in Fig. 9 where transparent filling patterns are superimposed: Only five cells (in light grey) at the top show a difference in filling.

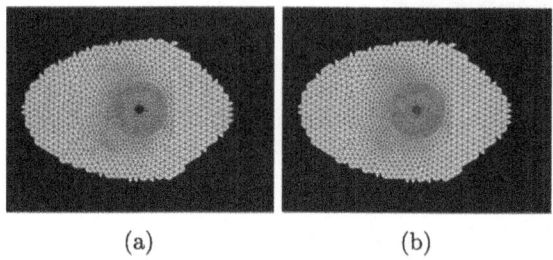

(a)   (b)

Fig. 8. Filling for test case 1 after 120 s: (a) LCMsim and (b) LCMsim v2.

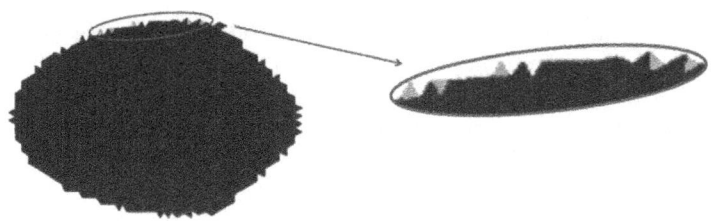

Fig. 9. Superposition of filled areas simulated with LCMsim and LCMsim v2, respectively, for test case 1 after 120 s. Light grey cells represent cells with different filling fraction.

3.2   Extending the Scope of Simulation

If a fast curing thermoset is used instead of oil as test fluid, curing of the resin must be considered during mold filling. The resin will undergo a chemical change due to the cross-linking reaction and will then form a highly cross-linked solid.

The degree-of-cure α is a variable used to describe the state of cure of the resin. It ranges from 0 to 1. When the degree of cure is 0, the resin is uncured, and when the degree of cure is 1, the resin is completely cured.

Depending on the resin system used and the type of chemical reactions there are several kinetic models available in literature for the cure reaction of thermosetting resins [5]. The most widely used and the simplest is the nth-order reaction model.

The degree-of-cure equation with convection which has to be implemented in LCMsim v2 is:

$$(\varepsilon\gamma\alpha)_t + \mathbf{u} \cdot (\gamma\alpha) = S_\alpha \tag{6}$$

with porosity ε, filling fraction γ, superficial velocity $\mathbf{u}$, source function $S_\alpha = k_{\mathrm{doc}}(1 - \alpha)^{n_{\mathrm{doc}}}$ and modeling parameters k_{doc} and n_{doc}. Similar to the implementation of the filling fraction in RTMsim [13] the equation is written in conservative form as:

$$(\varepsilon\gamma\alpha)_t + \nabla \cdot (\gamma\alpha\mathbf{u}) - \gamma\alpha\,(\nabla \cdot \mathbf{u}) = S_\alpha \tag{7}$$

The viscosity change with the degree of cure is described by:

$$\mu = \mu_0 e^{k_\mu \alpha} \tag{8}$$

with uncured dynamic viscosity μ_0 and modeling parameter k_μ.

In total, three additional parameters for the modeling equations (k_{doc}, n_{doc} and k_μ) and two parameters for the initial and boundary conditions ($\alpha_{\mathrm{init}} = 0$ and $\alpha_{\mathrm{a}} = 0$) are required.

A branch doc_eq was created on the GitHub repository where all the changes, described in Appendix A were implemented. The demonstrated procedure can be generalized for modifying existing or adding new conservation laws, for example a temperature equation.

Results for Filling Simulation with Degree-of-Cure. The considered test case is the same as in the previous section. Figure 10 shows the flow front after 200 s with (i) constant viscosity and (ii) degree-of-cure dependent viscosity. Parameters used are $k_{\mathrm{doc}} = 0.007$, $n_{\mathrm{doc}} = 1.2$ and $k_\mu = 0.0$ and 1.5 respectively, for case (i) and (ii). The parameters are chosen such that $\gamma > 0.5$ after 120 s and the flow front slows down significantly.

As expected, the filling for (i) is equal to the filling from the previous section. One can see numerical diffusion in the degree-of-cure at the flow front. This could be avoided by plotting the product of the binary filling fraction times degree-of-cure instead of the degree-of-cure. By plotting the product one gets a non-zero degree-of-cure only if already filled which is reasonable. The filling with degree-of-cure dependent viscosity shows a slower flow front propagation as expected.

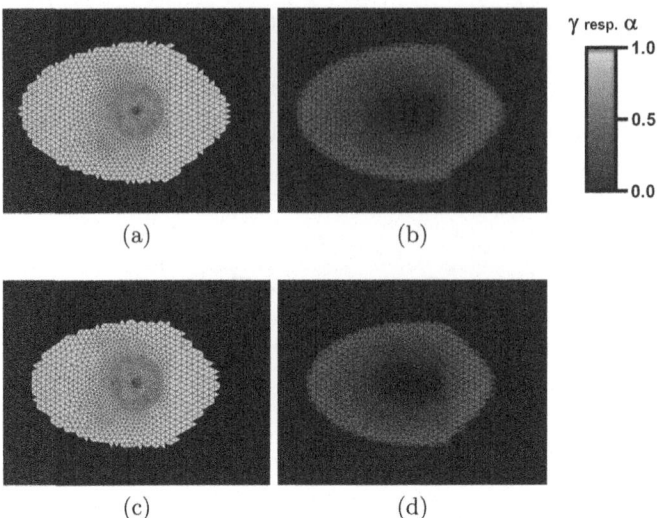

Fig. 10. Filling fraction γ (left) and degree-of-cure α (right) with constant (top) and variable (bottom) viscosity. The increase of the degree-of-cure caused by the chemical cross-linking reaction causes an increase in fluid viscosity and thus, a deceleration of flow propagation.

4   Conclusion and Outlook

The work at hand describes code modifications when a software engineer meets self-fabricated engineering software. The Julia package LCMsim for filling simulations in liquid composite molding was analyzed and requirements for a refactoring step were formulated. After refactoring, the required computational time for a typical benchmark problem is reduced by a factor of 17. Furthermore, adding supplementary equations describing phenomenological mechanisms was facilitated. The refactored version is available on a Github repository [14] where in a separate branch all described changes for adding a supplementary equation are shown in detail.

As a next step, new equations (for example the temperature equation) and new functionalities (for example for vacuum-assisted resin infusion) will be implemented. Improved performance together with the opportunity to customize the modelled physical and chemical processes constitute a very useful tool for engineers and scientists in the field of composite manufacturing.

A   Code Preparation

The following code modifications are necessary to solve an additional equation:

– The state vector will be extended for the variable doc (degree of cure).

– A new model type with ID 4 is created. The structure Model_4 extends Model_2_3. All structures have additional parameters used to describe the changes in the degree of cure and the viscosity (k_doc, n_doc, k_mu, doc_init, doc_a). Otherwise the states cannot be initialized.
– The csv-file with the simulation parameters optionally (in order to be backward compatible with the already existing model types) can have three additional parameters (k_doc, n_doc, k_mu; the values for doc_init=0, doc_a=0 need not be read).
– The functions save_model and load_model must be extended for model type 4 and for the five additional parameters in all model types.
– The function save_state must be extended for the variable doc.
– The function create_initial_state must be extended for the variable doc.
– In file abstract_model.jl, create a dummy function update_doc with returned value 0. and function update_viscosity must have an additional input argument doc_new.
– The file model_4.jl is created as copy of model_2.jl. Appropriate functions for the degree-of-cure update_doc and the viscosity update doc_new must be created.
– Include model_4.jl in file LCMsim_v2.jl.
– In function solve:
    • Create a vector for the old state doc_old and deep copy the values of the values from the vector with the old into the vector with the new state doc_new.
    • Inside the time evolution loop, call the functions for the numerical flux calculation and the degree-of-cure update, add the additional argument to the viscosity update function and assign the new values to the old values of the degree-of-cure at the end of one time step.
    • Add the degree of cure to the return of the final state.
– The numerical scheme itself is implemented in the functions update_doc and aggregate_neighbour_flux_doc which calls the functions numerical_flux_function_doc and numerical_flux_function_boundary_doc. The functions with the appendix _doc are modified copies of the corresponding functions without this appendix.

References

1. Refactoring: Improving the Design of Existing Code. Addison-Wesley Longman Publishing Co., Inc., USA (1999)
2. Barandun, G., et al.: Development of a next generation composite - annulus filler for rolls royce jet engines. In: SEICO 13 SAMPE EUROPE 34th International Conference and Forum (2013)
3. Bezanson, J., Edelman, A., Karpinski, S., Shah, V.B.: Julia: a fresh approach to numerical computing. SIAM Rev. **59**(1), 65–98 (2017)
4. Courant, R., Friedrichs, K., Lewy, H.: Über die partiellen Differenzengleichungen der mathematischen Physik. Math. Ann. **100**, 32–74 (1928)

5. Dodiuk, H., Goodman, S.: Handbook of Thermoset Plastics, vol. 254, January 2014
6. ESI Group: PAM-COMPOSITES: Composites Simulation Software, https://www.esi-group.com/products/pam-composites
7. Ferziger, J.H., Perić, M.: Computational methods for fluid dynamics. Springer (2002)
8. Martin, R.: Agile Software Development: Principles, Patterns, and Practices. Alan Apt series, Pearson Education (2003)
9. Niu, C., Niu, M.: Composite airframe structures: practical design information and data. Adaso Adastra Engineering Center (1992)
10. Obertscheider, C.: GUI for refactured LCMsim (v2). https://github.com/LCMsim/LCMsim_GUI (2024)
11. Obertscheider, C., Fauster, E.: RTMsim - a Julia module for filling simulations in resin transfer moulding. https://github.com/obertscheiderfhwn/RTMsim (2022)
12. Obertscheider, C., Fauster, E.: LCMsim - a Julia module for filling simulations in Liquid Composite Moulding. https://github.com/obertscheiderfhwn/LCMsim (2023)
13. Obertscheider, C., Fauster, E.: RTMsim - a Julia module for filling simulations in Resin transfer Moulding with the finite area method. J. Open Source Soft. 8(84), 4763 (2023)
14. Obertscheider, C., Fauster, E., Heber, L.: LCMsim_v2.jl - Julia module for the refactured LCMsim (v2). https://github.com/LCMsim/LCMsim_v2.jl (2024)
15. Obertscheider, C., Fauster, E., Stieber, S.: Experimental validation of a new adaptable LCM mold filling software. Adv. Manuf. Polymer Compos. Sci. 9(1), 2282310 (2023)
16. Reddy, J.N.: An Introduction to Continuum Mechanics. Cambridge University Press, 2 edn. (2013)
17. Stieber, S., Heber, L., Obertscheider, C., Reif, W.: Control of composite manufacturing processes through deep reinforcement learning. In: 2023 International Conference on Machine Learning and Applications (ICMLA), pp. 17–22 (2023)
18. Stieber, S., et al.: Inferring material properties from FRP processes via sim-to-real learning. Int. J. Adv. Manuf. Technol. 128, 1–17 (2023)
19. Strong, A.: Fundamentals of Composites Manufacturing, Second Edition: Materials. Society of Manufacturing Engineers, Methods and Applications (2008)
20. Versteeg, H., Malalasekera, W.: An Introduction to Computational Fluid Dynamics: The Finite Volume Method. Pearson Education Limited (2007)

Self-powered Embedded Systems: The Role of Non-volatile Memory Technology in IoT Devices

Stefan Wildermann$^{(\boxtimes)}$ [ID], Nils Wilbert [ID], Tobias Häberlein,
and Jürgen Teich [ID]

Friedrich-Alexander-Universität Erlangen-Nürnberg, Erlangen, Germany
{stefan.wildermann,nils.wilbert,tobias.haeberlein,juergen.teich}@fau.de

Abstract. Modern embedded systems are characterized by a steady progression towards greater efficiency, complexity, and autonomy. A key aspect of this evolution is the incorporation of self-* properties, enabling embedded systems that adapt and optimize performance autonomously. While self-* properties are fundamental for building *autonomous* embedded systems, they alone are not sufficient to realize *autarkic* embedded systems. A truly autarkic system should be able to operate self-sufficient by harvesting energy from its environment, eliminating the need for external power sources or frequent battery replacements, which is of particular interest in the Internet of Things. However, the possibility of unexpected power outages and the consequent loss of data and system states necessitates a fundamental re-evaluation of how modern embedded computer architectures are designed.

This article emphasizes the role of the memory system in this context. Here, non-volatile memory (NVM) technologies can retain data even in time intervals of power shutdown or intermitting power outages. In this paper, we present the fundamentals of intermittent computing with unreliable power supply as well as the benefits, design considerations, and challenges associated with integrating NVM into the memory hierarchy of self-powered systems. Finally, we discuss future research directions for enhancing the resilience and efficiency of using NVM in the context of self-powered IoT systems.

Keywords: Internet of Things · Self-sufficiency · Non-Volatile Memory

1 Introduction

The Apollo Guidance Computer (AGC) is widely considered to be one of if not the first embedded system. It was developed in 1965 for NASA's Apollo space

This paper is dedicated to Wolfgang Reif at the occasion of his 65th birthday. The work is supported in parts by the German Research Foundation (DFG) as part of the priority program "SPP 2377: Disruptive Memory Technologies" under project HYPNOS (project number 502213043).

G. Ernst et al. (Eds.): Wolfgang Reif Festschrift, LNCS 15765, pp. 155–177, 2025.
https://doi.org/10.1007/978-3-031-92196-4_8

mission. An embedded system is a computer system that has been designed for a particular technical context and performs specific tasks within this context. In case of the AGC, the tasks were monitoring the on-board sensors, navigation, and controlling the spacecraft's operations. With the development of the first microcontrollers and the growing demand for automation, embedded systems became increasingly widespread. Particularly, the emergence of ubiquitous computing and the Internet of Things has driven the seamless integration of computer systems into everyday objects and environments, making technology available everywhere and at any time.

Embedded systems that are deployed in modern applications like autonomous vehicles, smart grids, and the Internet of Things are characterized by a steady progression towards greater autonomy, efficiency, and complexity as systems must operate reliably in dynamic and often unpredictable environments. A key aspect of this evolution is the incorporation of self-* properties [15,24], such as self-configuration, self-optimization, self-healing, and self-protection. These properties allow embedded systems to adapt to changing conditions, optimize their performance, recover from faults, and protect themselves from threats without human intervention.

While self-* properties are fundamental for building *autonomous* embedded systems, they alone are not sufficient to realize *autarkic* embedded systems that are additionally capable in recuperating energy for their operation from their environment. While the power demands of computer systems are constantly growing, so is their energy consumption. Here, particularly for battery-powered systems, self-* properties can be utilized to provide smart power management for prolonging battery life. However, the capacity of the battery determines an upper bound on a system's availability. A truly autarkic system should be able to operate self-sufficient with respect to power supply even with only small battery capacities or completely without battery. They must be able to derive the necessary electrical energy from their environment via energy harvesting of, e.g., solar, wind, thermal or kinetic energy.

In this article, we take a closer look at such self-powered embedded systems. They must cope with the uncertainty and fluctuations of these sources, like the varying availability of solar energy during the day and night. Dynamic energy management techniques are usually deployed in this context to deal with this algorithmically. However, the risk of unforeseen power outages and the resulting loss of data and system states also calls for a fundamental rethinking of how to design the computer architecture of modern embedded processors deployed in this context. Emerging NVM technology [17] offers the potential for improving computer systems, including higher memory density, lower power consumption, and persistence.

Traditional computer systems today already employ a combination of non-volatile disks and volatile components like main memory, caches, and registers. These systems rely on file systems and operating system support to implement the mechanisms and data structures necessary for ensuring data persistence and consistency. However, the volatility of the CPU and main memory often introduces significant overheads in terms of latency and energy consumption. NVM technologies constitute a new approach for designing self-powered computer systems by bringing persistence closer to the CPU core. As NVM is able to retain data even without power supply, it offers the potential to design systems that are more robust against frequent and unforeseen power outages. We focus on the memory hierarchy of self-powered embedded systems. We discuss the design space of building such systems and how to program and operate them in the presence of power outages. This article particularly focuses on architectures with non-volatile main memory and also future systems that even employ non-volatile caches. We compare these designs both by evaluating their latency and energy overhead on the example of a programm intermittently filling data into a B+ tree data structure by employing a cycle-accurate system simulator [4].

The article is structured as follows. Section 2 provides more detail on the background of self-powered embedded systems. Section 3 takes a closer look at disruptive memory technology and how to utilize it in the memory hierarchy of embedded processors. Section 4 discusses the challenges and future research directions of using such computer architectures in self-powered IoT devices, before concluding this article in Sect. 5.

2   Background on Self-powered Embedded Systems and Intermittent Computing

Self-powered embedded systems are designed to operate using energy harvested from their environment instead of solely relying on traditional power sources like wired power, batteries or supercapacitors. Such systems are typically used in scenarios where recharging or replacing the battery is impractical or where long-term operation with minimal maintenance is required. These requirements particularly exist in various application domains from the Internet of Things. For example, wearable devices, i.e., electronic components that are attached to the human body, enable real-time monitoring of health data in medical and sports scenarios. While have to be small and provide continuous monitoring, they are operated in contexts that provide plenty of energy sources that can be harvested, ranging from environmental sources like solar energy to thermal and mechanical energy produced by the human body [2]. Moreover, many Internet of Things scenarios aim at collecting, processing, and analyzing sensor data. In wireless sensor networks, low-power embedded systems are deployed to collect sensor data. More and more of processing is already performed on these embedded systems before transferring relevant data to a backend (usually a cloud server). Particularly, systems used to perform remote monitoring (e.g., in areas like agriculture and biodiversity monitoring [33]) often relying on energy harvesting techniques due to the lack of other power supplies.

Figure 1 provides a high-level overview of the system architecture of self-powered embedded systems. The main purpose of such systems is to collect and process input data (e.g., sensor data) and produce an output (e.g., filtered and preprocessed data to be transmitted via a wireless interface, or control

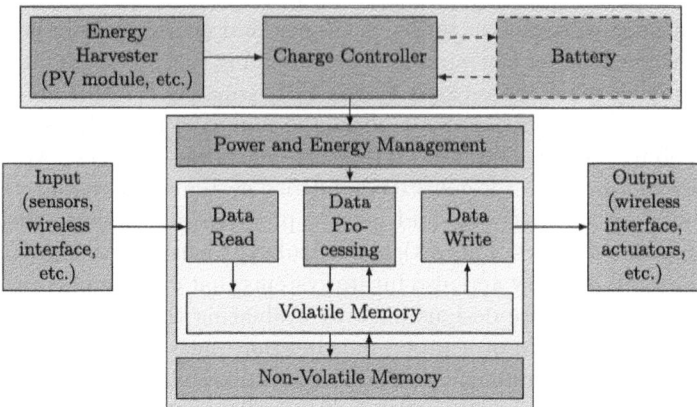

Fig. 1. System architecture of self-powered embedded systems.

signals to operate attached actuators). A central aspect of self-powered systems is their power supply. An energy harvester is responsible for generating power out of a renewable energy source. A charge controller distributes the power to the processing system. In case of battery-backed systems, the power can also be used to recharge the battery. The battery can be discharged in cases where insufficient energy can be harvested to operate the system. In contrast, battery-less systems do not contain such a backup and can only operate when energy is harvested. In both cases, the available energy is limited and may fluctuate during operating. An energy manager must monitor available energy and determine the timing of data processing as well as the appropriate response to power outages.

Intermittent computing [32] is a computing paradigm designed for environments where a continuous power supply is unreliable or unavailable, which is typically the case when operating on harvested energy. These systems are designed to perform useful computations in short bursts, aligning their operations with the availability of power. The major issue is that the complete or parts of the computational state may be stored in volatile memory. Then, in case of a power outage, the information contained in volatile memory is lost. For being able to make computational progress in the face of power outages, intermittent computing systems [32] thus have to be able to save their computational state and resuming operations once power is restored. Figure 2 illustrates three options for intermittent operation by means of an example.

Figure 2(a) shows the energy over time available to a system in this example, which is exposed to two power outages. A common approach to handle power outages is proactive *checkpointing* [32], i.e., saving the system's state at specific points in time, as illustrated in Fig. 2(b). This allows the system to restore the saved state and continue processing after power is restored. Also, persistent data structures (e.g., [34]) can be counted to this category. They are programmed such that they always either represent a consistent state in a non-volatile memory or a state from which the previous consistent state can be restored. Another con-

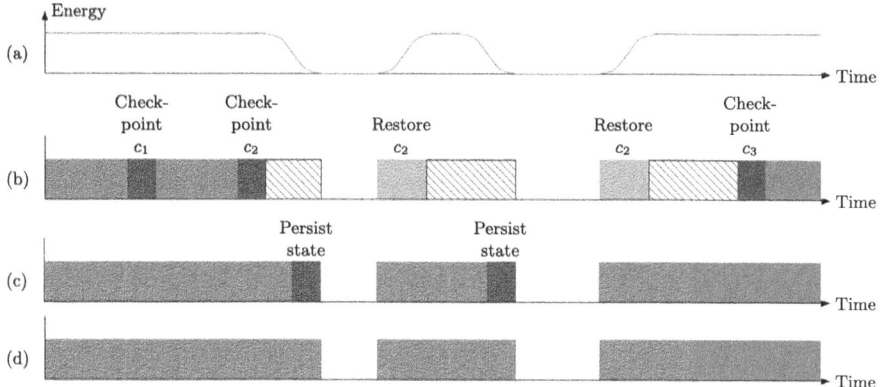

Fig. 2. Illustration of intermittent computing and the role of checkpointing. (a) Example of energy harvested over time. System operation (b) with regular checkpointing and restoring last checkpoint after power outage, (c) with power-outage-aware checkpointing, and (d) of a completely non-volatile system.

cept is logging [32]: Information about the computation (such as the current state) is stored in a log placed in non-volatile memory. When the system loses power and later resumes, the logged data allows the system to recover its state and continue processing from where it left off. However, these checkpointing concepts have several drawbacks. First of all, frequently performing checkpointing induces a performance and energy overhead, as the computation cannot be proceeded until the checkpoint is generated. In addition, the restoration of the state from the saved checkpoint can be energy- and time-consuming. Finally, the forward progression of computation may severely suffer from frequent power outages as all the computation after the last checkpoint is lost. This is exemplified in Fig. 2(b) by the operation filled with a hatched pattern. This operation is interrupted by two power outages. Each time, the computation has to start over from checkpoint c_2 before it can finally be finished.

Another approach is to *persist the system state* early enough before the power is fully gone (outage), as shown in Fig. 2(c). Once power is back, the system can resume immediately with the computation. Finally, the ideal case is a completely *non-volatile system*. In this case, no information is lost during power outages. The system thus neither needs explicit checkpointing, persisting, nor restoration, as illustrated in Fig. 2(d).

NVM technologies offer a new approach to designing intermittent computing systems by bringing data persistence closer to the CPU core. In the following section, we thoroughly discuss the design space of using NVM technology in computer architectures tailored to self-powered intermittent computing systems.

3   The Role of Disruptive Memory Technology for Intermittent Computing

Next-generation non-volatile memory technology opens new perspectives for designing computer architectures. In this section, we give an overview of the NVM technology and the design options for computer architectures. We further-more discuss how the design of the memory hierarchy influences which of the intermittent computing scenarios presented in Fig. 2 can actually be supported.

3.1   Non-Volatile Memory Technology

Besides the major advantage of persistence of the data, non-volatile, usually byte-addressable memory technologies also exhibit a better scalability in terms of memory density in comparison to DRAM. This is because it is becoming increasingly difficult to reduce the size of the capacitors used in DRAM tech-nology so that they can continue to reliably hold a charge [30]. Two of the most promising NVM technologies are presented below.

Phase-Change Memory (PCM) [10] is a non-volatile memory technology that uses the phase-changing properties of certain materials to store information. The material in the memory can switch between crystalline and amorphous phases when heated, which results in significant changes to its optical and electrical properties. In the amorphous phase, the atoms are disordered, leading to low optical reflectivity and high electrical resistance. In the crystalline phase, the atoms are arranged in an orderly fashion, resulting in higher reflectivity and lower resistance. PCM relies on the difference in electrical resistance between these two phases to store a bit. To write data, a small current pulse heats the material to just above its crystallization point for a logical 1, while a short, higher current pulse melts and rapidly cools it into the amorphous phase for a logical 0. Data is read by applying a small current and measuring the cell's resistance.

Spin-Transfer Torque RAM (STT-RAM) [19] is a variation of Magneto-Resistive RAM (MRAM), which utilizes quantum mechanical effects to store information by changing electrical resistance, similar to PCM. STT-RAM employs a magnetic tunnel junction (MTJ), composed of two ferromagnetic lay-ers separated by an insulating tunnel barrier. The lower layer, called the *reference layer* has a fixed magnetic orientation, while the upper *free layer* can have its magnetic orientation altered by applying a strong current. When the free layer's magnetic alignment matches the reference layer's, the MTJ exhibits low resis-tance, representing a logical 0. When their alignments are opposite (antiparallel), the resistance is high, corresponding to a logical 1. Writing data involves chang-ing the magnetic orientation of the free layer by controlling the current flow through the tunnel junction. Similar to PCM, STT-RAM requires a high cur-rent for write operations, and data is read by measuring the electrical resistance through the application of a small voltage.

PCM is considered a good candidate for main memory because of its high memory density and non-volatility. While STT-RAM does not reach the memory

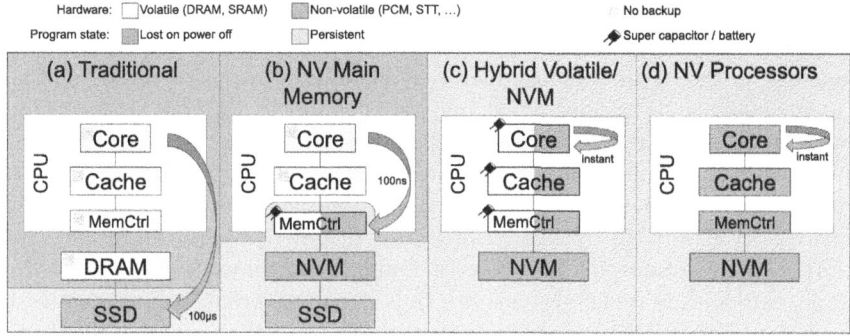

Fig. 3. Traditional (a), and hybrid (mixed volatile/non-volatile) computer memory architectures (b)-(d) (based on [17])

density of PCM, both read and write latencies are significantly lower with STT-RAM compared to phase-change memory. SST-RAM is thus considered a good candidate for cache memory.

To store information in NVM, write operations typically change the physical state of the used material for which they require a lot of time and energy. To read the information, the resistance of the material is then measured, which is usually much faster and consumes less power. The resulting asymmetry between read and write operations is a major disadvantage of these non-volatile memory technologies.

3.2 Design Space of Computer Architectures for Intermittent Computing

Figure 3 [17] gives an overview of different design options for computer system architectures with a focus on the volatility of the memory components. Traditional systems (a) only have a non-volatile disk while all other memory components are volatile. (b) First systems are commercially available, which also include non-volatile main memory (NVMM). Examples are Intel Optane[1] for the server domain and Texas Instrument MSP430FR microcontrollers[2] from the embedded domain. Despite the non-volatility of the main memory, these systems still contain volatile caches and registers, which has to be taken into account when using them in intermittent computing scenarios. (c) Systems that combine volatile and non-volatile memory components throughout the memory hierarchy are recently gaining more and more research attention. The idea is to equip the system with small super-capacitors or batteries which should provide sufficient energy for writing data from the volatile memory components to non-volatile ones as soon as a deeming power outage is detected. The

---

[1] https://www.intel.com/content/www/us/en/content-details/756238/intel-optane-technology-tiered-memory-white-paper.html.

[2] https://www.ti.com/tool/MSP-EXP430FR5739.

state is thus persisted, and the system can resume computation as soon as the power supply is regained. (d) Finally, non-volatile processors make use of NVM technology in the entire memory hierarchy, including processor registers. Thus, data and state is always persistent and never gets lost throughout power outages. Whereas the concept of (d) should allow the fastest progress of processing a given workload, unfortunately, non-volatile processors can only be operated with low clock frequencies due to the energy- and time-consuming write operations on NVM and an by several orders of magnitude lower endurance than SRAM. Design options (b) non-volatile main memory systems and (c) hybrid volatile/non-volatile cache systems are thus the most promising candidates for self-powered intermittent computing systems. In the following, we evaluate and compare both options.

3.3   Non-Volatile Main Memory (NVMM) Systems

In non-volatile main memory (NVMM) systems, the main memory (usually provided by DRAM) is replaced by a NVM technology (PCM is particularly suitable for this purpose). As all data in main memory is persistent, disk storage becomes obsolete in such systems. However, using NVMM correctly presents significant challenges. One of the main difficulties is ensuring that the data in the non-volatile main memory is in a consistent state, so that it is possible to recover from power failures. This requires understanding how and when writes are propagated to main memory, which is complicated in the presence of multiple volatile caches between the CPU and non-volatile main memory. As the content of volatile caches gets lost in case of power failures, inconsistent states can otherwise occur. In the following, we discuss the major issues and concepts how to handle them.

Challenge 1 – Reordering of memory operations: Processor performance has advanced faster than the performance of memory over recent decades. Thus, memory operations often have higher latency than other CPU operations, especially when data needs to be fetched from main memory rather than from a fast cache. By reordering memory operations, the processor can better utilize its execution units and parallelize operations, reducing idle time and increasing throughput. Moreover, modern caches and main memories are often divided into multiple banks, each of which can be accessed independently. Processors and memory controllers may reorder memory operations to perform multiple memory operations in parallel, improving memory bandwidth and reducing contention [28]. Essentially, this means that memory operations do not have to be executed in the order in which they are defined by the program.

Challenge 2 – Visibility $\neq$ persistence of memory operations: Processors may not arbitrarily reorder memory operations. Memory *consistency* models define rules in which order memory operations become visible in the system [27]. The reordering performed by the processor must comply with these rules. However, the order of *persistence* is not the same as the order of *visibility*. Due to the cache hierarchy between CPU and main memory, writes may not be propagated to main memory in the order they were issued by the processor. This can lead to unexpected outcomes. For example, after executing a simple program like

```
1                      x  :=  1;
2                      y  :=  1;
```

the cache may contain the correct values of x and y. However, a crash might leave the main memory with y = 1 and x = 0, when the write to x was cached but has not reached the non-volatile memory before the crash. This means that even though a memory operation was issued, its effect may not have been persisted. More and more effort is laid on defining proper *memory persistency models* for being able to specify valid orders in which data is written to persistent memory, see [25, 27].

Challenge 3 – Atomic persist granularity: Atomic persist granularity refers to the smallest unit of data that can be written to non-volatile memory in a single, indivisible operation, ensuring data consistency and durability even in the event of power failures or system crashes [25]. At this granularity, the system guarantees that either the entire memory operation is completed and persisted or none of it, preventing partial writes or corrupt data. This granularity is usually 8 Byte in modern computer architectures. However, the unit of data that is transferred between processor and memory corresponds to one cache line, which is usually 64 Bytes. This constitutes the granularity mismatch problem [18] so that in case of power outages during a data transfer, only a fraction of the cache line might get persisted.

Due to the challenges, intermittent computing on NVMM architectures requires the application of pro-active checkpointing, as discussed in Sect. 2 and illustrated in Fig. 2(b). Although NVMM offers byte-addressability and persistence, programming efficient data structures that survive power outages is extremely difficult. It is not enough to simply store an in-memory data structure in main memory. Rather, it is necessary to think about the ordering of memory operations and also make use of special instructions to enforce memory barriers and cache flushes, as briefly summarized below:

- *Memory barriers* (also called fences) are special instructions supported by many modern CPUs today to counter *Challenge 1*. They enforce an order of instruction execution in out-of-order processor pipelines by ensuring that all memory operations before the barrier are completed before processing any memory operation after it.
- *Cache flushes* are also special instructions supported by many modern CPUs to counter *Challenge 2*. They are used to explicitly write back dirty cache lines to main memory.
- Finally, x86 architectures include instructions for moving 64 Bytes as *direct-store* with 64-byte write atomicity to counter *Challenge 3*. However, most other instruction set architectures of CPUs (particularly those popular in the embedded domain) do not provide sufficient counterparts. As a workaround,

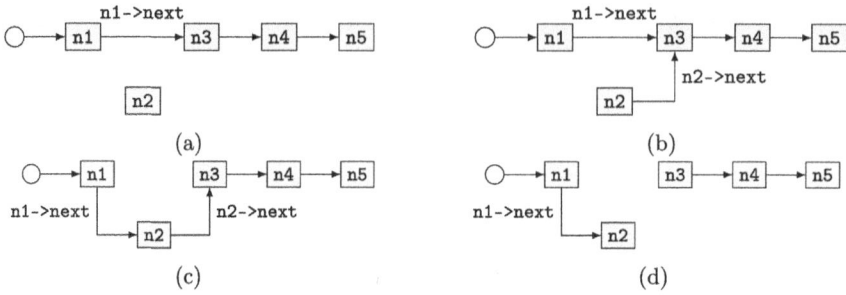

Fig. 4. Illustration of adding node n2 to a linked list. Steps (a) to (c): Linked list is always in a consistent state reachable from start node (white circle). (d) Potential situation after a power outage when memory operations get persisted out of order: All nodes starting from n3 would be unreachable from start node.

data structures would have to be programmed such that a consistent state is written back to main memory whenever changes at atomic persistence granularity took place. However, it can be assumed that the controller of the main memory buffers the data in its write buffer and only persists it when having obtained all data (i.e., the complete cache line). Under such assumptions, the granularity mismatch problem can be ignored.

Example of programming a persistent data structure: Let us illustrate how to program a persistent data structure by means of an example also shown in Fig. 4. Let an element n2 to be inserted into a linked list between connected elements n1 and n3 (see Fig. 4(a)). The procedure is to first link n2 with n3 (see Fig. 4(b)), and then linking n3 with n2 (see Fig. 4(c)). The list of linked elements reachable from the initial start node is thus always consistent: Up until the successful inclusion of n2, the linked list is always consistent with respect to its initial state. Now, several of the above-mentioned problems could arise when implementing this behavior as follows:

```
1              void append(node *n2, node *n1, node *n3) {
2                     n2->next = n3;
3                     n1->next = n2;
4              }
```

First, embedded processors like ARM processors use relaxed memory consistency models so that they do not guarantee total-store order. This means that it is possible for such processors to reorder store operations that do not have dependencies (see Challenge 1). This is in fact the case with the two instructions in lines 2 and 3, so that the processor could issue n1->next = n2 before n2->next = n3. If the system crashed before the second operation is executed, this would result in a system state with an inconsistent linked list as depicted in Fig. 4(d). It is therefore necessary to add a memory barrier between both instructions.

Second, the same inconsistent state could result even when both instructions are executed in the correct order but have only reached the cache. Although they are now globally visible in a cache-coherent system, the modifications may not have reached the non-volatile memory (see Challenge 2). If now n1->next is written back to main memory but the system crashes before writing back n2->next, the result would again be an inconsistent linked list. It is therefore necessary to add a cache flush after each instruction. To ensure that this cache flush instruction is not reordered by the processor, each has to be followed by a memory barrier.

The resulting code for programming the append function ensuring consistent states is thus as follows:

```
1        void append(node *n2, node *n1, node *n3) {
2            n2->next = n3;
3            FLUSH(n2->next);
4            BARRIER();
5            n1->next = n2;
6            FLUSH(n1->next);
7            BARRIER();
8        }
```

After each operation that modifies a next pointer, there must be a memory barrier to ensure the correct order of executing the instructions. Moreover, a cache flush must be included before each barrier to make sure the changes are written back to the non-volatile main memory.

Programming persistent data structures for NVMM systems can thus be tedious and error-prone. In order to avoid such errors, researchers have developed new concepts, often in the form of failure-atomic sections (FASEs). Such FASEs can be identified by compilers by automatically partitioning code into idempotent sections and using techniques like checkpointing [37] or logging [20] to provide consistency even after power failures. Other techniques rely on hardware transactional memory [14, 21] or keeping two replicas of variables in memory [13] (many of these techniques in fact rely on hybrid volatile/non-volatile main memories). Moreover, formal verification techniques can be applied to proof correct transformations for ensuring data durability in NVMM systems [5].

3.4   Hybrid Volatile/Non-Volatile Cache Systems

The idea of hybrid volatile/non-volatile cache systems as depicted in Fig. 3(c) is different. NVM technology, in general, and STT-RAM, in particular, have already been recognized as a promising technology for cache implementation [31, 38]. The concept of hybrid volatile/non-volatile caches has recently gained increasing attention in the research community [8, 9, 35, 36, 40], particularly due to its potential for intermittent-computing embedded systems. This technology provides the benefits of low read energy, high-memory density, and persistence. The latter is particularly of importance to overcome the challenges of NVMM systems discussed above. However, NVM at cache level introduces additional

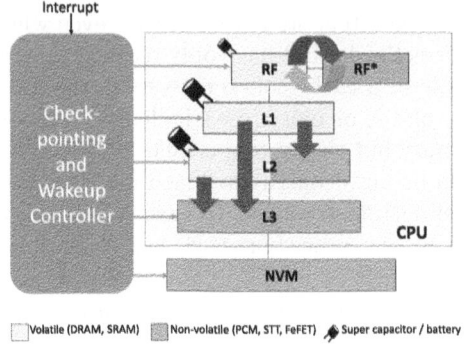

Fig. 5. Illustration of checkpointing (red) and recovery procedure (green). An interrupt issued by power-failure detection triggers the checkpointing controller to save the system state. Once power is regained, the saved register file is restored and the processor can resume with the next instruction. (Color figure online)

overheads compared to conventional SRAM in terms of increased write latencies and higher write energy consumption. As a result, fully non-volatile cache hierarchies, or even non-volatile processors (NVPs) [23] (depicted in Fig. 3(d)), are most often impractical. Consequently, hybrid caches that integrate both volatile and non-volatile memory technologies to balance performance and overhead have emerged as an active area of research, see, e.g., [1,38].

Intermittent computing on such architectures [32] follows the power-outage-aware checkpointing approach, as discussed in Sect. 2 and illustrated in Fig. 2(c). As the memory hierarchy still contains volatile components, the system state has to be persisted once a power outage is detected. An essential aspect is therefore the addition of a *reactive hardware-based checkpointing* procedure that is backed by a supercapacitor or a small battery. The system is equipped with a power-failure detection component [12] which issues an interrupt once the input-voltage level drops below a threshold. This interrupt is triggering the checkpointing controller (see Fig. 5). For checkpointing, the processor is blocked to fetch any new instructions, while continuing to process the rest of the processor pipeline. Afterwards, the CPU registers have to be saved in non-volatile shadow registers. In addition and parallel to saving the registers, all non-persisted dirty lines of non-volatile caches have to be written to non-volatile components of the memory hierarchy, following a selective writeback mechanism. This step can only start after draining the processor pipeline as otherwise instructions currently in the pipeline might still modify registers or cache lines. The procedure has to sequentially move along the memory hierarchy, starting at the cache closest to the CPU, since data located in multiple levels of the hierarchy the data closest to the CPU is guaranteed to be the most current version. It has to be taken into account that a write request that is no longer in the pipeline can still potentially have caused a cache miss that still has to be handled. A cache controller stores such pending requests in a dedicated register called Miss Status Holding

Register (MSHR). Thus, all MSHR entries have to be completed before writing back dirty cache lines. A writeback has to be performed for every volatile modified data. All writebacks are buffered in the cache's write buffer. Once the write buffer has run empty, signalling all writebacks at this level have been performed, the checkpointing controller can move on to the next cache level, with the main memory assumed to be non-volatile, thus not requiring any special handling. After harvesting enough energy to resume the execution, the saved registers (including the program counter) are restored from the shadow registers. The program can simply continue at the next instruction that is specified by the backed up program counter.

Dirty volatile cache lines have to be written back to a non-volatile memory in case of a power failure. In the following, we therefore assume that while one cache level can be hybrid, combining a volatile and a non-volatile section (e.g., L2 in Fig. 5), all memory below this level (including lower-level caches and main memory) are non-volatile. Moreover, we assume that the cache is inclusive, which means that the data stored in higher-level caches (such as L1) is also present in lower-level caches (such as L2 or L3).

The size of the backup supercapacitor/battery has to adhere to the worst case where all data stored in non-volatile section of caches are modified. The dimension of the backup capacitor is thus determined by the architecture itself rather than any type of replacement policy or the expected power outage frequency. In general, given an n-level cache hierarchy with the number of volatile cache lines at level i denoted as vCL_i, the maximum number of writebacks W_{max} to be issued during a power outage can be determined by the following equation:

$$W_{\mathrm{max}} = \sum_{i=1}^{n} vCL_i. \tag{1}$$

The higher the ratio of NVM in the cache hierarchy, the lower this number will be. It is furthermore directly proportional to the energy required to perform the checkpointing, and thus the capacity of the backup supercapacitor/battery.

3.5 Comparative Evaluation of Design Options

Now, how do both design options compare when deployed for intermittent computing? We compare both by evaluating their latency and energy overhead for executing an application that receives and stores key-value pairs. For this purpose, we make use of a B+ tree [6], as depicted in Fig. 6, with each node consisting of a list of ordered key-value pairs. However, only leaf nodes store the actual data associated with the key. The keys at inner nodes are the maximum key of the respective subtree and are used when searching for a specific key. Each leaf node contains information on the number of entries stored in the node, as well as a pointer towards the next leaf node. Leaf nodes are split in case they exceed a maximum number of entries and merged in case the number of entries falls below a lower bound.

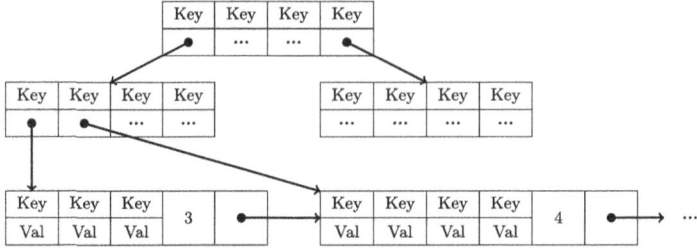

Fig. 6. Exemplary B+-Tree depicting a root and two inner nodes containing pointers. Two leaf nodes containing actual data, their number of entries and a pointer towards the neighbouring leaf node are further displayed.

Programming B+ Tree for NVMM System: In order to keep the data structure consistent for NVMM systems with power outages being a possibility, the B+ tree has further been adapted to a variant called the Crab tree [34]. Here, depending on the number of entries to be moved when inserting a new key-value pair, one of two different strategies is applied: Nodes are (a) either copied, the insertion performed on the copy and the copy replaces the original node, or (b) all entries of a node are shifted in place requiring less additional space and write transfers, yet more overhead in terms of additional cache flushes to guarantee a consistent state in case of a power outage. We have empirically determined that the best results are obtained when applying shift in place as long as less than 70% of the elements in the leaf node have to be shifted, and else copying the node.

Programming B+ Tree for Hybrid Cache System: Due to reactive hardware-based checkpointing, neither cache flushes nor memory barriers are required. As a direct consequence, this implementation always applies shift in place because, in contrast to the NVMM implementation, there is no overhead for explicitly persisting changes. Consequently, new key-value pairs are always directly inserted into the node and not into copies.

For evaluating both design options, we used the cycle-accurate system simulator gem5 [4], coupled with NVMain 2.0 [26] to simulate non-volatile main memories. We have extended the gem5 simulation by hybrid caches consisting of volatile and of non-volatile cache lines per cache set. We furthermore used the NVSim [11] to estimate the static power consumption of the caches.

Our evaluation architectures are based on embedded systems like the ARM Cortex-M85. We use gem5's generalized O3CPU model, based on the ARM ISA, with a 5-stage pipeline that issues instructions in-order and can execute them out-of-order. The simulated single-core CPU is clocked at 480 MHz, with a system clock of 240 MHz. Following the Cortex-M85, we implement a single-level cache hierarchy with a 32 KB 4-way associative data cache and a 32 KB 2-way associative instruction cache. In case of the *NVMM system*, the data cache is completely provided by volatile SRAM. In case of *hybrid caches*, three ways of each set in the data cache are volatile SRAM and the remaining one is non-volatile STT-RAM. The instruction cache remains fully non-volatile. In both cases, the main memory is a 4 GB non-volatile PCM, modeled after the specifications in [10]. Table 1 summarizes the parameters of the considered memories.

Table 1. Characterized read/write latencies and average read/write access energies of considered memories.

	Read Latency	Write Latency	Read Energy (per access)	Write Energy (per access)
SRAM Cache	2 Cycles @240 MHz	2 Cycles @240 MHz	0.009 nJ	0.009 nJ
STT-RAM Cache	2 Cycles @240 MHz	8 Cycles @240 MHz	0.007 nJ	0.056 nJ
PCM Main Memory	48 Cycles @400 MHz (tRCD)		0.081 nJ	1.685 nJ

In the following tests, 1000 key-value pairs are inserted into the tree. The execution involves a mixture of memory read operations and write operations. After that, a power outage is stimulated to occur and (in case of NVMM systems), the data structure has to be restored. Table 2 summarizes the results obtained via simulation. The overhead in terms of latency and energy required for NVMM systems is immense. After a power outage, the data structure has to be validated for consistency and reconstructed in case of unfinished insertions, which introduces additional overhead. The hybrid system, on the contrary, has relatively low execution overhead and no restoration overhead at all, as the consistency of the data structure is given at all times. The static energy consumption (while not shown in the table) is also much lower for the hybrid system. Not only because of the lower latency, but also because the static power consumption of the hybrid cache is lower than that of the volatile cache (see Table 2). The reason is that the leakage power of STT-RAM (which constitutes 25% of the hybrid data cache) is much lower than that of SRAM (the other 75%). Overall, the results show the dominance of NVM technology in combination with hardware-based checkpointing over NVMM systems.

4   Open Research Questions

Hybrid volatile/non-volatile cache systems represent the future for implementing CPUs of self-powered embedded systems. Their benefits were highlighted in the previous section. However, there are still a lot of open research questions. A few of them are discussed in the following.

4.1   Hybrid Cache Design Space

Novel NVM technologies integrated at the cache level offer significant advantages over traditional SRAM, such as lower energy consumption for read operations and reduced backup overhead during power outages in intermittent computing scenarios. Despite these benefits, write operations to NVMs still consume more

Table 2. Latency and dynamic energy for inserting 1000 key-value pairs into a B+ tree and for restoring a consistent state after a power outage, comparing non-volatile main memory (NVMM) and hybrid data cache with 25% of cache lines being non-volatile. The table also includes the static power consumption of the caches.

	NVMM	Hybrid 25%NV
Static Power (Cache) [mW]	54	43
Insertion		
CPU Cycles	23,944,823	1,293,984
Dyn. Energy (Cache) [nJ]	16,572	4,138
Dyn. Energy (Main Memory) [nJ]	968,845	17,401
Restore		
CPU Cycles	87,157	0
Dyn. Energy (Cache) [nJ]	42	0
Dyn. Energy (Main Memory) [nJ]	1,708	0

energy compared to conventional SRAM. Combining both NVM and SRAM in the cache hierarchy offers the potential to balance performance by reducing NVM write overhead while taking advantage of NVM's read efficiency and non-volatility. Ideally, read-intensive data that is frequently read should be kept in non-volatile memory, while write-intensive data that is frequently written in volatile memory. However, this raises the questions of how to partition the memory into volatile and non-volatile sections and how to decide where to place each piece of data.

Figure 7 sketches a hybrid cache architecture, also outlining the structure of a cache set. Each cache set is divided into volatile and non-volatile ways (cache lines). The concrete ratio is a design decision, denoted by *degree of non-volatility*. Hybridization of caches also complicates the choice of an appropriate *cache replacement policy*, as conventional policies like least recently used (LRU) do not account for the distinct characteristics of hybrid caches. Specifically, due to NVM's higher write overhead, architecture-aware policies should avoid placing write-intensive data in the non-volatile sections. Additionally, few existing policies support intermittent computing, making it crucial to explore how both the degree of non-volatility and the replacement policy affect performance based on the application and power supply stability.

In [35], we have analyzed the impact of these design decisions on the energy and latency for executing three different applications: one read-intensive and one write-intensive application, and one with a mixture of reads and writes. We investigated three different cache replacement policies: an architecture-agnostic policy (LRU), a hybrid-cache-aware policy [1], and a hybrid-cache-aware policy tailored to intermittent computing [3]. Our investigations involved scenarios without power outages and with frequent outages.

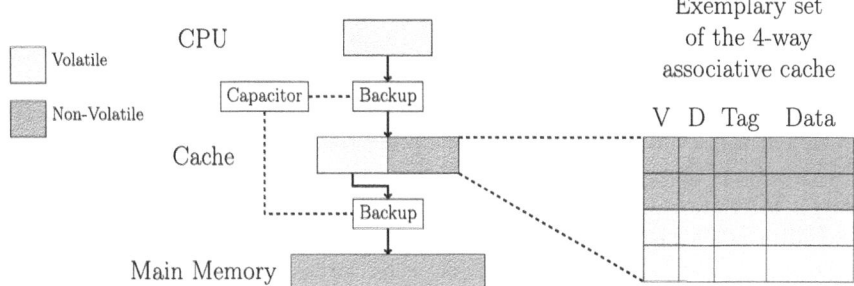

Fig. 7. Schematic outline of hybrid cache architecture including a non-volatile main memory based on [35]. An exemplary cache set of the hybrid cache is shown on the right including valid (V) and dirty (D) flags, tag and data. The cache set is divided into volatile and non-volatile cache lines.

The results have shown that, depending on the application, even with a stable power supply, the hybrid cache configurations we evaluated showed reduced latencies and energy consumption compared to traditional SRAM caches. However, the results also show that the interaction between application characteristics, architecture features, and cache replacement policies can lead to unpredictable effects, both positive and negative, on overall performance. Particularly, the memory access behavior of the application and the presence or absence of power outages strongly influence which design decisions dominate with respect to energy and latency. For each of the examined applications, the set of the best (non-dominated) design points was always only a small subset out of all design points. Neither a clearly dominant degree of volatility nor a clearly dominant policy could be identified.

While architecture-aware replacement policies have proven effective in our evaluations, it is to be expected that developing new replacement policies is not the most critical area for future research. Instead, the focus should shift toward better understanding and characterizing these complex cache effects. This is especially relevant for intermittent computing, where to balance the NVM overhead and the benefit of retaining non-volatile data close to the CPU.

The ultimate goal is to develop automated design tools that help to explore the design space and determining the application-specific hybrid memory system. These tools would optimize cache hierarchies based on identifiable application characteristics, user annotations, known architecture traits, and expected power stability.

4.2   Closing the Hardware/Software Gap

When the mix of applications deployed on an embedded system is dynamic and exhibits diverse memory access patterns, performing design-time exploration to determine the best static system configuration is only one part of the solution.

In addition to that, it is necessary to close the gap between the software and hardware.

Section 3.3 has presented the challenges for programming consistent data structures in the presence of non-volatile main memory and volatile caches. Hybrid caches avoid this by providing persistence at the granularity of single instructions by applying power-outage-aware checkpointing. However, a gap between software and hardware still remains: When placing new data in the cache after a cache miss, cache replacement policies have to apply heuristics to guess whether the data will be used write-intensive or read-intensive (see [1,3]). This may in fact lead to suboptimal placement decisions, as shown in [35].

A remedy could be a hardware/software co-design so that the cache controller provides an interface for the replacement policy to be fed with additional information by the software. Such additional information could be provided via programming extensions, e.g., by annotations in the application code (a concept known as *hints* from resource-aware programming [16]). Even a compiler could perform a static code analysis to derive such hints automatically. Hints may help the cache controller to, e.g., identify read-intensive cache lines, which benefit best from NVM characteristics, or to retain critical data with persistence constraints in non-volatile section.

Another problem is that the degree of non-volatility is fixed once determined in the design phase. The *volatile* sections offer low latency and energy overheads for write operations. The *non-volatile* sections offer low latency and energy overheads for read operations and, in addition, exhibit a lower static power consumption. Depending on the memory access pattern of the executed applications and the cache replacement policy, the utilization of volatile and non-volatile sections may differ. Adaptive hardware [7] may help to dynamically reconfigure the cache by powering off parts of the cache currently not needed, reducing the static power consumption, while providing cache-lines in the technology best suited for the current access pattern can reduce the dynamic energy consumption. Dynamically powering down NVM (or hardware components that depend on it) is feasible even when the memory holds important data, due to its non-volatile nature. This enables even more potential for energy savings [22].

4.3   Endurance, Accidental Writes, and Retention

Furthermore, NVM technologies face the challenge of significantly lower write endurance compared to their volatile counterparts and the exposure to various reliability threads due to failures and noise. Write endurance refers to the finite number of write cycles before memory degrades. Failures and noise (particularly thermal noise) can impact the data integrity and lead to accidental writes [29]. Countermeasures (see [17]) range from using error correction techniques, minimizing writes to NVM, and wear-leveling. While techniques like wear-leveling can be applied to main memory, implementing such strategies at the level of registers or lower-level caches would be prohibitively costly. However, as discussed above, the usage of hybrid caches can reduce this issue by placing write-intensive data in the volatile segments of the memory hierarchy.

Interestingly, there is a further tradeoff of using NVM technology that can be explored, namely, the tradeoff between write overhead and retention time: Fast write modes provide low latency and write energy, but data have a short retention time [39]. Consequently, they would require to be frequently refreshed as else data integrity can get corrupted over time.

4.4   How to Forget?

In hybrid cache systems, conceptionally, everything is always persistent. While this is a property that is generally desired in intermittent computing, there are also cases in which this can turn out to be a problem.

One of these cases are software errors (e.g., an infinite loop or deadlock) mostly caused by software bugs. A traditional approach to recover from such errors is to reset or reboot the system, which effectively loses all volatile state. The presented systems retain data even after a power outage, thus all effects caused by bugs or errors in software are persisted. Unlike volatile memory, where a system reboot clears the memory state and provides a fresh start, NVM errors can survive and continue affecting the system, leading to ongoing issues.

A similar problem arises in applications in which temporal consistency of data must be guaranteed. A simple example is the following program for calculating the average value of n consecutive samples read from a sensor. The program code could look like this:

```
1        float sum = 0;
2        for (int i = 0; i < n; i++) {
3             float value = sensor_read();
4             sum += value;
5        }
6        average = sum/n;
```

Let's assume the power outage occurs after $m < n$ iterations and the system is then down for multiple minutes or even hours before power supply is restored. This would lead to temporal inconsistency as there can be a huge time gap between the first m and the remaining $n - m$ samples.

One consequence could be to re-introduce the concept of checkpointing according to Fig. 2(b) in order to be able to mark consistent states and to be able to roll back to them in the event of an error or power outage. Another option would be that a program could separate its state logically into a volatile scratch space where intermediate data is kept and a (larger) persistent state. Only when reaching a consistent state will the program merge the intermediate state into the persistent state. This could make further adaptations on the hardware level necessary, like introducing volatile scratchpad memory or hardware transaction memory to hold the intermediate state. Via dedicated instructions, a program could then explicitly commit the intermediate state into the persistent memory or abort the computation/transaction.

5   Conclusion

An evolution of autonomous systems was introduced as autarkic systems that are completely self-sustained. While computer systems used in many areas are too power hungry, IoT devices are just small enough that they can be operated based on typical energy harvesting sources. However, the strong power fluctuation of such sources makes it necessary that systems can cope with frequent power outages. The main problem is that the system should always stay in a consistent state. While this paper has shown that NVM provides the foundation to build computer architectures for self-powered systems, we have also discussed that there are still a lot of research questions left before we can successfully deploy self-powered IoT devices. Looking ahead, we envision the emergence of autonomous computer systems that can correctly operate infinitely long without any human intervention.

References

1. Ahn, J., Yoo, S., Choi, K.: Prediction hybrid cache: an energy-efficient STT-RAM cache architecture. IEEE Trans. Comput. **65**(3), 940–951 (2016). https://doi.org/10.1109/TC.2015.2435772
2. Ali, A., Shaukat, H., Bibi, S., Altabey, W.A., Noori, M., Kouritem, S.A.: Recent progress in energy harvesting systems for wearable technology. Energ. Strat. Rev. **49**, 101124 (2023). https://doi.org/10.1016/j.esr.2023.101124
3. Badri, S., Saini, M., Goel, N.: Efficient placement and migration policies for an STT-RAM based hybrid L1 cache for intermittently powered systems. Des. Autom. Embed. Syst. (2023). https://doi.org/10.1007/s10617-023-09272-w
4. Binkert, N., Beckmann, B., Black, G., Reinhardt, S.K., Saidi, A., Basu, A., Hestness, J., Hower, D.R., Krishna, T., Sardashti, S., Sen, R., Sewell, K., Shoaib, M., Vaish, N., Hill, M.D., Wood, D.A.: The gem5 simulator. SIGARCH Comput. Archit. News **39**(2), 1–7 (2011). https://doi.org/10.1145/2024716.2024718
5. Bodenmüller, S., Derrick, J., Dongol, B., Schellhorn, G., Wehrheim, H.: A fully verified persistency library. In: Verification, Model Checking, and Abstract Interpretation: 25th International Conference, VMCAI 2024, London, United Kingdom, 15–16 January 2024, Proceedings, Part II, pp. 26–47. Springer-Verlag, London, United Kingdom (2024). https://doi.org/10.1007/978-3-031-50521-8_2
6. Braginsky, A., Petrank, E.: A lock-free B+tree. In: Proceedings of the Twenty-Fourth Annual ACM Symposium on Parallelism in Algorithms and Architectures. SPAA 2012, pp. 58–67. Association for Computing Machinery, Pittsburgh, Pennsylvania, USA (2012). https://doi.org/10.1145/2312005.2312016
7. Chen, Y.-T., et al.: Dynamically reconfigurable hybrid cache: an energy-efficient last-level cache design. In: 2012 Design, Automation & Test in Europe Conference & Exhibition (DATE), pp. 45–50 (2012). https://doi.org/10.1109/DATE.2012.6176431
8. Chen, Y.-C., et al.: Modeling and simulating emerging memory technologies: a tutorial. arXiv (2025)
9. . Choi, J., Zeng, J., Lee, D., Min, C., and Jung, C.: Write-light cache for energy harvesting systems. In: Proceedings of the 50th Annual International Symposium on Computer Architecture. ISCA '23. Association for Computing Machinery, Orlando, FL, USA (2023). https://doi.org/10.1145/3579371.3589098

10. Choi, Y., et al.: A 20 nm 1.8V 8Gb PRAM with 40MB/s program bandwidth. In: 2012 IEEE International Solid-State Circuits Conference, pp. 46–48 (2012). https://doi.org/10.1109/ISSCC.2012.6176872
11. Dong, X., Xu, C., Xie, Y., Jouppi, N.P.: NVSim: a circuit-level performance, energy, and area model for emerging nonvolatile memory. IEEE Trans. Comput. Aided Des. Integr. Circuits Syst. **31**(7), 994–1007 (2012). https://doi.org/10.1109/TCAD.2012.2185930
12. Eichler, C., Hofmeier, H., Reif, S., Hönig, T., Nolte, J., Schröder-Preikschat, W.: Neverlast: towards the design and implementation of the NVM-based everlasting operating system. In: 54th Hawaii International Conference on System Sciences, HICSS 2021, Kauai, Hawaii, USA, 5 January 2021, pp. 1–10. ScholarSpace (2021)
13. Friedman, M., Petrank, E., Ramalhete, P.: Mirror: making lock-free data structures persistent. In: Proceedings of the 42nd ACM SIGPLAN International Conference on Programming Language Design and Implementation, PLDI 2021, pp. 1218–1232. Association for Computing Machinery, Virtual, Canada (2021). https://doi.org/10.1145/3453483.3454105
14. Genç K., Bond, M.D., Xu, G.H.: Crafty: efficient, HTM-compatible persistent transactions. In: Proceedings of the 41st ACM SIGPLAN Conference on Programming Language Design and Implementation, PLDI 2020, pp. 59–74. Association for Computing Machinery, London, UK (2020). https://doi.org/10.1145/3385412.3385991
15. Güdemann, M., Nafz, F., Ortmeier, F., Seebach, H., and Reif, W.: A specification and construction paradigm for organic computing systems. In: 2008 Second IEEE International Conference on Self-Adaptive and Self-Organizing Systems, pp. 233–242 (2008)
16. Hannig, F., Roloff, S., Snelting, G., Teich, J., Zwinkau, A.: Resource-aware programming and simulation of MPSoC architectures through extension of X10. In: Proceedings of the 14th International Workshop on Software and Compilers for Embedded Systems, SCOPES 2011, pp. 48–55. Association for Computing Machinery, St. Goar, Germany (2011). https://doi.org/10.1145/1988932.1988941
17. Henkel, J., e tal.: Special session - non-volatile memories: challenges and opportunities for embedded system architectures with focus on machine learning applications. In: Proceedings of the International Conference on Compilers, Architecture, and Synthesis for Embedded Systems, CASES 2023 Companion, pp. 11–20. Association for Computing Machinery, Hamburg, Germany (2024). https://doi.org/10.1145/3607889.3609088
18. Hwang, D., Kim, W.-H., Won, Y., and Nam, B.: Endurable transient inconsistency in byte-addressable persistent B+-tree. In: Proceedings of the 16th USENIX Conference on File and Storage Technologies, FAST 2018, pp. 187–200. USENIX Association, Oakland, CA, USA (2018)
19. Imani, M., Patil, S., and Rosing, T.: Low power data-aware STT-RAM based hybrid cache architecture. In: 2016 17th International Symposium on Quality Electronic Design (ISQED), pp. 88–94 (2016). https://doi.org/10.1109/ISQED.2016.7479181
20. Izraelevitz, J., Kelly, T., Kolli, A.: Failure-atomic persistent memory updates via JUSTDO logging. SIGARCH Comput. Archit. News **44**(2), 427–442 (2016). https://doi.org/10.1145/2980024.2872410
21. Jeong, J., Hong, J., Maeng, S., Jung, C., Kwon, Y.: unbounded hardware transactional memory for a hybrid DRAM/NVM memory system. In: 2020 53rd Annual IEEE/ACM International Symposium on Microarchitecture (MICRO), pp. 525–538 (2020). https://doi.org/10.1109/MICRO50266.2020.00051

22. Karim, A., Falk, J., Schmidt, D., Teich, J.: Self-powering dataflow networks – concepts and implementation. In: 22nd ACM-IEEE International Symposium on Formal Methods and Models for System Design (MEMOCODE), pp. 69–74 (2024). https://doi.org/10.1109/MEMOCODE63347.2024.00013
23. Ma, K., et al.: Nonvolatile processor architectures: efficient, reliable progress with unstable power. IEEE Micro **36**(3), 72–83 (2016). https://doi.org/10.1109/MM.2016.35
24. Müller-Schloer, C., Schmeck, H., Ungerer, T.: Organic computing—a paradigm shift for complex systems. Springer Science & Business Media (2011)
25. Pelley, S., Chen, P.M., Wenisch, T.F.: Memory persistency. In: Proceeding of the 41st Annual International Symposium on Computer Architecuture. ISCA 2014, pp. 265–276. IEEE Press, Minneapolis, Minnesota, USA (2014)
26. Poremba, M.,et al.: NVMain 2.0: a user-friendly memory simulator to model (Non-) volatile memory systems. IEEE Comput. Archit. Lett. **14**, 140– 143 (2015)
27. Raad, A., Wickerson, J., Vafeiadis, V.,et al.: Weak persistency semantics from the ground up: formalising the persistency semantics of ARMv8 and transactional models. Proc. ACM Program. Lang. 3(OOPSLA), 140–143 (2019). https://doi.org/10.1145/3360561
28. Rixner, S., Dally, W.J., Kapasi, U.J., Mattson, P., Owens, J.D., et al.: Memory access scheduling. SIGARCH Comput. Archit. News **28**(2), 128–138 (2000). https://doi.org/10.1145/342001.339668
29. Sayed, N., Mao, L., Bishnoi, R., Tahoori, M.B., et al.: Compiler-assisted and ProfilingBased analysis for fast and efficient STT-MRAM on-chip cache design. ACM Trans. Des. Autom. Electron. Syst. **24**(4), 128–138 (2019). https://doi.org/10.1145/3321693
30. Shiratake, S.,et al.: Scaling and performance challenges of future DRAM. In: 2020 IEEE International Memory Workshop (IMW), pp. 1–3. Association for Computing Machinery, New York, NY, USA (2020). https://doi.org/10.1109/IMW48823.2020.9108122
31. Sun, G., Dong, X., Xie, Y., Li, J., Chen, Y.,et al.: A novel architecture of the 3D stacked MRAM L2 cache for CMPs. In: 2009 IEEE 15th International Symposium on High Performance Computer Architecture, pp. 239–249. Association for Computing Machinery, New York, NY, USA (2009). https://doi.org/10.1109/HPCA.2009.4798259
32. Umesh, S., Mittal, S., et al.: A survey of techniques for intermittent computing. J. Syst. Architect. **112**(4), 101859 (2021). https://doi.org/10.1016/j.sysarc.2020.101859
33. Wägele, J., et al.: Towards a multisensor station for automated biodiversity monitoring. Basic Appl. Ecol. **59**(4), 105–138 (2022). https://doi.org/10.1016/j.baae.2022.01.003
34. Wang, C., Chattopadhyay, S., Brihadiswarn, G., et al.: Crab-tree: a crash recoverable B+-tree variant for persistent memory with ARMv8 architecture. ACM Trans. Embed. Comput. Syst. **19**(5), 105–138 (2020). https://doi.org/10.1145/3396236
35. Wilbert, N., Wildermann, S., Teich, J.,et al.: hybrid cache design under varying power supply stability - a comparative study. In: 10th International Symposium on Memory Systems, pp. 105–138. Association for Computing Machinery, New York, NY, USA (2024). https://doi.org/10.1145/3695794.3695819
36. Wilbert, N., Wildermann, S., Teich, J.,et al.: To keep or not to keep - the volatility of replacement policy metadata in hybrid caches. In: Proceedings of the 2nd Workshop on Disruptive Memory Systems. DIMES 2024, pp. 17–24. Association for

Computing Machinery, Austin, TX, USA (2024). https://doi.org/10.1145/3698783.3699381

37. Woude, J.V.D., Hicks, M.,et al.: Intermittent computation without hardware support or programmer intervention. In: 12th USENIX Symposium on Operating Systems Design and Implementation (OSDI 16), DIMES 2024, pp. 17–32. USENIX Association, Savannah, GA (2016). https://doi.org/10.1145/3698783.3699381

38. Wu, X., Li, J., Zhang, L., Speight, E., Xie, Y.,et al.: Power and performance of read-write aware hybrid caches with non-volatile memories. In: 2009 Design, Automation & Test in Europe Conference & Exhibition, DIMES 2024, pp. 737–742. USENIX Association, Savannah, GA (2009). https://doi.org/10.1109/DATE.2009.5090762

39. Zhang, M., Zhang, L., Jiang, L., Chong, F.T., Liu, Z., et al.: Quick-and-dirty: an architecture for high-performance temporary short writes in MLC PCM. IEEE Trans. Comput. **68**(9), 1365–1375 (2019). https://doi.org/10.1109/TC.2019.2900036

40. Zhou, Y., Zeng, J., Jeong, J., Choi, J., Jung, C.,et al.: SweepCache: IntermittenceAware cache on the cheap. In: Proceedings of the 56th Annual IEEE/ACM International Symposium on Microarchitecture, MICRO 2023, pp. 1059–1074. Association for Computing Machinery, Toronto, ON, Canada (2023). https://doi.org/10.1145/3613424.3623781

Overview of Bounded Model Checking for Stack-Based Virtual Machines

Matthias Güdemann[(✉)]

University of Applied Sciences Munich (HM), Lothstrasse 64, 80335 Munich, Germany
matthias.guedemann@hm.edu

Abstract. Bounded model checking has become a powerful approach to analyzing software, particularly for bug detection. Despite the undecidability of proving arbitrary properties of a program, model checking can detect bugs and generate a witness demonstrating how the faulty behavior can be reached. This fully automated process makes it an attractive technique for continuous integration in industrial settings. Modern SAT and SMT solvers enable the analysis of large state spaces, even though the underlying problems are $\mathcal{NP}$-hard.

In this work, we explain the details of bounded model checking of a stack-based virtual machine with the example of the Java Virtual Machine bytecode instructions. These virtual machines are commonly used as intermediate languages for various front-ends. In theory, the ability to analyze programs in such a virtual machine enables the analysis of multiple source languages.

1 Introduction

Model checking has emerged as an economically viable method for verifying software and hardware [27]. Its primary benefits lie in the full automation of the verification process, rendering it conducive for integration into continuous integration/continuous deployment (CI/CD) systems. Additionally, the method enables the conduction of precise bit-level analyses, especially relevant in cases involving IEEE754 floating point operations. In contrast, other automated techniques, such as abstract interpretation [8], adopt abstraction as the primary mechanism, thereby sacrificing analysis precision. Interactive theorem proving techniques, while capable of tackling undecidable problems, necessitate expert manual intervention. Despite theoretical claims of efficacy in handling very large state spaces and bit-precise analyses, actual implementation often presents practical challenges. Nonetheless, each verification strategy exhibits its unique mix of advantages, disadvantages, and application domains.

In this article, we aim to demonstrate the utilization of bounded model-checking to verify properties of stack-based virtual machines. The objective is to enhance the capacity for automated bug detection and coverage analysis, which are key strengths of model-checking. Stack-based virtual machines operate at a low-level and serve as the foundation for various contemporary programming

G. Ernst et al. (Eds.): Wolfgang Reif Festschrift, LNCS 15765, pp. 178–194, 2025.
https://doi.org/10.1007/978-3-031-92196-4_9

languages. Examples of stack-based virtual machines include the Python Virtual Machine (PVM), the Java Virtual Machine (JVM) utilized for Java, Kotlin, and Clojure, as well as WebAssembly (Wasm) targeted by multiple languages through the LLVM framework.

This article focuses on the Java Virtual Machine (JVM) and demonstrates the practical implementation of the approach using the Java Bounded Model Checker (JBMC) [7]. JBMC, which is built upon the CProver Framework, shares its foundation with the C Bounded Model Checker (CBMC) [17]. Analogous to the flexibility provided by a compiler framework like LLVM for various back-end implementations, employing a common formal framework allows for different property instrumentation and analyses.

The rest of the article is structured as follows: Sect. 2 gives the necessary background for BMC, satisfiability based reasoning and SMT solving, Sect. 3 explains how JVM bytecode can be transformed into an model-checking problem that can be analyzed using an SMT solver, Sect. 4 explains some concrete applications for model-checking bytecode, Sect. 5 describes some of the practical challenges of the approach and Sect. 6 concludes the article.

2   Background

We give a general view of some background on BMC. The illustration uses the JVM as an example, but similar approaches are possible for other languages. The principles employed in the construction of a formal model and its solution are very similar.

2.1   Related Work

The approach described here is implemented in a similar way in analysis tools like CBMC [17], JBMC [7], LLBMC [21], ESBMC [10] or Kani [26]. The description here generally outlines the principles underlying these tools, each one implements additional features that can make specific use cases more efficient.

These tools generally provide an intermediate representation of programs and use different front-ends to transform different languages into this representation. CBMC provides a front-end for C and the SystemC C++ subset, JBMC provides a front-end for JVM class files, LLBMC uses the LLVM IR as its input language, ESBMC provides front-ends for C, a subset of C++, Python [9] as well as Solidity [25] and Kani provides a Rust front-end. JBMC, CBMC, LLBMC, ESBMC and Kani all use GOTO programs as their intermediate representation.

Other tools which use similar techniques but a different intermediate representation are CPAChecker [6] and Ultimate Automizer [13]. The Software Verification Competition (SV-COMP) [5] is an annual competition for comparing the efficiency of such automated analysis tools on different benchmarks. The language for the benchmark problems is mostly C, but there is also a Java track.

ByteBack, as described in [22], is a Java verification tool that leverages Java bytecode for its foundation. It specializes in verifying programs with exceptions and employs a translation mechanism from Java bytecode to the Boogie verification language [18]. Similar to JBMC, this approach offers the benefit of analyzing other languages compiled to Java bytecode.

The JJBMC approach [4] offers a compelling combination of techniques. It translates Java programs leveraging annotations written in the Java Modeling Language (JML) into native Java code, which then undergoes analysis using the JBMC model checker. This method benefits from JML's expressiveness, enabling verification that encompasses loop invariants and function contracts to achieve more comprehensive results.

2.2   Bounded Modelchecking

Bounded Modelchecking effectively transforms a program into a mathematical model which can be analyzed using efficient automated solvers. This mainly deals with reachability properties or invariants of programs. The mathematical model corresponds to a set of constraints, fulfilling the constraints corresponds to a program execution.

A property P in the program is analyzed by converting the program into a set of constraints and conjoining this model with $\neg P$, the negation of the property to analyze. If this extended set of constraints has a solution, then it is possible to execute the program and reach $\neg P$. If no solution to the extended set of constraints exists, then $\neg P$ is not reachable, and P is invariant.

The bounded aspect of BMC is the fact that loops and recursive calls have to be bounded in order to get a finite set of constraints when constructing the mathematical model. While this limits the properties that can be analyzed, bounded analysis is often possible to conduct in pure reachability analysis, e.g., for coverage. In domains where real-time behavior is necessary, all loops have to be bounded in the program and therefore these bounds can be used in the analysis.

2.3   SAT-Based Reasoning

Formal verification based on satisfiability solvers is a very powerful approach for bit-precise reasoning. While the supported properties are mostly limited to invariants or reachability, the size of the problems can be quite large despite the theoretical hardness of the underlying problem. Algorithmic advances are fostered via different yearly SAT competitions where different tools participate against each other. While SAT solving is $\mathcal{NP}$-hard, real world problems are now analyzed using these tools, as captured in the following quote:

> SAT is now the inexpensive, easy-to-solve workhorse for really hard problems. People still have it in their heads that SAT equals $\mathcal{NP}$-hard, therefore difficult to solve or impossible to solve. But for us, it's the lowest entry point. On top of SAT, we build algorithms for solving problems that are way harder [15].

In the most general form, SAT solvers can only answer satisfiable (SAT) or unsatisfiable (UNSAT) for a given propositional logic formula. This reasoning is used in the following way, if for two propositional logic formulas A and B we have $SAT(A) \wedge UNSAT(A \wedge \neg B)$ then $A \rightarrow B$.

In the context of BMC, A corresponds to the program and B to the property. $SAT(A)$ means that the program can be executed, $UNSAT(A \wedge \neg B)$ means that no execution can cause the invariant B to be falsified.

If $A \wedge \neg B$ is SAT, then there exists an execution that can lead to the invariant being violated. In the case of SAT, most solvers also generate a model for the formula. In the case of BMC this corresponds to a witness for which A is executed and $\neg B$ does not hold, this can be translated into input parameters for the program which lead to the violated property.

2.4   SMT Solving

An approach built on-top of SAT solvers is satisfiability modulo theory (SMT). SMT solvers are powerful tools for solving problems that benefit from higher-level reasoning. By utilizing diverse logics, SMT solvers can reason more efficiently and on a more abstract level, leading to solutions that would otherwise be challenging to obtain, in particular for bit-level reasoning. Just as SAT solvers, SMT solvers are now routinely used to solve difficult real-world problems [24].

The SMT-LIB2 [3] contains the specification of different logics that can be used in SMT solving. For program analysis the most important ones are bitvectors, floating point and arrays. These allow the expression of a large part of program semantics of commonly used programming languages.

An SMT solver effectively uses *purification* to separate a formula into sub-formulas which belong to exactly one logic. These separated formulas are solved separately using a variant of the basic Nelson-Oppen decision procedure combination [16, p. 229 ff.]. In the case one is UNSAT, the whole formula is unsatisfiable. In case all are SAT, new information in the form of equivalences is propagated to the other local formulas and then rechecked. Once no new information can be found and all parts remain SAT, the whole problem is satisfiable.

The advantage of SMT solving in comparison with pure SAT solving is that the theories can use higher-level abstraction and mathematical reasoning. In the Nelson-Oppen decision procedure combination, deciding UNSAT in one of the local formulas is enough to decide UNSAT for the whole formula. It is therefore desirable to find the *easiest* logic and decision procedure to decide UNSAT if possible.

The example [11] in Listing 1 for BMC illustrates this point. Here, the goal is to show that the values $p = prod(a, b)$ and $q = prod(c, d)$ are equal. As a is equal to c, b is equal to d and `prod` has no side-effects, this is actually the case.

```
1   int prod(int a, int b) {
2       int p = 1;
3       for (int i = 0; i < 3; i++) {
4           p = p * (a + b);
5       }
6       return p;
7   }
8   void property(int a, int b) {
9       int c = a;                      // set c to the value of a
10      int d = b;                      // set d to the value of b
11      assert(prod(a, b) == prod(c, d));   // property
12  }
```

Listing 1: Example for Uninterpreted Functions

The function prod contains integer multiplication which creates a rather complex propositional logic conjunctive normal form (CNF) formula for bit-level verification when unrolled and bit-blasted. The problem is not necessarily the size of the resulting CNF but the dependencies between the variables.

Using an SMT solver which supports uninterpreted functions (UF) logic, this is very easy to prove via Shostak's congruence closure [16, p. 85]. The key factor here is that the prod function is pure, i.e., has no side effects and will always produce an equivalent output in case of equivalent inputs.

Showing the unsatisfiability of the negated property prod(a, b) != prod(c, d) using Shostak's procedure works as follows here: a and c are equivalent because of the assignment, analogously b and d. Therefore, in the congruence step, prod(a, b) is also considered to be equivalent to prod(c, d) as the respective arguments are equivalent. As this is a contradiction to the negated property, the problem is UNSAT and the original property holds.

Executing the congruence closure decision procedure here is close to trivial. In contrast, bit-blasting the integer multiplications gives rise to a relatively complex propositional SAT problem. While the problem is not very big in terms of the number of variables and clauses, due to the nature of multiplication there is a lot of internal interconnectedness and even current state-of-the-art SAT solvers require a substantial amount of time to solve the decision problem.

Similar effects can be seen with solving IEEE754 floating point instances. Bit-blasting floating point problems can also lead to very complex propositional formulas which are difficult to solve. For some IEEE754 floating point problems, higher-level mathematical reasoning is possible and can be much more efficient. Often it is difficult to tell beforehand which solver backend will perform best and it makes sense to use a portfolio approach which is easy to parallelize.

3   Checking JVM Bytecode

The Java Virtual Machine (JVM) defines a stack-based VM with a well-defined semantics for the bytecode instructions [1]. The bytecode mnemonics are prefixed with their type. Java being an object-oriented language also requires bytecodes for object casting etc.

Basing the analysis on the bytecode and not the Java language directly has the advantage that other languages that use the JVM can also be supported. Examples of the most widely used such languages are Kotlin and Clojure, but there exist numerous other languages that use the JVM, including a Java version for security critical smartcards.

```
1    int f(int n) {
2        int sum = 0;
3        for (int i = 0; i < n; i++) {
4            sum += i;
5        }
6        return sum;
7    }
```

Listing 2: Example Program computing $\sum_{i=0}^{n} i$ for $n \geq 0$

3.1 Example Program in Java Bytecode

The Listing 3 shows the mnemonics of JVM bytecode instructions from the compiled example function shown in Listing 1. The format is `byte offset: mnemonic`. Every non-static function has the object instance as its implicit first argument, named `this`. The integer argument n to the function corresponds to the local variable number 1. These local variables are valid from byte 0 to byte 20 of the function which comprises all bytecode instructions. The local variable 2 corresponds to the integer `sum` and local variable 3 to the loop counter `i`.

```
1    int f(int);
2      Code:
3         0: iconst_0
4         1: istore_2              // store value 0 in sum
5         2: iconst_0
6         3: istore_3              // store value 0 in i
7         4: iload_3               // loop start
8         5: iload_1               // load parameter n
9         6: if_icmpge    19       // if i >= n go to ---------------\
10        9: iload_2               //                                |
11       10: iload_3               //                                |
12       11: iadd                  //                                |
13       12: istore_2              // store value of sum + i in sum  |
14       13: iinc         3, 1     // increment i                    |
15       16: goto         4        // go to loop start               |
16       19: iload_2               //                     <----------/
17       20: ireturn               // return value of sum
```

Listing 3: Disassembled Java Bytecode

3.2 Bytecode Conversion into Intermediate Imperative Code

In order to analyze a program with BMC, it is transformed into a simple C-like imperative language without an explicit stack. For JBMC, this representation is called GOTO program. The transformation of the above bytecode follows the algorithm outlined in [19]. This approach can easily be adapted to other stack based virtual machines. The basic algorithm works in the following steps:

1. **Build the Control Flow Graph (CFG)** The control flow graph specifies the possible execution flow in a program. Linear flows are combined into *basic blocks* which form the nodes of the graph, edges between the nodes correspond to conditional or unconditional jumps.
2. **Compute the Imperative Program from the CFG** Using a breadth first search (BFS) through the CFG, compute the output expressions for each instruction based on its input expression. Doing this according to BFS along the CFG guarantees that the necessary input is always computed as long as the compiled bytecode instruction control flow is valid, i.e., correctly generated by a compiler.

Listing 4 shows the imperative program that corresponds to the bytecode in GOTO format. The loop is expressed as a conditional branch in line 6 and the backwards GOTO in line 10.

```
1    loop.f(int) /* java::loop.f:(I)I */
2        DECL sum : signedbv[32]
3        DECL i : signedbv[32]
4        ASSIGN sum := 0
5        ASSIGN i := 0
6    1: IF !(i >= n) THEN GOTO 2
7        GOTO 3
8    2: ASSIGN sum := sum + i
9        ASSIGN i := i + cast(1, signedbv[32])
10       GOTO 1
11   3: ASSIGN return_value := sum
12       END_FUNCTION
```

Listing 4: Imperative GOTO Program after Conversion from JVM Bytecode

3. **Loop unrolling** Bounded checking limits the number of loop iterations in the execution and analysis of a program. The elimination of the loops is done via loop unrolling unto a given bound. In the general case this bound cannot be deduced automatically but is specified manually.

Listing 5 shows the program after loop unrolling with a bound of 3. In this case the loop body is executed at most 3 times. The CFG does not contain any backwards edge and therefore the code does not contain any loop. This is required for bounded modelchecking to make the problem decidable as loop invariants and variants cannot be computed automatically.

The number of times a loop gets unwound before the program is transformed into a constraint system is variable. The bound can be chosen specifically for each loop of the program or the same global value can be used. It is also possible to use incremental unwinding, i.e., iteratively increasing the loop bound from 1 until a problem is found or the time budget is exhausted. Which strategy works best depends on the task at hand.

```
1   loop.f(int) /* java::loop.f:(I)I */
2       DECL sum : signedbv[32]
3       DECL i : signedbv[32]
4       ASSIGN sum := 0
5       ASSIGN i := 0
6    1: IF !(i >= n) THEN GOTO 2
7       GOTO 3
8    2: ASSIGN sum := sum + i
9       ASSIGN i := i + cast(1, signedbv[32])
10      IF !(i >= n) THEN GOTO 4
11      GOTO 3
12   4: ASSIGN sum := sum + i
13      ASSIGN i := i + cast(1, signedbv[32])
14      IF !(i >= n) THEN GOTO 6
15      GOTO 3
16   6: ASSIGN sum := sum + i
17      ASSIGN i := i + cast(1, signedbv[32])
18   3: ASSIGN return_value := sum
19      END_FUNCTION
```

Listing 5: Example Program After Loop Unrolling With a Bound of 3

4. **Conversion to SSA form** Using standard algorithms [23, pp. 23] the static single assignment (SSA) form of the imperative code is constructed. In SSA form, each variable is assigned *exactly once*. This is realized via *versioning* of the variables. At convergence points in the CFG, φ-functions are inserted to discern the values depending on which basic block the flow is coming from.

Listing 6 shows the Single Static Assignment (SSA) form of the loop-less program above. In the SSA form, each variable is assigned exactly once. This is achieved by creating a new *version* each time a variable gets assigned a new value. Each conditional branch either continues executing the loop body or branches to basic block number 3 (l. 18), indicating wither a violated loop condition or reaching of the last of the unwound iterations.

When reaching basic block 3, the final assignment to the last value of sum uses the φ function which selects which value is on the right hand side of the assignment depending on which basic block the control flow is coming from.

```
1   loop.f(int) /* java::loop.f:(I)I */
2       DECL sum : signedbv[32]
3       DECL i : signedbv[32]
4       ASSIGN sum0 := 0
5       ASSIGN i0 := 0
6   1: IF !(i0 >= n) THEN GOTO 2
7       GOTO 3
8   2: ASSIGN sum1 := sum0 + i0
9       ASSIGN i1 := i0 + 1
10      IF !(i1 >= n) THEN GOTO 4
11      GOTO 3
12  4: ASSIGN sum2 := sum1 + i1
13      ASSIGN i2 := i1 + cast(1, signedbv[32])
14      IF !(i2 >= n) THEN GOTO 6
15      GOTO 3
16  6: ASSIGN sum3 := sum2 + i2
17      ASSIGN i3 := i2 + cast(1, signedbv[32])
18  3: ASSIGN sum4 := phi(1: sum0, 2: sum1, 4: sum2, 6: sum3)
19      ASSIGN return_value := sum4
20      END_FUNCTION
```

Listing 6: Example Program After SSA Transformation

5. **Linearization of the Branches** The SSA form is transformed into a linearized version, where the branches are represented as Boolean guards. The assignment using a φ-function in the SSA form is then replaced via a conditional assignment using the respective truth values of the guards.
6. **Generation of Verification Conditions** In the unrolled, linearized SSA form, every variable is assigned exactly once. Because of this, every assignment can be formulated as a constraint, stating that the lefthand side equals the righthand side. The whole program then forms a set of constraints which are fulfilled if a correct run of the program is executed. This forms the verification conditions of the program.

Listing 7 shows the generated verification conditions. Effectively, one assumes that the conditions modeling the program are fulfilled and checks whether this conjoined to the negation of the property is satisfiable. The φ function in the SSA form is transformed into a conditional equivalence using the ternary conditional operator.

```
1   sum0 = 0,
2   i0 = 0,
3   sum1 = sum0 + i0,
4   guard0 = !(i0 >= n),
5   i1 = i0 + 1,
6   sum2 = sum1 + i1,
7   guard1 = !(i1 >= n),
8   i2 = i1 + 1,
9   sum3 = sum2 + i2,
10  guard2 = !(i2 >= n),
11  i3 = i2 + 1,
12  sum4 = !guard0 ? sum0 : (!guard1 ? sum1 : (!guard2 ? sum2 : sum3))
13  |------------------------------------------------------------------
14  !(sum4 < 2)
```

Listing 7: Verification Conditions (VCCs) for the property $sum < 2$

Such a logical deduction can be solved by transforming it into an SMT-LIB2 constraint problem and running an SMT solver as described in Sect. 2.4. Here, the sort of the logic variables is 32-bit signed bitvectors and the operations are bitvector addition. The conditional equivalence (l. 12) is translated into a sequence of SMT-LIB2 *ite* (if-then-else) constraints.

The value of n is free to choose for the solver. In order to falsify the conclusion that sum4 is less than 2, the solver has to find a satisfying assignment to all variables. With a value of $n = 3$, the whole constraint system including the negated property is falsifiable and therefore $sum < 2$ is not invariant. This can directly be translated into a test-case which calls the function f with a value of 3 for the parameter n.

3.3   Special Case: Direct Variable Modification

The approach to verify stack based bytecode as outlined above works if every update is done via the stack. Unfortunately in real virtual machines like the JVM, this is not always the case, as there exist bytecode instructions that modify local variables directly.

This problem can be seen in the Listing 8 illustrated with JVM bytecode instructions. Here, first the value of local variable 1 is pushed onto the stack, then the iinc (integer increment) bytecode instruction increments local variable 1 by 1 and then the original value of the local variable is used to access an array.

In such a case, if the value of a variable v is put on the stack, this creates an expression that contains v, if then v is directly modified and the value on the stack is used afterwards, the value on the "stack" has also changed because reading the value is deferred to accessing the variable. Therefore, direct variable modification violates the assumption that underpins the approach to build expressions from the operations on the stack.

```
1          ...
2     14: iload_1                  // push value of local variable 1
3     15: iinc        1, 1         // direct increment of local variable 1
4     18: iaload                   // access an array at the index on the stack
5          ...
```

Listing 8: Direct Variable Modification

From a point of view of the stack machine, no problem exists. The value of the local variable is saved on the stack, therefore the content of the local variable can safely be changed. Due to the nature of the translation into imperative code without an explicit stack, this leads to a modification of the content of the built expression. Because of this, it is necessary to treat instructions that modify variables directly in a different way than the rest. In particular, one needs to create a temporary variable to which the original value is assigned. This temporary variable is then used to build the expression that models the stack content.

Such direct modifications exist to optimize common code constructs like incrementing of variables. Therefore, similar direct modifications exist in other stack-based virtual machine bytecode, too, and must be treated in a similar way.

4 Applications

Bounded Model-checking can be used in different applications. While its expressive power is somehow limited, it can be executed in a fully automated way. Because of the use of SMT solvers which support bit-vector logic, bit-precise analyses are possible. This is in particular beneficial for reachability analysis, e.g., automated bug finding or coverage analysis.

4.1 Run-Time Error Checking

Checking for potential run-time errors is one of the areas where this analysis approach can be used. Examples of such run-time errors include division-by-zero, array-out-of-bounds, null pointer dereference or null object use[1].

Additional properties that can be analyzed are for example IEEE754 floating point computations that contain unintended division by zero[2] or potential not-a-number (NaN) results. Strictly speaking these properties are not run-time errors because no exception is caused, but this makes them even more difficult to detect without bit-precise analysis.

---

[1] Abstract interpretation is able to show the absence of these problems, not the presence.

[2] In IEEE754 floating point, division by zero does not cause an exception and is sometimes used to generate the constant ∞.

Such errors can be checked by automated instrumentation of the converted program. For example, before an integer division, a modelchecker can insert an assertion that the quotient is non-null after the computation of the imperative program. Just as explicitly added assertions, the resulting program can be analyzed as described in Sect. 3.2.

If a violation to the assertion is found, the checker generates a witness which corresponds to a function input that will cause a run-time error. Once fixed, this witness can be added to the testsuite to prevent a reintroduction of the problem.

4.2   Coverage Checks and Test Generation

Another application which is well-suited to bounded model-checking is coverage checking and test generation. Coverage checking effectively corresponds to reachability analysis on the CFG while test generation corresponds to translate the result of the reachability analysis into an executable test case.

A simple reachability analysis instruments each basic block with a wrong property, i.e., `false`. BMC then tries to find input parameters in such a way that this property gets reached when executing the program. If such an input parameter is found, then this input parameter can be used to generate a test case which validates that the corresponding basic block is actually covered when the test case is run.

There exist different criteria for coverage apart from simple branch coverage as outlined above. The principle always stays the same. Of particular interest in industry applications is the modified condition/decision coverage (MC/DC) which is often required in safety critical applications when developing according to a given standard like ISO26262 in automotive [14].

The validation of the coverage using generated test cases can be tricky. This is because standard coverage analysis tools often use a very different notion of coverage e.g., coverage measured in lines on the source code instead of the coverage on the generated bytecode. It is therefore beneficial to use coverage analysis tools which use the exact same coverage metric as the coverage analysis and test generation, e.g., BlueCov [12] and JBMC for Java.

5   Challenges for Modern Programming Languages

Traditional languages like C can be challenging to verify due to things like unspecified or undefined behavior. Modern programming languages pose different challenges for formal verification. This is despite the fact that the languages themselves are generally defined in a more precise way and often have a single implementation effectively giving an implementation as specification.

5.1   Strings

Nowadays, strings are often a primitive datatype in modern programming languages. In C, strings are simply realized as null-terminated byte arrays. Most modern languages support Unicode and many different string operations. In Java for example, Strings are represented with an array of 16 Bit wide UTF-16 characters and explicit length information.

Efficiently solving string constraints in SMT formulas requires specialized solvers or at least specialized theories that support string operations. Because of the ease of use, strings are used ubiquitously in many programs.

```
1   int f(String s) {
2     int r = 0;
3     if(s.subStringOf("longString")) {
4       r = 1;
5     }
6     assert(r != 1);
7     return r;
8   }
```

Listing 9: Branch Dependent on String Values

Listing 9 shows a program where the property in line 6 depends on whether the string parameter "s" is a substring of a given string. Solving this or similar constraints, e.g., matching regular expressions require approaches like [20] which iteratively creates constraints that describe the string functions.

5.2   Standard Library

Unlike C, most modern languages come with an extensive standard library that offers efficient data structures and algorithms. These standard libraries are widely used, which means that any verification approach also has to support the use of these.

In theory, it is straightforward to simply create a mathematical model of the code in the standard library. For Java this would mean to read the corresponding bytecode from the runtime jar file and create the model as described in Sect. 3. In practice, this approach does not scale. The two main reasons for this are

1. Most parts of the standard library are not used in a program
2. The implementation of data structures and algorithms is often made as efficiently as possible which generates rather complex models.

The first challenge can be solved by only lazily providing the parts of the library which are actually used. Any non-references classes are not generated, any unused method is simply stubbed and its code only generated if a call to the method is actually found.

The second challenge can be more difficult to solve. While in theory it is possible to simply include the implementation of the data-structures and algorithms into the formal model, this leads to scalability problems. Generally, the standard library implementations are well optimized for execution but not for verification. An approach to solve this challenge is to abstract the implementation of certain algorithms and data-structures with a specification and make this specification as efficient as possible for verification.

This approach is shown in Listing 10 for the square root implementation for Java. The real implementation in the standard library uses an efficient approach to calculate the square root. In this specification, the square root of a `double` a is defined as a number $sqrt$ for which

$$|sqrt \cdot sqrt - a| < \epsilon$$

holds for a pre-specified `double` $\epsilon > 0$. This is represented via the function calls to `CProver.nondetDouble` which creates a random double value and `CProver.assume` which creates constraints for this value, specifying the required property for the square root.

This is an approximation for the following reason: these constraints can be true for multiple values of $sqrt$ so the result might not be identical to the result of an algorithm. It is important to note here that an exact square root for a given arbitrary `double` value is not even guaranteed to exist. Therefore whether this is a viable approximation has to be decided for each application. A more exact specification would be possible but would lead to a more complex model which potentially requires more resources to find a solution for the generated constraints.

```
1    double EPS = ... ;
2    public static double sqrt(double a) {
3        if (Double.isNaN(a) || a < 0.0) {
4            return Double.NaN;
5        }
6        if (a == Double.POSITIVE_INFINITY || a == 0.0) {
7            return a;
8        }
9        double sqrt = CProver.nondetDouble();
10       CProver.assume(sqrt >= 0);
11       CProver.assume(abs(sqrt * sqrt - a) < EPS);
12       return sqrt;
13   }
```

Listing 10: Model for Floating Point Square Root Function

6   Summary and Outlook

This paper presented an approach to translate a stack-based virtual machine language into a formal model to be analyzed using bounded model-checking.

Using analyses based on virtual machine bytecode instructions has the advantage that different input languages which use this language can be analyzed.

The approach removes the explicit stack by creating expressions that model the stack content, unroll the resulting imperative program and use SSA form to create a finite model that can be solved using SMT solvers.

Users can specify properties via assertions in the code. In BMC, the negation of such properties is conjoined with the model of the program and then checked for satisfiability. If the negated property is reachable, then the result of the analysis being a counterexample of a property or the witness for a test case. If the negated property is not reachable, it is proven to be invariant.

Standard properties, e.g., overflow checks, array bounds, division by zero, NULL pointer dereference, or NaN results can be added and checked automatically. This makes the integration of BMC into a CI/CD process interesting in software development. It is also possible to instrument the intermediate representation with reachability properties that model coverage in the program. In this way, it is possible to automatically generate additional test cases to increase coverage or to prove part of a program as being unreachable.

The outlined approach for BMC of Java bytecode can be used as a template to create the possibility to analyze other stack-based virtual machine code, e.g., Webassembly (Wasm) [2]. Wasm has some additional challenges like vector instructions and blockwise branching which have to be translated correctly into SMT constraints.

Another promising area for future work is to extend the analysis with the possibility to annotate loops with invariants and variants and then verify those. While this generally requires manual annotations, this could be an interesting area for the application of generative models as the results can easily be validated using solvers.

References

1. The Java® Virtual Machine Specification. https://docs.oracle.com/javase/specs/jvms/se23/html/index.html
2. WebAssembly Specification - WebAssembly 2.0 (Draft 2024-11-12). https://webassembly.github.io/spec/core/
3. Barrett, C., Fontaine, P., Stump, A.: The SMT-LIB Standard
4. Beckert, B., Kirsten, M., Klamroth, J., Ulbrich, M.: Modular verification of JML contracts using bounded model checking. In: Margaria, T., Steffen, B. (eds.) ISoLA 2020. LNCS, vol. 12476, pp. 60–80. Springer, Cham (2020). https://doi.org/10.1007/978-3-030-61362-4_4
5. Beyer, D.: State of the art in software verification and witness validation: SV-COMP 2024. In: Finkbeiner, B., Kovács, L. (eds.) Tools and Algorithms for the Construction and Analysis of Systems, pp. 299–329. Springer, Cham (2024). https://doi.org/10.1007/978-3-031-57256-2_15
6. Beyer, D., Keremoglu, M.E.: CPACHECKER: a tool for configurable software verification. In: Gopalakrishnan, G., Qadeer, S. (eds.) CAV 2011. LNCS, vol. 6806, pp. 184–190. Springer, Heidelberg (2011). https://doi.org/10.1007/978-3-642-22110-1_16

7. Cordeiro, L., Kesseli, P., Kroening, D., Schrammel, P., Trtik, M.: JBMC: a bounded model checking tool for verifying Java bytecode. In: Chockler, H., Weissenbacher, G. (eds.) CAV 2018. LNCS, vol. 10981, pp. 183–190. Springer, Cham (2018). https://doi.org/10.1007/978-3-319-96145-3_10
8. Cousot, P.: Abstract interpretation. ACM Comput. Surv. **28**(2), 324–328 (1996). https://doi.org/10.1145/234528.234740
9. Farias, B., Menezes, R., De Lima Filho, E.B., Sun, Y., Cordeiro, L.C.: ESBMC-python: a bounded model checker for python programs. In: Proceedings of the 33rd ACM SIGSOFT International Symposium on Software Testing and Analysis, pp. 1836–1840. ACM, Vienna, Austria (2024). https://doi.org/10.1145/3650212.3685304
10. Gadelha, M.R., Monteiro, F., Cordeiro, L., Nicole, D.: ESBMC v6.0: verifying C programs using k-induction and invariant inference. In: Beyer, D., Huisman, M., Kordon, F., Steffen, B. (eds.) TACAS 2019. LNCS, vol. 11429, pp. 209–213. Springer, Cham (2019). https://doi.org/10.1007/978-3-030-17502-3_15
11. Güdemann, M., Riedl, K.: Level-up - from bits to words. In: Lima, L., Molnár, V. (eds.) Formal Methods: Foundations and Applications. LNCS, pp. 124–142. Springer, Cham (2022). https://doi.org/10.1007/978-3-031-22476-8_8
12. Güdemann, M., Schrammel, P.: BlueCov: integrating test coverage and model checking with JBMC. In: Proceedings of the 38th ACM/SIGAPP Symposium on Applied Computing, pp. 1695–1697. ACM, Tallinn, Estonia (2023). https://doi.org/10.1145/3555776.3577829
13. Heizmann, M., et al.: Ultimate automizer with SMTInterpol. In: Piterman, N., Smolka, S.A. (eds.) TACAS 2013. LNCS, vol. 7795, pp. 641–643. Springer, Heidelberg (2013). https://doi.org/10.1007/978-3-642-36742-7_53
14. ISO: Road vehicles – Functional safety (2011)
15. Kroening, D.: Automated reasoning at Amazon: a conversation. https://www.amazon.science/blog/automated-reasoning-at-federated-logic-conference-floc. Accessed 19 Apr 2023
16. Kroening, D., Strichman, O.: Decision Procedures. Springer, Cham (2016)
17. Kroening, D., Tautschnig, M.: CBMC – C bounded model checker. In: Ábrahám, E., Havelund, K. (eds.) TACAS 2014. LNCS, vol. 8413, pp. 389–391. Springer, Heidelberg (2014). https://doi.org/10.1007/978-3-642-54862-8_26
18. Leino, K.R.M.: This is boogie 2 (2008)
19. Leroy, X., Rocquencourt, I.: Java bytecode verification: algorithms and formalizations
20. Li, G., Ghosh, I.: PASS: string solving with parameterized array and interval automaton. In: Bertacco, V., Legay, A. (eds.) HVC 2013. LNCS, vol. 8244, pp. 15–31. Springer, Cham (2013). https://doi.org/10.1007/978-3-319-03077-7_2
21. Merz, F., Falke, S., Sinz, C.: LLBMC: bounded model checking of C and C++ programs using a compiler IR. In: Joshi, R., Müller, P., Podelski, A. (eds.) VSTTE 2012. LNCS, vol. 7152, pp. 146–161. Springer, Heidelberg (2012). https://doi.org/10.1007/978-3-642-27705-4_12
22. Paganoni, M., Furia, C.A.: Reasoning about exceptional behavior at the level of Java bytecode. In: Herber, P., Wijs, A. (eds.) Integrated Formal Methods, pp. 113–133. Springer, Cham (2024). https://doi.org/10.1007/978-3-031-47705-8_7
23. Rastello, F., Bouchez Tichadou, F. (eds.): SSA-Based Compiler Design. Springer, Cham (2022). https://doi.org/10.1007/978-3-030-80515-9
24. Rungta, N.: A billion SMT queries a day (invited paper). In: Shoham, S., Vizel, Y. (eds.) Computer Aided Verification. LNCS, vol. 13371, pp. 3–18. Springer, Cham (2022). https://doi.org/10.1007/978-3-031-13185-1_1

25. Song, K., Matulevicius, N., de Lima Filho, E.B., Cordeiro, L.C.: ESBMC-solidity: an SMT-based model checker for solidity smart contracts. In: Proceedings of the ACM/IEEE 44th International Conference on Software Engineering: Companion Proceedings, ICSE 2022, pp. 65–69. Association for Computing Machinery, New York (2022). https://doi.org/10.1145/3510454.3516855
26. VanHattum, A., Schwartz-Narbonne, D., Chong, N., Sampson, A.: Verifying dynamic trait objects in rust. In: Proceedings of the 44th International Conference on Software Engineering: Software Engineering in Practice, ICSE-SEIP 2022, pp. 321–330. Association for Computing Machinery, New York (2022). https://doi.org/10.1145/3510457.3513031
27. Vardi, M.Y.: Program verification: vision and reality. Commun. ACM **64**(7), 5–5 (2021). https://doi.org/10.1145/3469113

How to Drawjectory? - Trajectory Planning Using Programming by Demonstration

Leonhard Alkewitz🆔, Timo Zuccarello🆔, Alexander Raschke^(✉)🆔,
and Matthias Tichy🆔

Institute of Software Engineering and Programming Languages, Ulm University,
89081 Ulm, Germany
{leonhard.alkewitz,timo.zuccarello,alexander.raschke,
matthias.tichy}@uni-ulm.de

Abstract. A flight trajectory defines how exactly a quadrocopter moves in the three-dimensional space from one position to another. Automatic flight trajectory planning faces challenges such as high computational effort and a lack of precision. Hence, when low computational effort or precise control is required, programming the flight route trajectory manually might be preferable. However, this requires in-depth knowledge of how to accurately plan flight trajectories in three-dimensional space.

We propose planning quadrocopter flight trajectories manually using the *Programming by Demonstration (PbD)* approach – simply drawing the trajectory in the three-dimensional space by hand. This simplifies the planning process and reduces the level of in-depth knowledge required. We implemented the approach in the context of the *Quadcopter Lab* at Ulm University.

In order to evaluate our approach, we compare the precision and accuracy of the trajectories drawn by a user using our approach as well as the required time with those manually programmed using a domain specific language. The evaluation shows that the *Drawjectory* workflow is, on average, 78.7 s faster without a significant loss of precision, shown by an average deviation 6.67 cm.

Keywords: Quadrocopter · (Semi-Automatic) Trajectory Planning · Programming by Demonstration

1 Introduction

Life is full surprising coincidences. Wolfang Reif twice had – in totally unrelated circumstances – a profound impact on the Institute of Software Engineering and Programming Languages at Ulm University.

First, Wolfgang Reif was professor at the institute – then called Software Engineering and Compiler Construction – from 1994 to 2000 (before moving on to heading the Institute of Software and Systems Engineering at the University of Augsburg) working on formal verification and the KIV system. He put Ulm on the international map as a renowned place for formal

G. Ernst et al. (Eds.): Wolfgang Reif Festschrift, LNCS 15765, pp. 195–219, 2025.
https://doi.org/10.1007/978-3-031-92196-4_10

methods research in the software engineering community.

*Second, Wolfgang Reif gave me (Matthias Tichy) in 2010 the first oppor-
tunity for independent research and teaching by offering me the position as
acting professor for Self-Organizing Systems at the University of Augsburg
for one year. While my goal was then to leave academia to join industry,
seizing that opportunity allowed me to grow as a person and as an academic and put me on my academic path.*

*Finally, those two unrelated circumstances surprisingly met when i was
offered to become head of the Institute of Software Engineering and Compiler Construction at Ulm University in 2015. Impressed by the research of
Wolfgang Reif's group in the area of autonomous quadrocopters, I decided
to build a quadrocopter lab as a research pillar in Ulm as well. Wolfgang
Reif and his group, particularly, Andreas Angerer and Alwin Hoffmann,
have been a tremendous help in building up that lab and our research.
Hence, we chose to present one of our lab's recent outcomes in this paper
to commemorate Wolfgang Reif's 65th birthday.*

Quadrocopters have a wide range of applications, from automated parcel
delivery to search and rescue operations. Therefore, it is not surprising that the
topic of planning flight routes or trajectories for drones is becoming increasingly
important [4,19].

Generally speaking, trajectory planning "consists in assigning a time law to
the geometric path" [15]. Trajectory planning distinguishes two planning strategies: manual and automatic planning. When considering automatic trajectory
planning, there are several promising approaches, such as genetic algorithms,
artificial neural networks, or simple A* algorithms [4,15].

However, some tasks require high precision and specific points of interest to
be visited. Thus, the mentioned strategies may not always be ideal for trajectory
planning [36]. For instance, an old building in danger of collapse might need to
be screened by a drone with the drone having to check certain rooms. In this
case, the creation of a route through the building, encompassing all points of
interest, can be a particularly time-consuming and error-prone endeavor, particularly when the objective is to implement an automatic planning process, as
this approach does not permit any form of intervention in the selection process
of the points to be visited.

One possible solution to ensure that all points of interest are visited is manual trajectory planning, whereby the drone's flight path is not determined automatically but by hand. There are alternative approaches to manual planning
including i) programming the path (mostly using low-code approaches) and ii)
creating interactive points in a 3D environment on a PC [20,34]. However, these
methods often require knowledge of either how to program or how to interact
with the points, making planning unintuitive.

Accordingly, there is a desire for a user-friendly and still accurate method
for manual trajectory planning, such as demonstrating the trajectory in the
real world. This concept of "transfer[ring] new skills to a machine by relying on
demonstrations from a user" [10] is called *Programming by Demonstration*. Due

to its high intuitiveness, it is ideal for domain experts in a certain topic who need the support of quadrocopters, despite their potential lack of experience in controlling quadrocopters [26], which is why programming by demonstration has seen significant growth [31].

Based on the aforementioned problem and goals, we formulate the following research questions:

RQ 1 How to automatically plan a trajectory by demonstrating the flight path once?

RQ 2 How does trajectory planning by demonstration compare against trajectory planning by manual programming using a domain specific language in terms of accuracy and effort?

We developed an approach for planning trajectories from a user's demonstration, thus applying the *programming by demonstration* paradigm answering RQ1. Specifically, we track the demonstration, i.e., the point sequence of the desired flight path. Subsequently, a selection of points, designated as waypoints, is made from the recorded point sequence. These waypoints are then used to interpolate a trajectory using natural cubic splines.

In principle, our approach could also be applied to other domains (walking robots, industrial robots, etc.). The trajectory planning must be adapted to the domain and, if necessary, be able to take other restrictions into account, e.g. degrees of freedom in the movement, obstacles on the ground, or stairs.

To answer RQ2, we evaluate the accuracy and the planning effort by systematically using both approaches to plan three different classes of trajectories: planar geometric figures, planar non-trivial figures, 3D figures. We measure the time to plan those trajectories as well as the accuracy of the planned trajectory compared to the intended trajectory. The evaluation shows that the *Drawjectory* workflow is, on average, 78.7 seconds faster without a significant loss of precision, shown by an average deviation 6.67 cm compared to using a domain specific language to manually program the trajectories.

Section 2 and Sect. 3 introduces the quadrocopter lab setting as well as relevant foundations of trajectory planning. Thereafter, Sect. 4 presents our trajectory planning by demonstration approach and the proof-of-concept implementation. Section 5 presents the results of the evaluation. After discussing related work in Sect. 6, we conclude in Sect. 7 by summarizing the work and addressing limitations as well as discussing potential future work.

2   Overview of *Quadcopter Lab*

The proposed *Drawjectory* workflow and its proof-of-concept implementation are incorporated into the quadrocopter laboratory (*Quadcopter Lab*) at the Institute of Software Engineering and Programming Languages at Ulm University. The quadrocopter laboratory is a spacious room intended for research and students' software engineering projects in the context of quadrocopters. The lab supports, e.g., trajectory planning with collision avoidance for multiple quadrocopters by providing a domain specific language and an augmented reality (AR) system to visualize important information for the quadrocopters.

The setup includes a flight area ($9 \times 5 \times 4$ m), a motion-capturing system supported by a total of 16 cameras on a traverse near the ceiling and a Linux machine running all necessary ROS2 infrastructure. The setup is shown in Fig. 1.

Fig. 1. The motion-capturing system in the *Quadcopter Lab*

The Robot Operating System 2 (ROS2) is an open-source "software platform for developing robotics applications" [24] with widespread applications in both academic and industry settings [9]. Despite its name, ROS2 is not a standalone operating system, it is rather an ecosystem or collection of software designed to create modular, distributed and asynchronously communicating units called *nodes*, that usually encapsulate one specific function [24]. Running *nodes* can communicate with each other and exchange data asynchronously using different methods. Another part of the ROS2 ecosystem is *RViz* [3], a 3D visualizer and *RQt*, a framework for custom user interfaces [2].

An *OptiTrack*-system [1] precisely tracks the position and orientation, the so-called *pose*, of any marked objects within an area of approximately $6 \times 4 \times 3$ m. A marked object refers to an object with special reflective markers attached that can be tracked by the cameras using infrared light. The *OptiTrack*-system utilizes these markers to precisely determine the position of an object in three-dimensional space relative to a reference "world" frame. The system continuously sends a stamped pose to the Linux machine at a frequency of nearly 100 Hz. A stamped pose includes the timestamp at which the point was tracked as well as the points' cartesian coordinates and a quaternion for the orientation[1] of the corresponding object.

The quadrocopters can either be controlled manually by a controller or automatically by so-called mission scripts. The mission scripts are programmed using a simple Lua-based domain-specific language (DSL) presented in [34].

The DSL includes several commands like `takeoff` or `land`. The most important is `moveTo(x',y',z',`ψ`)` that directs the quadrocopter to fly from its current position (x, y, z) to the given position (x', y', z'), while adjusting its orientation by the angle ψ. The (artificial) coordinate system of the quadcopter laboratory has its origin in the centre of the area on the ground and is measured in metres. (0,0,1.5) therefore corresponds approximately to the centre of the room, so that there is 3 m to the front and back, 2 m to the left and right and 1.5 m to the top and bottom.

For this work, we extended the original language presented by Witte et al. [34] by convenient functions for creating waypoints for arcs. For example, the function `arcLeft(n, x, y, z, `ϕ`, angle, forward, lateral)` generates n equidistant waypoints from (x, y, z) to the target, which is determined by the left-hand elliptical arc with the `angle` of an ellipse with the radii `forward` and `lateral`. ϕ is the angle between the initial direction of the arc and the current orientation of the drone.

3   Foundations

In this section, we briefly formalize terms used throughout this work. Firstly, we define *paths* and *trajectories* and their properties, then we formalize their respective planning process and give a definition for *cubic splines* as a way of trajectory interpolation. Lastly, definitions for different kinds of errors used in the evaluation are given.

We define the *Geometric Path* as follows,

Definition 1 ((Geometric) Path). *A path can be defined as a function*

$$P : [0, 1] \to C$$
$$s \mapsto p(s)$$

devoid of any timing information, which satisfies the following constraints [30]:

$$p(0) = p_{start}$$
$$p(1) = p_{goal}$$

where p_{start} denotes the first point of the path and p_{goal} the last point of the path.

---

[1] https://docs.ros2.org/galactic/api/geometry_msgs/msg/PoseStamped.html, last accessed 02/03/2025.

where C is called the configuration space, which—in the context of this work—is set to $C = \mathbb{R}^3$ as we disregard the orientation of the quadrocopter. Thus, leaving only the spatial three of the usual six degrees of freedom of a quadrocopter.

Then, by adding a time parameterization s that maps the time to a specific section of the path, defined by

$$s : [t_0, t_{max}] \rightarrow [0, 1]$$
$$t \mapsto s(t)$$

with $s(t_0) = 0$ and $s(t_{max}) = 1$, a trajectory is defined as following [30]:

Definition 2 (Trajectory). *A trajectory T is a path P endowed with a time parameterization s:*

$$T : [t_0, t_{max}] \rightarrow \mathbb{R}^3$$
$$t \mapsto p(s(t))$$

More descriptively, "a trajectory [can be] defined as a sequence of time-stamped locations" [22].

In order to retrieve the velocity and the acceleration of the trajectory at each point in time t, the trajectory T can simply be derived [27]. Then the velocity is defined as $v(t) = \dot{T} = \frac{d}{dt}T$ and the acceleration as $a(t) = \ddot{T} = \frac{d^2}{dt^2}T$. In addition, we assume for the trajectory planning that the velocity and acceleration at the first and the last point of the trajectory is set to 0:

$$v(t_0) = v(t_{max}) = 0$$
$$a(t_0) = a(t_{max}) = 0$$

Definition 3 (Trajectory smoothness). *Given a trajectory without excessive or fast-changing accelerations—meaning the so-called jerk is low, which is defined as the derivative of the acceleration—this trajectory is called* **smooth** [15].

A low or at least limited *jerk* also has the advantage that the trajectory can be executed faster and with higher accuracy [15], but a low jerk also increases the time required to reach cruise speed and hence, in some scenarios, may actually increase trajectory time. This means that the jerk threshold needs to be determined depending on the individual case. In our scenario, we focus on a high trajectory smoothness.

Based on the definitions of a path and a trajectory, the terms *path planning* and *trajectory planning* are defined as:

Definition 4 (Path planning). *Path planning is the process of generating a geometric path (see Definition 1) from a start to a goal point, passing through pre-defined waypoints [15].*

Definition 5 (Trajectory planning). *Trajectory planning is the process of endowing a geometric path (see Definition 1) with time information* [15].

As the geometric path is already defined in the demonstration, further consideration of path planning is unnecessary. In practice, different approaches for trajectory planning are employed focusing on speed, efficiency, or minimal jerk [15], however, in this work, the focus is on generating a smooth and precise trajectory. Therefore, the (natural) cubic spline is presented as this is a simple, smooth, and accurate trajectory planning method. Cubic splines are used to create smooth curves fitting through several waypoints and are defined as follows:

Definition 6 (Cubic spline). *Given a set* $X = \{(x_1, y_1), \ldots, (x_n, y_n)\}$ *of point-pairs, a cubic spline is a function* $S : [x_1, x_n] \mapsto \mathbb{R}$ *defined piece-wise by cubic functions* [25]:

$$S(x) = \begin{cases} s_1(x) & x_1 \le x \le x_2 \\ \ldots \\ s_{n-1}(x) & x_{n-1} \le x \le x_n \end{cases}$$

where s_i *is a polynomial of third degree given by:*

$$s_i(x) = a_i(x - x_i)^3 + b_i(x - x_i)^2 + c_i(x - x_i) + d_i$$

A cubic spline S *is a natural cubic spline, if* $s_1''(x_1) = s_{n-1}''(x_n) = 0$.

From this, we can derive the following properties for a cubic spline function [25]:

1. $\forall x_i \in X : S(x_i) = y_i$, i. e. S interpolates all (way)points y_i exactly
2. $S \in C^2$, i. e. S is twice continuously differentiable meaning $S(x)$, $S'(x)$, $S''(x)$ are continuous on $[x_1, x_n]$

Using cubic splines to interpolate trajectories can lead to errors with regard to the demonstration—the so-called *interpolation error*. A typical method to determine this error is to calculate the position error vectors [20,22,23]. The position error vector for a specific point in time t is defined as

$$e_{pos}(t) = P(t) - T(t) \tag{1}$$
$$= [P_x(t) - T_x(t), P_y(t) - T_y(t), P_z(t) - T_z(t)]^T \tag{2}$$

where $P(t)$ describes the point of the demonstrated flight path at time t and $T(t)$ the corresponding point of the trajectory at the same time t.

In addition, there may be an error between the theoretically planned trajectory and the actually flown trajectory in the real world, potentially caused by external influences, such as a ventilation, but as it is not directly related to the trajectory planning process, this type of error is not considered further in this work. We also limit the calculation of errors to the planned trajectory as opposed to the edited trajectory, as the latter would not necessarily represent an error of the trajectory planning process itself.

These previously mentioned position error vectors are the foundation of further error analysis, wherefore the root-square-mean-error (RSME) [14,20,36] and the mean-absolute-error (MAE) are commonly used. The advantage of the MAE over RSME is that it emphasizes outliers less and can be interpreted directly. The RSME and MAE can be calculated as follows[2] [11]:

$$RSME = \sqrt{\frac{1}{n} \sum_{i=1}^{n} ||e_{pos}(t_i)||_2^2} \tag{3}$$

$$MAE = \frac{1}{n} \sum_{i=1}^{n} ||e_{pos}(t_i)||_2 \tag{4}$$

where n is the number of recorded points making up the demonstrated flight path and t_i is the time of the i-th recorded point.

However, a non-zero interpolation error is not necessarily problematic: Even though during trajectory planning physical limitations are taken into account and jitter in demonstrations is ignored as much as possible, the errors are still calculated between the raw demonstration and planned trajectory.

4   Concept and Implementation of the *Drawjectory* Workflow

To solve the problem of a precise and intuitive manual trajectory planning for quadrocopters, we propose the *Drawjectory* workflow. This workflow comprises four distinct phases, depicted in Fig. 2, that build upon each other:

1. Track the desired flight path of the quadrocopter using a gesture wand.
2. Plan a smooth, feasible trajectory based on the demonstration.
3. Edit the planned trajectory, by shifting, rotating or linearly scaling it, or by moving certain points.
4. Process trajectory to control the quadrocopter.

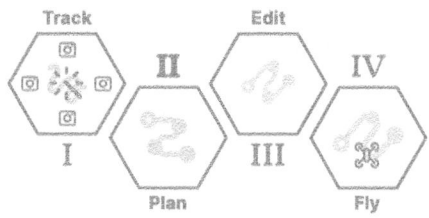

Fig. 2. The *Drawjectory* phases

The objective of this workflow is to demonstrate the desired trajectory intuitively in the first phase. The adapted trajectory planning in phase two ignores the noise from the user's demonstration by creating a smooth, feasible trajectory. To address the issue of the limited space in the quadrocopter laboratory,

---

[2] In both Eq. 3 and Eq. 4, $|| \bullet ||_2$ denotes the Euclidean norm.

or of slightly deviating demonstrations, we introduce the third phase. This phase provides the user with the option to edit the trajectory, by either moving certain points or shift, scale or rotate the trajectory relative to the first point of the trajectory[3].

As motivation for this phase, our programming by demonstration approach is restricted to the space covered by the tracking system. Consequently, it is necessary to enlarge the planned route at a later stage and, if necessary, to rotate or move it relative to a calibration point. If the user later realizes that parts of the flight were unintentional or too inaccurate, moving the waypoints is useful.

In order to test the *Drawjectory* workflow, we created a RQt plugin called `Drawjectory Control Panel` that serves as a proof-of-concept implementation. More details on the architecture, the implementation and integration of the `Drawjectory Control Panel` into the *Quadcopter Lab* can be found in [7].

4.1   Tracking of the Gesture Wand

The first phase of the workflow is the *demonstration of the flight path*, consisting of the following steps: i) Tracking an object that determines the flight path, and ii) the saving, editing and visualizing of the demonstrated flight path.

To demonstrate the flight path of the quadrocopter, we use a custom gesture wand. As shown in Fig. 3, the wand is a stick with three reflective markers attached to it, aligned with the axes of a Cartesian coordinate system, that is tracked by the *OptiTrack*-system. During the demonstration, the physical limits of the drone, such as acceleration limits, should be considered, as the speed of the demonstration will later be transferred one-to-one to the drone. It should be noted that the demonstration may be imprecise, or not deter-

Fig. 3. The gesture wand with grey reflective markers

mined at all, as the tracking cameras are mounted on a truss near the ceiling and can only observe a limited area within the laboratory. However, this area is marked on the floor and working solely in this area reduces the likelihood of poor quality tracking (see Fig. 1).

The position of the gesture wand is sent to the `Drawjectory Control Panel` only approximately every 10 *ms*, although this is not a serious issue as points are selected to interpolate a trajectory later. These recorded points make up the demonstrated flight and are the basis for later trajectory planning.

---

[3] In the *Quadcopter Lab* it is not possible to scale the trajectory arbitrarily, as the quadrocopters need a connection to the *OptiTrack*-system and Linux machine.

In addition, the *Drawjectory* workflow includes the option to trim and visualize the recorded flight path, exemplified in Fig. 4. "Trimming" the flight path is the process of redefining the actual start and end points of the flight path so that all points before the start point and all points after the end point are ignored when planning the trajectory. One advantage of trimming is that the user is allowed to cut out parts at the beginning or end of the demonstration that contained errors rather than restarting the tracking, or cut out a section of the demonstrated flight path that is only necessary, for instance, if the flight path is only to extend an automatically planned trajectory.

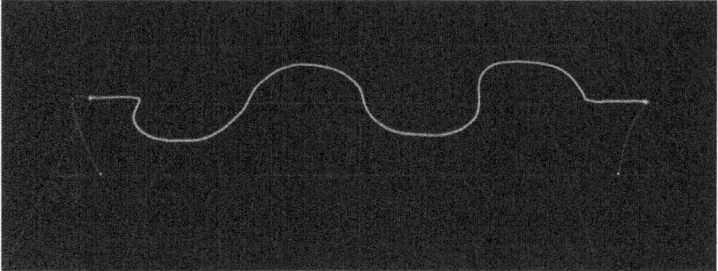

Fig. 4. Visualization of the demonstrated, trimmed flight path in *RViz*, whereas the two yellow spheres mark the start and end point of the trimmed flight path.

It is advisable to allow the user of the *Drawjectory* workflow to save these recorded points in order to reload them later or elsewhere if necessary, e.g. to proceed the workflow in a different environment. Therefore, the `Drawjectory Control Panel` provides methods to save and (re-)load a list of n recorded points in an unambiguous format.

4.2   Adapted Trajectory Planning

Based on the truncated demonstrated flight, the next phase of the *Drawjectory* workflow is *to plan or reconstruct the trajectory*. This is typically achieved by sampling points from the demonstration and interpolating a path passing through the sampled points, with the aim of including the omitted points as well [5,21,26,38]. The reason for the sampling is to reduce the noise of the demonstration and create smoother curves by interpolation.

Selection of the Waypoints. A number of sampling strategies exist to extract waypoints from a sequence of recorded points. These include random sampling, equidistant sampling and corner detection [21], whereas random and equidistant sampling are techniques developed in-house. In the proof-of-concept implementation of the workflow, we employed random and equidistant sampling, as random sampling is a simple and fast-to-implement approach, while equidistant sampling regarding the spatial distance is commonly used for obtaining waypoints for the interpolation [8]. The principal disadvantage of these methods is the lack of focus on "significant" points, such as points in sharp curves, which is why the corner detection algorithm exists. Nevertheless, this algorithm is too complex for a proof-of-concept implementation and reduces noise to little [21].

Algorithm 1 Equidistant waypoint selection

Require: $n \geq 2$
$\quad W \leftarrow \{\}$
$\quad d \leftarrow round(\frac{|P|-n}{n-1})$
$\quad i \leftarrow 0$
$\quad$**while** $i < |P| - 1$ **do**
$\quad\quad W \cup p_i$
$\quad\quad i \leftarrow i + d + 1$
$\quad$**end while**
$\quad W \leftarrow \{p_l\}$

In general, the strategies used in the *Drawjectory* workflow are required to include i) the first and last point P_0 and P_l of the trimmed demonstrated flight path P and ii) a specified number of waypoints n. Random sampling retrieves a point p_i from the sequence of recorded points P (except the already chosen points) as long as the number of desired waypoints is not reached. An alternative approach is to select the waypoints in an equidistant manner. Algorithm 1 introduces a technique of selecting the waypoints in such a way that there are approximately the same number of omitted points between each adjacent waypoint.

Assuming that the time interval between each recorded point p_i is perfectly equal, this technique is equivalent to selecting the waypoints regarding a certain time interval, so use equidistant sampling in terms of time. Consequently, the "distance", i.e. the number of omitted points between two waypoints and not the spatial distance, must be determined by the following formula:

$$d = \frac{\text{number of points in sequence} - \text{number of waypoints}}{\text{number of waypoints} - 1}$$

It describes the points to be omitted ($|P| - n$) are split into ($n - 1$) sections, as there are this many sections between the waypoints. The rationale behind this sampling technique is based on the assumption that a user demonstrates more complex sections, such as a sharp curve, of the trajectory at a slower pace and simpler ones at a faster pace. Consequently, there is a higher density of points in sections that are more complex (to interpolate), which in turn leads to a greater number of waypoints in these sections, which are used for interpolation. This sampling technique therefore produces results that are comparable to those obtained through corner detection, but is more likely to reduce noise, provided that not too many waypoints are selected.

Interpolation of the Trajectory. For interpolating a trajectory based on given waypoints, there exist a couple of methods, such as Bezier curves or NURBS, as presented in Sect. 6. The most basic idea, however, is to linearly interpolate, i.e. connect the waypoints with straight lines. The method's principal advantage is its straightforward implementation and the fact that it creates exact lines. However, this method also has significant drawbacks, including its inefficiency as it creates infinite curvature at the waypoints, which necessitates the quadrocopter's halt at the waypoints to precisely adhere to the trajectory [27].

Another commonly used approach to planning the trajectory are polynomial piecewise functions, such as natural cubic splines, which were introduced in Sect. 3. Cubic splines offer the advantage of a simple and expedient calculation, while their smoothness remains quite high and they pass exactly through the waypoints in contrast to B-splines [33]. As cubic splines are only capable of interpolating two-dimensional functions and a trajectory requires the interpolation of a four-dimensional function, the trajectory cannot be interpolated directly but need to be decoupled. This entails interpolating the trajectory not as a whole, but separately for the x, y and z coordinates, and subsequently coupling these coordinates [17].

Given the sequence of waypoints, denoted as $W := p_0, ..., p_n$, where each point p_i was recorded at time t_i, three sets of time-coordinate pairs, one for each of the spatial axes x, y and z, are defined: $X := \{(t_0, p_{0,x}), ..., (t_n, p_{n,x})\}$, $Y := \{(t_0, p_{0,y}), ..., (t_n, p_{n,y})\}$ and $Z := \{(t_0, p_{0,z}), ..., (t_n, p_{n,z})\}$. These time-coordinate pairs determine the cubic splines S_x, S_y that in turn define the trajectory T as follows:

$$T(t) = [S_x(t), S_y(t), S_z(t)]^T$$

To fly the trajectory, it is necessary to obtain control points from the trajectory, as well as the velocity and acceleration at certain points. Therefore, the splines and their derivatives are evaluated at timestamps $t \in [t_0, t_{max}]$.

Feasibility Constraints for Trajectories. As either the demonstration or the interpolation can result in a trajectory that is not flyable, also called *not feasible*, it is necessary to fulfill certain constraints before the trajectory is flown.

The first constraint is the size of the *Quadcopter Lab*, or more precisely the area in which the quadrocopters are allowed to fly. As previously stated in Sect. 2, the area in question has a size of 6 × 4 x 3 m. Consequently, each point p_i of the trajectory T must be within this area, i.e.

$$\forall p_i \in T : 0 \le p_{i,x} \le 6 \wedge 0 \le p_{i,y} \le 4 \wedge 0 \le p_{i,z} \le 3$$

otherwise the demonstration must be repeated.

The controllers of the quadrocopter impose the second constraint, as they limit the velocity and acceleration of the quadrocopter [28]. In general, the velocity $v(t)$ and acceleration $a(t)$ must fulfill the following constraints:

$$||v|| = \sqrt{v_x^2 + v_y^2 + v_z^2} \le v_{max}$$

$$||a|| = \sqrt{a_x^2 + a_y^2 + a_z^2} \le a_{max}$$

In the context of the *Quadcopter Lab*, experiments with different velocity and acceleration limits revealed that $v_{max} = 1.5 \frac{m}{s}$ is appropriate for the *Bebop Parrot 2* drones used in it. Additionally, it is sufficiently low to prevent the quadrocopter from reaching a critical acceleration a_{max}, which would result in a deviating flown trajectory.

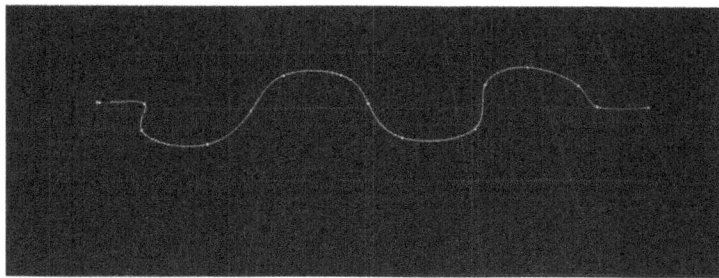

Fig. 5. Visualization of a trajectory, colored according to the interpolation error (Color figure online)

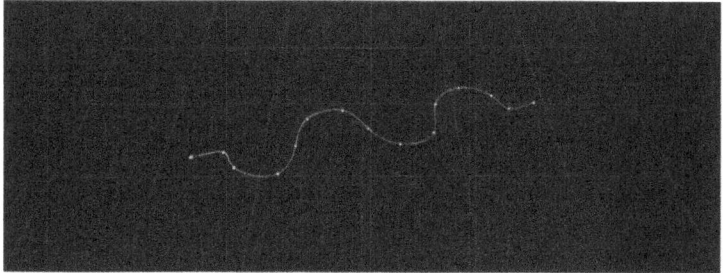

Fig. 6. Visualization of an shifted, scaled, rotated trajectory with moved waypoints. (Color figure online)

Visualization of Trajectory. In the case of the proof-of-concept implementation, a line visualizes the computed trajectory in a manner analogous to the demonstrated flight path. This visualization is extended by yellow spheres representing the waypoints. The trajectory is colored with a color gradient, ranging from green (minor deviation) to red (major deviation), indicating the interpolation error (see Fig. 5).

Editing Planned Trajectory. Once the trajectory planning phase is complete, the user may *edit the trajectory*. There are several operations, which may or may not be applied and which may be combined, and which therefore allow a trajectory to be edited arbitrarily, as long as these operations do not produce a trajectory control point that violates the feasibility constraints (cf. Sect. 4.2).

One operation is to manipulate the waypoints of the trajectory within a three-dimensional virtual environment, such as *RViz*. Additionally, the trajectory may be shifted by an offset, rotated by an angle around the first point in the xy-plane or linearly scaled by a factor relative to the first trajectory point[4].

Figure 6 illustrates an example of an edited trajectory. As can be seen, several parts of the trajectory are red, indicating that these section deviate the most

---

[4] See [13] for more information about how these operations can be defined.

from the demonstration. However, as the lines are now relatively close to half circles, these adaptions probably match the intended flight path more closely.

5   Evaluation

To evaluate the *Drawjectory* workflow, we applied the workflow to a couple of scenarios and measured the planning time as well as the accuracy. This experiment should provide some insight into whether the *Drawjectory* workflow is faster - i.e. easier to use - and at what cost, namely loss of accuracy compared to programming the trajectory, this possible acceleration comes. We provide a package online with the material used for the evaluation, namely the raw programming, demonstration and trajectory data, the evaluation scripts and detailed scenario descriptions [6].

Unit Testing and Test Coverage. As the correctness of the trajectory planning process is important too, this evaluation is complemented by unit tests. Therefore, 45 test cases with multiple conditions have been written that test the implementation of the *Drawjectory* workflow automatically. These test cases achieve a statement coverage of 60.2% and excluding the parts responsible for the GUI, it increases to 95.5%.

5.1   Experiment Setup

In general, the experiment (setup) consists of three phases: i) the definition and specification of the scenarios, ii) conducting the experiment by realizing the scenarios using two input modalities and measuring the planning time as well as the accuracy, iii) evaluation and analysis of the measurements. The experiment was conducted by one author[5], an experienced user of both input modalities. While programming and demonstrating the trajectories, the author was allowed to use the scenario specification to read the instructions. Corrupted demonstrations have been repeated until no more corruption could be detected. Demonstrations could be corrupted by missing data from the *OptiTrack* system, for example because the cameras' view of the tracking markers was obscured.

Scenarios. We designed a total of 16 scenarios, divided into three categories of increasing complexity (cf. Table 1):

- *Planar geometric figures* (5), such as a square or an ellipse
- *Planar non-trivial figures* (6), such as simulating the inspection of rooms
- *3D figures* (5), such as a spiral or staircase

---

[5] Additionally, the workflow was evaluated in three selected scenarios by five participants regarding its usability. Details on the experimental setup, results, and potential expansions are provided in the accompanying thesis [7].

In general, the scenarios were designed to represent complete (second and third group) or at least components (first group) of possible real-world scenarios. However, the scenarios were intended to be solvable in time, meaning they should not be too complicated. The different scenarios should test, how well trajectories containing sharp turns, smooth curves, long lines or alternating heights can be planned.

Each scenario is precisely specified and described, complemented by a visualization to illustrate the description, as exemplary shown in Fig. 7.

Input Modalities (IV). The scenarios were realized using two different input modalities (independent variable):

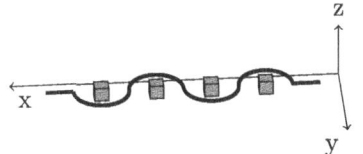

- **Programming** using an extended version of the Domain-specific language (DSL) of the *Hybrid Editor* [34], already mentioned in Sect. 2.

Fig. 7. Visualization of the scenario *Virtual slalom*, whose realization was shown in Sect. 4.

- **Demonstration** of the trajectory in the real-world, in this case in the *Quadcopter Lab* with the *OptiTrack* system.

Since the *Hybrid Editor* does not directly support the planning and visualization of trajectories, the programmed flight path was sent to the `Drawjectory Control Panel` as substitute of the tracking data. Hence, the further planning process, including trimming the flight path, determining the number of waypoints and starting the trajectory planning is identical, only the input method is different. The resulting trajectory is displayed in *RViz* to evaluate whether the demonstration or programming was correct, complete and meets the specification.

Measurements (DV). To indirectly evaluate the usability of the *Drawjectory*, we measured the time required to create a trajectory according to the scenario description, as the required time is an indicator for the usability. The planning time was measured automatically by pressing two buttons on the `Drawjectory Control Panel`. Since the planning time does not include the time required to actually plan the trajectory, the time is not dependent on the system on which it is determined.

In order to asses the accuracy of the demonstrated trajectories, we also measured the interpolation error as well as a collection of widely used trajectory similarity measurements. As the demonstrated and the programmed trajectory are not necessarily the same length or duration, we cannot simply calculate the RSME or MAE of the demonstrated trajectory compared to the programmed trajectory. That is why we analyzed the demonstrated trajectories using measurements that can determine the similarity of two different long trajectories. As a comparison value for the trajectory calculated on the basis of the demonstration, we use the programmed flight path, which represents the intended flight path.

The *Hausdorff* distance measures the maximum point-to-trajectory distance, similar to the *Discrete Frechet* distance that is defined as "the maximum distance between any matched point pairs". The *Dynamic Time Warping* (DTW) algorithm finds an optimal matching between the points of two trajectories by minimizing the sum of point-to-point distances, which is called DTW distance [32]. Since the DTW distance depends on the length of the two trajectories, we also consider a normalized DTW distance, which is determined by dividing the DTW distance by the maximum length of two trajectories. This measure overestimates the average distance between two matched trajectory points [12]. In general, the smaller the distances, the more similar the trajectories are.

5.2   Results

Interpolation Error: Measuring the interpolation error from both inputs, the demonstration and the programmed flight path, revealed that the interpolated trajectory is quite similar to the input, as can been see in Fig. 8.

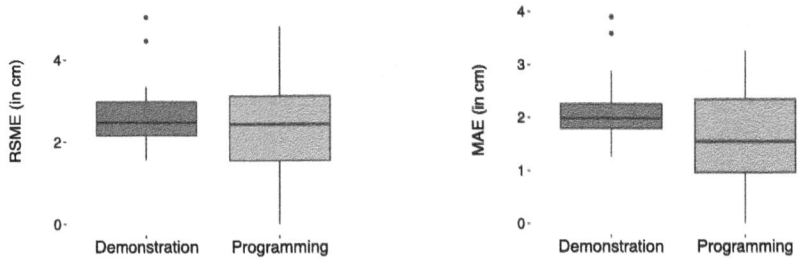

Fig. 8. Interpolation errors RSME (left) and MAE (right) for the two input modalities *Demonstration* and *Programming*.

The average RSME of trajectories planned from demonstrations is 0.027 (SD = 0.0093), in comparison to 0.0232 (SD = 0.014) for programmed flights. Regarding the MAE, the results are very similar: 0.0217 (SD = 0.0073) for the demonstration and 0.0158 (SD = 0.0098) for the programming. This can be interpreted as an average deviation of 2.17 cm respectively 1.58 cm from the trajectories to their input flight path. These results show that demonstrating the trajectories also generates sections that cannot be flown directly or that contain noise. However, it can be seen that this is of the same order of magnitude as programming and that the two input modalities are therefore comparable.

Planning Time: Averaged over all scenarios, using the demonstration approach requires a significantly lower planning time of 75.2 s (SD = 18.39) compared to programming the trajectories with a planning time of 153.86 s (SD = 96.96).

As can be seen in Fig. 9, the required planning time for demonstrating a scenario does not show a strong increase for more complex scenarios (from 60.2 s for geometric figures to 79.0 s for complex 3D figures on average).

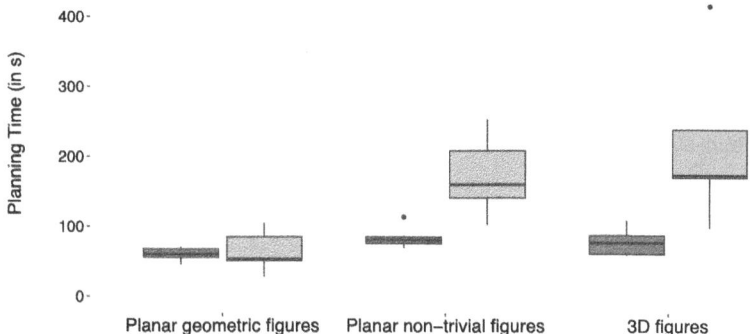

Fig. 9. The trajectory planning times for the three scenario groups.

On the other hand, the time required to program a trajectory increases as the scenarios become more complex, especially if the trajectory is not planar (from 64.7 s for geometric figures to 219.7 s for complex 3D figures on average). This indicates that, probably due to its greater intuitiveness, the *Drawjectory* workflow scales better and saves time, in particular, in more complex scenarios.

Trajectory Similarity: As mentioned in Sect. 5.1, we investigated whether the planning time saved comes at the expense of accuracy. Therefore, we measured the Hausdorff, Frechet, Dynamic Time Warping and normalized Dynamic Time Warping distances between the programmed flight path and the trajectory based on the demonstration for each scenario. The results are presented in Table 1 as well as visualized in Fig. 10.

Figure 10 clearly illustrates, that the more complex the scenarios become (from easy *Planar non-trivial figures* to complex *3D figures*), the greater the distances are. This means that the trajectory based on the demonstration is on average less similar to the programmed trajectory for more complex scenarios. As the Dynamic Time Warping distance depends on the trajectory length, the increasing distance for more complex scenarios is not surprising. The average Frechet distance is 1.644, but there are strong outliers, as evidenced by a standard deviation of 1.813.

In contrast, the average of all Hausdorff distances is with 0.1186 more meaningful, as the standard deviation is relatively lower with 0.048. This average

Table 1. Trajectory Similarity Measures for all scenarios

Scenario Name		Hausdorff	Frechet	DTW	avg. DTW
Line	—	0.076	1.989	30.525	0.077
Square		0.070	0.036	69.339	0.044
Triangle		0.075	0.052	38.424	0.063
Circle		0.075	0.081	50.056	0.039
Ellipse		0.085	0.084	100.091	0.064
Fly around obstacle (1)		0.090	2.969	41.854	0.055
Fly around obstacle (2)		0.093	2.998	102.798	0.075
Virtual slalom		0.120	4.888	90.120	0.059
Virtual maze		0.112	4.175	304.325	0.059
Room inspection		0.140	0.019	187.737	0.056
The house of Wolfgang Reif		0.138	1.000	125.505	0.073
Cuboid		0.206	0.064	393.892	0.090
Spiral		0.135	1.017	96.676	0.071
Staircase		0.078	1.434	29.034	0.034
3D slalom		0.208	5.015	229.538	0.138
Room inspection on two levels		0.195	0.489	437.796	0.070

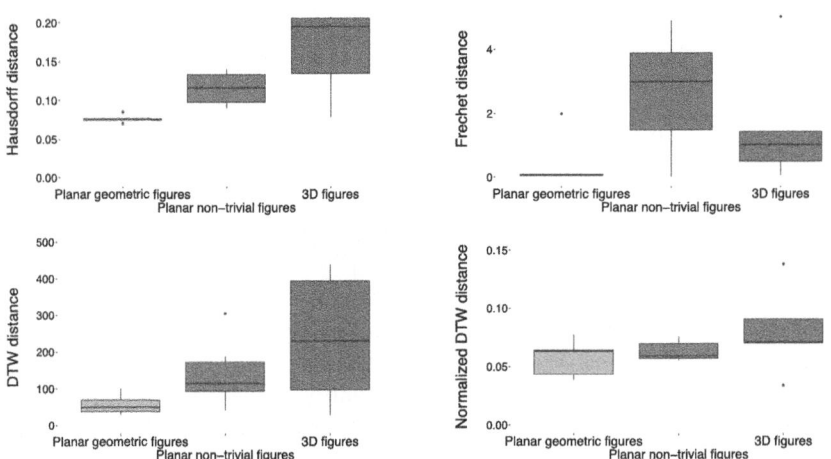

Fig. 10. Trajectory similarity of the demonstrated and programmed trajectory, distinguished by the scenario groups.

distance can be interpreted as, on average over all scenarios, a maximum distance from any point of the demonstrated trajectory to the programmed trajectory of about 11.8 cm. Furthermore, also averaged over all scenarios, the average distance between the demonstrated, optimally matched, and the related programmed trajectory is less than 6.67 cm (SD = 2.4) - called the normalized DTW distance.

In general, the results indicate that the *Drawjectory* workflow accelerates the trajectory planning process (-78.7 s on average). In particular, when planning more complex trajectories, the programming of these trajectories requires a more time ($+82.2$ s), while the demonstration exhibits a minimal additional overhead ($+21.6$ s). This acceleration comes at the expense of accuracy, as shown by the Hausdorff or normalized DTW distance increasing by 0.0876 and 0.0232 (in m), respectively, for longer and more complex scenarios.

To conclude, the *Drawjectory* workflow is a fast and intuitive approach to manual trajectory planning while maintaining a high level of precision. In particular, when used alongside the *Hybrid Editor*, that can be used for programming trajectories, the proposed workflow can be an efficient and effective method for manual trajectory planning.

5.3   Threats to Validity

As this study is an experiment, it is subject to a number of potential threats. Accordingly, we discuss these threats to *internal*, *external*, *construct* and the *conclusion* validity [35] and how we addressed them.

Internal Validity. One potential threat to the internal validity is the different time points at which the two input modalities were tested. To address this issue, both input modalities were tested without any time pressure, starting at the same time of day. Additionally, knowledge of the procedure could have a negative effect on the internal validity. However, as the test person knew all the scenarios exactly, had a lot of experience with the programs and had carried out a test run on two sample scenarios, there was no familiarity effect after testing the first input modality. In general, using the same scenarios and the same version of the programs (*Hybrid Editor* and *Drawjectory Control Panel*) increased the internal validity.

External Validity. The external validity is mainly threatened by the fact that the scenarios described in Sect. 5.1 were rather simple and are not necessarily comparable to potential real-world scenarios, but they do contain many base cases from which real-world scenarios could be built. Further factors threatening the external validity are:

1. The used tracking system in the *Quadcopter Lab*, as the *OptiTrack*-system cannot be applied in every situation, especially outside laboratories.
2. The experiment was conducted by a person experienced in both demonstrating and programming trajectories.

Although the *OptiTrack*-system is not be widely used in the private sector due to its expensive and specialized hardware, it is used in industry [1]. Furthermore, it could potentially be replaced by another, more accessible tracking system, as the entire workflow or proof-of-concept does not require a specific tracking system.

In general, the *Drawjectory* workflow on its own is only a concept, which can be implemented by any person in any situation and is therefore generalizable.

Construct Validity. To increase the construct validity we employed different measures for the trajectory similarity, which allowed us to cross-check the results.

Conclusion Validity. The employed measures for assessing the trajectory similarity between the demonstrated and the programmed trajectories are reliable and widely used in the literature [12,32]. Furthermore, the planning time was objectively measured, as it was automatically determined. As the experiment was conducted in the quadrocopter laboratory by one author, the results were not affected by any external disturbances.

6   Related Work

The programming of robots is a mature field of research, particularly in relation to industrial robots [18]. The difference to the drones considered here is that industrial robots typically have fixed axes and can therefore be positioned very precisely. Drones, on the other hand, hover in the air and therefore cannot actually be held in exact positions. Hence, in the following, we focus on related work that deals with the trajectory planning of drones according to the approaches used in this work.

In "A Hybrid Editor for Fast Robot Mission Prototyping" [34], Witte et al. introduce a high-level domain-specific language (DSL) that can be used to directly set waypoints the drone has to pass by. The programming is supported by a *RQt* plugin including a visualization of the calculated trajectory and the functionality that any move of the waypoints in the visualization is reflected in the source code. The authors claim that this editor allows novice and advanced programmers to quickly prototype flight routes because of visual feedback, but do not provide a user study. For this paper, we extended this language with additional constructs as described in Sect. 2 and evaluated how fast a trajectory can be created by using either the DSL or the demonstration approach.

Hoppe et al. [20] propose a framework for rapid prototyping of drone routines called *DronOS*. Similarly to our approach, the authors assess *DronOS* by using three different methods for planning a flight route for a quadrocopter:

- *Unity Scripting*: the flight route is planned using waypoints in a 3D environment using drag-and-drop. This process is analogous to the interactive waypoints presented in [34].
- *Vive Scripting*: Waypoints are set using a controller in the real world and can be edited later in a virtual environment.

– *Vive Realtime*: The user holds a controller and points to different targets. The quadrocopter follows these targets in realtime and records the used flight path.

The difference between the *Vive Realtime* method and our approach presented here is that we demonstrate the complete path and not just individual points. In Drawjectory, the path can be drawn without a drone having to fly it in realtime. However, the challenge in our approach is to automatically extract appropriate waypoints from the drawn trajectory. The presented methods were evaluated in a small user study with 12 participants, whose task completion time as well as the mental workload using a NASA-TLX questionnaire [16] were measured. Similar to our results, they conclude that real-time demonstration was the fastest method, while *Unity scripting* is the most accurate and *Vive scripting* is the most intuitive.

Yau et al. [37] try to improve flight route planning by introducing interaction models for controlling quadrocopters with one hand. Their challenge was to map the six degrees of freedom of a flying drone to the limited options of one hand controllers. They evaluated four different mappings by a user study with 32 participants with respect to the task completion time and mental load (NASA-TLX) for very simple task. Based on these results, the conducted a second user study for the best option for more complex tasks. The participants "found the device easy to use and easy to learn" [37].

These proposed input modalities concentrate on how to achieve the most practical quadrocopter live control method, rather than focusing on a precise and intuitive (pre-)planning of a trajectory, which is the main goal of our approach. In the following three programming by demonstration approaches, trajectories are not planned for drones but for classic robots. What all these approaches have in common is the handling of inaccuracies when demonstrating a trajectory, but with different techniques.

Melchior et al. [26] use a neighbor graph to identify correspondences between multiple imperfect demonstrations. Compared to Drawjectory, this process has the weakness that it takes several demonstrations to produce a trajectory close to the desired one.

Aleotti et al. [5] propose a workflow for learning robot trajectories based on single or multiple demonstrations. These demonstrations are tracked by a *CyberTouch VR glove*, which returns a set of points in three-dimensional space. Similar to *Drawjectory*, the user can change the position of selected waypoints at the end. The difference, however, is that we are not drawing the trajectory in a virtual space but in real space, which makes the input more intuitive.

The third presented work by Hwang et al. [21] is in several aspects closer to our approach: Firstly, the user draws the desired path with a finger or a stylus on a touchscreen. Then, an improved version of a corner detection algorithm extracts significant waypoints, which can be edited afterwards. Finally, a smooth trajectory is interpolated by using so-called piece-wise cubic Bezier curves (PCBC). The main drawback of this method is that it is designed for two-dimensional space and the PCBCs, in contrast to the natural cubic splines employed in our implementation, are not twice continuously differentiable at the waypoints [21,29]. However, this is necessary to ensure a smooth trajectory, which is the goal of the trajectory planning in the *Drawjectory* workflow.

7 Conclusion

Quadcopters are becoming an increasingly common part of our everyday lives, often controlled automatically thanks to ever-improving algorithms for automatic planning. In some scenarios, however, planning quadrocopter flight paths must still be done manually. Therefore, there arises a need for an intuitive but still accurate process to manually plan flight paths. To solve this problem, we propose the fast and intuitive *Drawjectory* workflow, which allows "programming" flight paths for a quadrocopter by demonstration.

The workflow consists of three main steps: i) demonstrating the flight path by tracking a gesture wand, ii) selecting points from this flight path and planning a trajectory based on these points, and iii) optionally editing the planned trajectory with input from the user.

The presentation of this workflow is supported by a proof-of-concept implementation in ROS2 that runs in the *Quadcopter Lab* of the University of Ulm. Based on a demonstration using motion capture, the implementation plans a trajectory from selected waypoints using a natural cubic spline interpolation. Both the demonstrated flight path and the planned trajectory are visualized in the *RViz* tool and can optionally be edited by shifting, moving, or scaling waypoints in the Drawjectory Control Panel.

Finally, we evaluate the *Drawjectory* workflow and its implementation. We find that while maintaining high precision the *Drawjectory* workflow intuitively accelerates the trajectory planning process greatly compared to programming, More specifically, the proposed workflow is, on average, 78.7 s faster than programming and excels especially when planning complex trajectories, but causes an average deviation of approximately 6.67 cm and an average maximum point-to-trajectory deviation of 11.9 cm.

However, during the development of the workflow and its proof-of-concept implementation, several issues arose that could not be solved within the scope of this work:

Recalling the goals of the workflow, in particular the requirement that this workflow should use only few technical aids, not all goals are achieved, due to the fact that the motion capture system is immobile and difficult to set up. This limitation could potentially be overcome by instead using VR devices requiring much less setup.

Furthermore, in this work, the orientation of the quadrocopter is not considered at all, as it is not necessary for flying along a path. However, in real world scenarios, for example when a camera is mounted on the quadrocopter, the orientation is important. The tracking and trajectory planning process could be adapted accordingly in future work.

Moreover, the waypoint selection strategy and the interpolation method for the trajectory are quite simple. It could be interesting to investigate how different strategies affect—for example—the accuracy in respect to the user's intentions.

Lastly, one potential follow-up work could investigate how the intention of the demonstration can be identified automatically. The user could then be suggested corrections based on the predicted intention.

References

1. Optitrack. https://www.optitrack.com. Accessed 11 Apr 2024
2. Overview and usage of RQT. https://docs.ros.org/en/foxy/Concepts/About-RQt.html. Accessed 22 Apr 2024
3. Rviz user guide. http://docs.ros.org/en/iron/Tutorials/Intermediate/RViz/RViz-User-Guide/RViz-User-Guide.html. Accessed 15 Apr 2024
4. Aggarwal, S., Kumar, N.: Path planning techniques for unmanned aerial vehicles: a review, solutions, and challenges. Comput. Commun. **149**, 270–299 (2020)
5. Aleotti, J., Caselli, S., Maccherozzi, G.: Trajectory reconstruction with NURBS curves for robot programming by demonstration. In: 2005 International Symposium on Computational Intelligence in Robotics and Automation, pp. 73–78. IEEE (2005)
6. Alkewitz, L.: How to Drawjectory? - trajectory planning using programming by demonstration (reproducibility package) (2024). https://doi.org/10.5281/zenodo.13992727
7. Alkewitz, L.: How to Drawjectory? Trajectory planning using programming by demonstration (2024). https://doi.org/10.18725/OPARU-54401. Accessed 01 Nov 2024
8. Bochkanov, S.: Spline interpolation and fitting. https://www.alglib.net/interpolation/spline3.php. Accessed 08 Oct 2024
9. Bonci, A., Gaudeni, F., Giannini, M.C., Longhi, S.: Robot operating system 2 (ROS2)-based frameworks for increasing robot autonomy: a survey. Appl. Sci. **13**(23), 12796 (2023)
10. Calinon, S.: Learning from demonstration (programming by demonstration). Encycl. Robot. 1–8 (2018)
11. Chai, T., Draxler, R.R.: Root mean square error (RMSE) or mean absolute error (MAE)?-arguments against avoiding RMSE in the literature. Geosci. Model Dev. **7**(3), 1247–1250 (2014)
12. Chang, Y., Tanin, E., Cong, G., Jensen, C.S., Qi, J.: Trajectory similarity measurement: an efficiency perspective. Proc. VLDB Endow. **17**(9), 2293–2306 (2024)
13. Deakin, R.E.: 3-D coordinate transformations. Surv. Land Inf. Syst. **58**(4), 223–234 (1998)
14. Foehn, P., Romero, A., Scaramuzza, D.: Time-optimal planning for quadrotor waypoint flight. Sci. Robot. **6**(56), eabh1221 (2021)
15. Gasparetto, A., Boscariol, P., Lanzutti, A., Vidoni, R.: Path planning and trajectory planning algorithms: a general overview. In: Motion and Operation Planning of Robotic Systems: Background and Practical Approaches, pp. 3–27 (2015)
16. Hart, S.G., Staveland, L.E.: Development of NASA-TLX (task load index): results of empirical and theoretical research. In: Advances in psychology, vol. 52, pp. 139–183. Elsevier (1988)

17. Hehn, M., D'Andrea, R.: Quadrocopter trajectory generation and control. IFAC Proc. Vol. **44**(1), 1485–1491 (2011)
18. Heimann, O., Guhl, J.: Industrial robot programming methods: a scoping review. In: 2020 25th IEEE International Conference on Emerging Technologies and Factory Automation (ETFA), vol. 1, pp. 696–703 (2020). https://doi.org/10.1109/ETFA46521.2020.9211997
19. Herdel, V., Yamin, L.J., Cauchard, J.R.: Above and beyond: a scoping review of domains and applications for human-drone interaction. In: Proceedings of the 2022 CHI Conference on Human Factors in Computing Systems, pp. 1–22 (2022)
20. Hoppe, M., Burger, M., Schmidt, A., Kosch, T.: DronOS: a flexible open-source prototyping framework for interactive drone routines. In: Proceedings of the 18th International Conference on Mobile and Ubiquitous Multimedia, pp. 1–7 (2019)
21. Hwang, J.H., Arkin, R.C., Kwon, D.S.: Mobile robots at your fingertip: Bezier curve on-line trajectory generation for supervisory control. In: Proceedings 2003 IEEE/RSJ International Conference on Intelligent Robots and Systems (IROS 2003) (Cat. No. 03CH37453), vol. 2, pp. 1444–1449. IEEE (2003)
22. Joachim Gudmundsson, N., Laube, A.P.: Computational movement analysis (2012)
23. Luis, C., Ny, J.L.: Design of a trajectory tracking controller for a nanoquadcopter. arXiv preprint arXiv:1608.05786 (2016)
24. Macenski, S., Foote, T., Gerkey, B., Lalancette, C., Woodall, W.: Robot operating system 2: design, architecture, and uses in the wild. Sci. Robot. **7**(66), eabm6074 (2022)
25. McKinley, S., Levine, M.: Cubic spline interpolation. Coll. Redwoods **45**(1), 1049–1060 (1998)
26. Melchior, N.A., Simmons, R.: Graph-based trajectory planning through programming by demonstration. In: 2012 IEEE/RSJ International Conference on Intelligent Robots and Systems, pp. 1929–1936. IEEE (2012)
27. Mellinger, D., Kumar, V.: Minimum snap trajectory generation and control for quadrotors. In: 2011 IEEE International Conference on Robotics and Automation, pp. 2520–2525. IEEE (2011)
28. Mueller, M.W., Hehn, M., D'Andrea, R.: A computationally efficient motion primitive for quadrocopter trajectory generation. IEEE Trans. Rob. **31**(6), 1294–1310 (2015)
29. Olkin, Z., Rogers, J.: Autonomous quadrotor trajectory planning and control for in-flight aerial vehicle capture. In: 2021 IEEE Aerospace Conference (50100), pp. 1–10. IEEE (2021)
30. Pham, Q.C.: Trajectory planning 52 (2015)
31. Ravichandar, H., Polydoros, A.S., Chernova, S., Billard, A.: Recent advances in robot learning from demonstration. Annu. Rev. Control Robot. Auton. Syst. **3**, 297–330 (2020)
32. Toohey, K., Duckham, M.: Trajectory similarity measures. Sigspat. Spec. **7**(1), 43–50 (2015)
33. Usenko, V., Von Stumberg, L., Pangercic, A., Cremers, D.: Real-time trajectory replanning for MAVs using uniform B-splines and a 3D circular buffer. In: 2017 IEEE/RSJ International Conference on Intelligent Robots and Systems (IROS), pp. 215–222. IEEE (2017)
34. Witte, T., Tichy, M.: A hybrid editor for fast robot mission prototyping. In: 2019 34th IEEE/ACM International Conference on Automated Software Engineering Workshop (ASEW), pp. 41–44. IEEE (2019)
35. Wohlin, C., Runeson, P., Höst, M., Ohlsson, M.C., Regnell, B., Wesslén, A.: Experimentation in Software Engineering. Springer (2012)

36. Xu, S., Ou, Y., Duan, J., Wu, X., Feng, W., Liu, M.: Robot trajectory tracking control using learning from demonstration method. Neurocomputing **338**, 249–261 (2019)

37. Yau, Y.P., Lee, L.H., Li, Z., Braud, T., Ho, Y.H., Hui, P.: How subtle can it get? A trimodal study of ring-sized interfaces for one-handed drone control. Proc. ACM Interact. Mob. Wearable Ubiquit. Technol. **4**(2), 1–29 (2020)

38. Zimmermann, M., Vu, M.N., Beck, F., Nguyen, A., Kugi, A.: Two-step online trajectory planning of a quadcopter in indoor environments with obstacles. IFAC-PapersOnLine **56**(2), 11002–11009 (2023)

Go Where Energy Can be Saved: A Vision for a Green Infrastructure Evaluation, Optimization, and Alignment System

Benjamin Weigell[(✉)] and Bernhard Bauer

Software Methodologies for Distributed Systems, University of Augsburg,
Augsburg, Germany
{benjamin.weigell,bernhard.bauer}@uni-a.de

Abstract. The Information and Communication Technology (ICT) sector contributes to the global Greenhouse Gas (GHG) mainly through its electric energy consumption. This consumption is predicted to increase significantly by 2030, with data centers being one of the largest energy consumers. Data centers are not the root cause of the high energy consumption; they are the visible part; the main drivers are the executed software and built IT infrastructures. However, current solutions address only parts of this problem by proposing solutions for efficient green routing, green schedulers, renewable energy sources, and determining the environmental impact of single applications. Therefore, we envision the development of the Green Infrastructure Evaluation, Optimization, and Alignment System (GIEOAS) that will transform existing IT infrastructures into energy-efficient systems. GIEOAS will achieve this by evaluating existing IT infrastructures against Key Performance Indicators (KPIs) and patterns while allowing users to define custom KPIs and patterns using a Domain-Specific Language (DSL). In addition, it will propose optimized infrastructure designs that meet the specified objectives. Finally, GIEOAS aspires to automatically align the current IT infrastructure with the optimized design by generating Infrastructure-as-Code (IaC) to achieve energy reduction. For now, our primary objective is to lay the basis for GIEOAS by defining five problem statements and presenting our envisioned solutions for each. The overarching goal of GIEOAS is to reduce the energy consumption of the ICT sector.

Keywords: Green IT · Software Architectures · DSL · Code Generation · Architecture Mining · Architecture Optimization

1 Introduction

The Information and Communication Technology (ICT) sector plays a vital role in modern society, from supporting the critical infrastructure sector to personal

This work was partially funded by the German Federal Ministry of Education and Research (BMBF) under reference number 01IS24070G.

entertainment purposes, but it comes with a significant environmental cost. The Greenhouse Gas (GHG) emission of the ICT sector contributed 1.4% to global GHG emissions and accounted for 4% of global electricity usage in 2020 [25]. A substantial portion of this emission, around 64%, originated from electrical energy consumption, with data centers depleting 24% of this energy [25]. Furthermore, [1] predicts that the ICT sector could account for 21% of the global energy usage by 2030 and data centers could be responsible for roughly a third of the energy consumption. Notably, these predictions did not consider potential additional energy demands from Artificial Intelligence (AI) and blockchain technologies.

In response to the growing environmental impact, data center providers have already developed several strategies to reduce energy consumption and provide clean energy for their data centers. The strategies include efficient green routing, green schedulers, smart cooling, and renewable energy sources [14].

While data centers are visible contributors to high energy consumption, a root cause of energy inefficiency lies within the software executed on them and the IT Infrastructure built upon them. According to [16], an IT Infrastructure, or in short infrastructure, is: "The system of hardware, software and service components that supports the delivery of business applications and IT-enabled processes.". Energy-inefficient IT infrastructures could be the result of poor design choices and a lack of a precise overview of the current IT infrastructure [5]. For example, poor design choices include the underutilization of rented Virtual Machines (VMs) [17] or the placement of software applications in geographically distant data centers despite high data exchange between them. A current example of a company's missing overview over their IT infrastructure, is Twitter's (X's) re-discovery of an idle GPU cluster [33].

Beyond contributing to GHG emissions, energy-inefficient IT infrastructure is also problematic for critical sectors such as hospitals. The World Health Organization (WHO) identifies the adequacy of hospitals to cope with disasters as a basic problem that plays an essential role in reducing disaster mortality rates. Hence, in the event of a power outage, a hospital's primary goal must be to maintain uninterrupted services for as long as possible despite the limited emergency power supply. As a result, energy-ineffective IT infrastructures would strain emergency resources, endanger the hospital's continuity of services, and increase mortality rates.

To tackle the problem of energy-inefficient IT infrastructures, we develop the Green Infrastructure Evaluation, Optimization, and Alignment System (GIEOAS). The GIEOAS aims to transform existing IT infrastructures into energy efficient ones by achieving the following goals

G1 **Resource-Aware Modeling:** Creating and maintaining an always up-to-date model of the current IT infrastructure, including resource utilization, to serve as a basis for the subsequent steps.

G2 **Evaluation:** Enable users to evaluate running IT infrastructures against pre-defined Key Performance Indicators (KPIs) and software patterns, and support the creation of custom KPIs and patterns via a Domain-Specific Language (DSL).

G3 **Optimization:** The created infrastructure model is optimized with regard to objectives such as low energy consumption, redundancy and low latency times applying suitable multi-criteria optimization methods.

G4 **Alignment:** Transformation of the current IT infrastructure by generating Infrastructure-as-Code (IaC) in a state that aligns with the proposed optimized IT infrastructure model.

G5 **Explainability:** Comprehensible visualization of the necessary adjustments to align the current IT infrastructure to the optimized model.

Therefore, to guide the development of GIEOAS and achieve these goals, the remaining part of the paper is organized as follows. In the next four sections, we address the first four goals in separate sections. Each section begins with a broader description of the goal, followed by a brief overview of the relevant State-of-the-Art (SOTA) approaches, and concludes with our concrete vision for solving the problem and achieving the overall goal. Additionally, aspects of G5 are integrated into these goals where applicable. After addressing all goals, we summarize the envisioned GIEOAS and map each goal to technical components introduced in the concrete visions. Finally, we conclude the paper.

2   Resource-Aware IT Infrastructure Mining

GIEOAS requires an up-to-date representation of the running IT infrastructure that captures all relevant infrastructure details to evaluate, optimize, and transform existing IT infrastructures into energy-efficient systems as stated in G1. Therefore, two subgoals must be achieved. Firstly, the information must be extracted from the running infrastructure, and secondly, the extracted data must be stored in a suitable format. The format should enable the subsequent goals of evaluating, optimizing, and aligning the current infrastructure via IaC generation. Both subgoals are addressed in dedicated parts in the following. The capturing is addressed in *Infrastructure Mining*, and the data format in *Infrastructure Modeling*.

2.1   Infrastructure Mining

Before creating an infrastructure model, all components must be extracted from a running infrastructure, including hardware servers, virtual machines, containers, and applications. Additionally, the relationships between components - such as "is executed on" and "communicates with" - need to be identified, as well as their resource usage (e.g., CPU, RAM, GPU utilization). The identified components, their relationships, and resource usage are essential for evaluating and optimizing the current infrastructure against objectives like energy consumption. Finally, configuration details, such as ports and endpoints, are derived from these components, which later support the accurate generation of IaC. Therefore, we

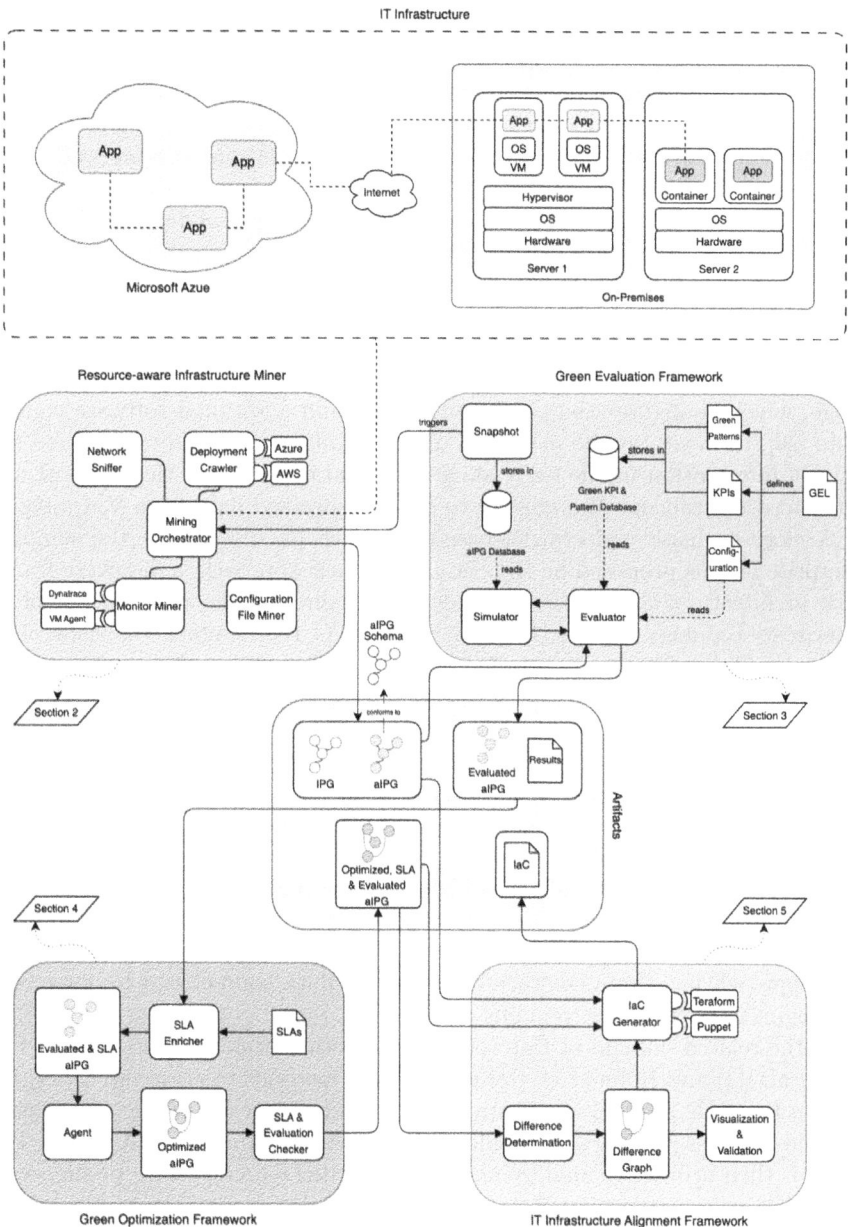

Fig. 1. Structure of GIEOAS and Exemplary IT Infrastructure

define the infrastructure mining task as the process of identifying and extracting all relevant components, their relationships, and resource usage from the running infrastructure, including configuration details.

Approches to Infrastructure Mining. One of the first approaches to the infrastructure mining task is presented by Machiraju et al. [24]. They introduce a model-driven approach to application auto-discovery, automatically capturing applications and their relationships in a format consumable by application management solutions. Therefore, Machiraju et al. [24] propose an application template model that allows the description of input for the auto-discovery engine, specifying the applications to be discovered and the output of discovered applications. During the discovery process, the auto-discovery engine employs different discovery agents that query the information directly from the application itself or indirectly from sources like the operating system.

With the transition to cloud-based technologies, Virtual Machines have gained importance in IT infrastructure. A Virtual Machines is based on a VM image, which configures the operating system and additional software components [26]. However, as the metadata of a VM image often did not include this detailed information in the early stages of cloud computing, Menzel et al. [26] introduced the concept of a crawler to extract specified data from VM images.

A more sophisticated crawler-based approach for discovering the complete enterprise IT was proposed by Binz et al. [5]. Their approach, a discovery framework architecture, automatically finds and maintains the components of an Enterprise Topology Graph (ETG) [5]. An ETG represents a snapshot of an enterprise IT, and contains detailed technical information about each component, such as processes, services, software, infrastructure, and their relationships and dependencies [6]. The core of the proposed discovery framework architecture is the discovery layer. In this layer, an ETG graph is built by iteratively crawling an existing IT infrastructure using plugins from a specified starting point. A plugin is a software component that retrieves information from a specific data source, such as a Tomcat web server. Additionally, a scheduler determines the plugins' execution order, and a reconciliation component refines the final ETG. The overall interaction with the discovery layer is managed by the discovery manager, allowing, for example, to specify the starting point of the crawling. This approach has the advantage that various data sources can be integrated via plugins and, therefore, remains expandable.

In the related domain of Enterprise Architecture Management (EAM), Farwick et al. [11] and Holm et al. [18] propose two methods to create an Enterprise Architecture (EA) model automatically. EAM aims to capture the relationship between business processes and underlying IT architecture in an EA model. This model is then utilized to analyze and optimize the IT architecture to align with the overall business strategy. However, keeping the EA models up-to-date is challenging due to an ever-evolving IT architecture [11]. To address this, Farwick et al. [11] propose creating and continuously updating EA models by obtaining information (semi)-automatically from humans and different technical interfaces, namely web services and business process management tools. Similar to Farwick et al. [11], Holm et al. [18] automate the creation of EA models by scanning the network. They divide the information obtained from a network scan into three categories: System information in the form of MAC and IP addresses; user information including active hosts; and software information such as the operating system, its version, applications running on it, and the applications' protocols, types, and versions.

In a recent approach to holistic application management, Harzenetter et al. [15] propose a method of automatically creating management workflows for running applications based on dynamically extracted instance models. An instance model is generated from a running infrastructure in three steps. First, the instance information retriever extracts information about the running applications by querying the APIs of the underlying deployment technologies. This retriever offers a plugin interface to support different deployment technologies and to be expandable. Each plugin extracts information for its specified deployment technology, such as Kubernetes or Puppet. Second, the technology-specific data are converted into a technology and vendor-agnostic model format, namely TOSCA. Finally, the model is further enriched with additional data using the crawler approach of [5]. The resulting model contains technology-agnostic deployment data and application-specific information that enable the generation of management activities, such as backup creation and dependency updates.

Vision: The Resource-Aware Infrastructure Miner. We propose the Resource-aware Infrastructure Miner to solve the previously defined infrastructure mining goal. This miner extracts the required information, i.e., components and their relationships, resource usage, and configuration details, from a running infrastructure by building upon and extending the concepts presented in [5], [15], and [18]. Our overall envisioned concept of the Resource-aware Infrastructure Miner consists of a central component, the Mining Orchestrator, and four specialized data extraction components: the Deployment Crawler, Network Sniffer, Monitor Miner, and Configuration File Miner and is visualized in the broader context of GIEOAS in Fig. 1.

The Mining Orchestrator serves as the central interaction and control unit of the Resource-aware Infrastructure Miner. It provides a UI for configuring the mining process and orchestrates the execution of the specialized data extraction components.

The Deployment Crawler, inspired by [15], and the Network Sniffer, motivated by [18], are primarily responsible for extracting information about the underlying deployment topology and identifying active components, such as servers, VMs, and applications. To further refine this coarse-grained information, both components are enhanced. The Deployment Crawler reuses the plugin concept introduced by Binz et al. [5] to extract more detailed information, including configuration specifics and runtime environment details. The Network Sniffer captures more in-depth network details, such as the number and size of packets exchanged between components.

As defined in this goal G1, resource usage information must be determined for active components within the IT infrastructure. These resource usage details shall be determined by our planned Monitor Miner that retrieves data from different monitoring services via plugins, and can also deploy monitoring agents if resource usage information is unavailable for a component. For example, a plugin may be used to retrieve CPU and memory usage metrics from an existing monitoring service, such as Dynatrace. In contrast, a monitoring agent would be deployed on an unmonitored VM to capture these metrics directly.

Since IaC has been widely adopted in DevOps practices [3,20], we introduce the Configuration File Miner, which will utilize IaC files as data sources. The Configuration File Miner combines specialized plugins with a Large Language Model (LLM) to extract information from IaC files. Each plugin is tailored to a specific technology, such as Terraform, and thereby enables precise information extraction. In contrast, utilizing an LLM offers the advantage of understanding nearly any configuration file. Thus, the LLM allows us to obtain the details from configuration files for which we do not possess a plugin.

2.2 Infrastructure Modeling

To utilize the information obtained from the infrastructure mining approach, the information has to be stored in a suitable data structure. This data structure should support the assessment of the evaluation objectives in a straightforward manner, such as through queries comparable to SQL statements or a DSL. In addition, the data structure should be adequate for optimization and IaC generation processes or alternatively be efficiently transformable into a proper representation. In summary, the model must be flexible enough to represent any infrastructure topology while still providing sufficient structure for the evaluation, optimization, and IaC generation processes.

Graph-based and Meta-Object Facility (MOF)-based approaches are commonly used to model and store details about an IT infrastructure. While Domain-Specific Languages (DSLs) such as Terraform and Ansible are employed to configure infrastructure, they are not considered here as they are less suitable for the required modeling, evaluation, and optimization capabilities.

In the following sections, we briefly explore each approach by providing relevant literature and evaluating their suitability in four key areas: (1) modeling arbitrary infrastructure topologies; (2) providing a foundation for the evaluation; (3) supporting the optimization process; and (4) enabling the IaC generation process.

Graph-Based Approaches. Building upon graphs is beneficial as they allow the application of established graph algorithms, can be stored in Graph Databases (GDBs), and can be queried by the Graph Query Language (GQL), thereby building the foundation for the evaluation capability. Additionally, graphs are appropriate for the optimization process as they serve as input for optimization techniques such as Graph Neural Networks (GNNs) [31] and Graph Reinforcement Learning (GRL) [27].

Furthermore, graphs are suitable for capturing and modeling an IT infrastructure, as shown by Binz et al. [6]. They [6] propose the Enterprise Topology Graph (ETG), a formal graph model designed to capture all IT components and their relationships. An ETG consists of nodes and edges, where both can hold arbitrary properties in the form of key-value pairs, similar to a property graph. In an ETG, nodes and edges have a type, which defines specific properties, for example, the type web server has a property named port. These types are not embedded directly within the graph but referenced and specified in an additional hierarchical tree structure, which makes the types extensible and allows low- and high-level analyses. Therefore, graphs represent a data structure that fulfills all our required capabilities.

MOF-based Approaches. The Meta-Object Facility (MOF), developed by the Object Management Group (OMG), provides a semi-formal approach to defining a metamodel and describing model transformations [28]. MOF consists of four hierarchy levels that build upon each other. Level 0 represents real-world objects, such as an existing IT infrastructure. At level 1, these real-world objects are described by a model that conforms to a metamodel, namely UML 2.0 or a custom metamodel defined at level 2. This metamodel complies with the MOF meta-metamodel at level 3, either the Complete MOF (CMOF) or the Essential Meta-Object Facility (EMOF). The Eclipse Modeling Framework (EMF) helps implement MOF and focuses on modeling and code generation by providing the Ecore metamodel conforming to EMOF. For IT infrastructure modeling, Chiari et al. [8] introduced the DevOps Modeling Language (DOML). DOML enables the technology-agnostic modeling of an IT infrastructure, which can then be translated into multiple IaC languages. Internally, DOML defines a three layered metamodel based on Ecore. The Application Layer defines the structure and composition of an application. The Abstract Infrastructure Layer specifies provider-agnostic infrastructure elements on which an application can be executed. Lastly, provider-specific components are described in the Concrete Infrastructure Layer. In order to generate provider-specific IaC, the abstract elements can be connected to concrete infrastructure elements.

However, like most MOF-based approaches, DOML is limited by a static metamodel. A static metamodel limits the modeling process to structures that the metamodel designers anticipated in advance. Consequently, unforeseen design structures or domain changes may be challenging to incorporate as they require adapting the existing metamodel. Therefore, this restricts the flexibility of MOF-based approaches. Furthermore, MOF-based approaches, primarily when implemented with EMF, do not offer inherent advanced analysis or optimization capabilities. Nevertheless, an Ecore model could be easily translated into a more suitable data structure for evaluation and optimization.

In summary, while MOF-based approaches combined with EMF effectively support IaC generation, they tend to be less flexible and require transformation for advanced analysis and optimization tasks.

Vision: The Two Layered Infrastructure Model. We propose to generate two models to combine the advantages of the graph-based and the MOF-based modeling approaches. The first model is a graph-based, technology-specific, and metamodel-agnostic representation that describes the current infrastructure in detail. The second model is a technology- and vendor-agnostic version of the first model, which is also graph-based but is compliant to a static metamodel.

For the first model, the infrastructure information extracted from the Resource-aware Infrastructure Miner is transformed into a Property Graph (PG), referred to as the Infrastructure Property Graph (IPG). A PG is a directed graph including nodes and edges that have labels and can store arbitrary properties as key-value pairs [2]. This generated IPG reflects the current state of the infrastructure and is stored in a Graph Database (GDB) to retrieve and query it efficiently. For example, nodes could represent an application, a container, or a VM; the node properties could specify the Operating System (OS), ports, or resource utilization; and edges could define where an application is executed or how many messages are exchanged between components.

The benefit of the IPG is that it does not limit the infrastructure capturing process through a static metamodel, the final model is well understood by end-users, and GDBs offer query languages that support advanced analysis.

However, as the IPG is vendor- and technology-specific, it is less suitable for the evaluation, optimization, and IaC generation processes. Therefore, to tackle this, we plan to introduce a second model type, the Abstract Infrastructure Property Graph (aIPG). The aIPG follows a fixed schema, that defines the structure of an infrastructure in a technology- and vendor-agnostic way. A PG Schema is comparable to a metamodel, as it describes the allowed structure for the PG, including node and edge types, connection constraints, property constraints, and general constraints [2].

An IPG is transformed into an aIPG in three steps. In the first step, the schema of the IPG is extracted. Next, a mapping between the extracted schema and the predefined aIPG Schema is automatically determined to the greatest possible extent and can be refined manually. In the final step, this mapping is employed to translate the IPG to an aIPG instance.

Overall, building upon property graphs and the combination of the IPG and the aIPG fulfill the desired capabilities. The IPG allows the flexible modeling of an arbitrary infrastructure topology, while the aIPG provides a solid, predefined structure for the subsequent task evaluation, optimization, and IaC generation.

3   Green Infrastructure Evaluation

Evaluating IT infrastructure in terms of greenness and thereby gaining insights into its optimization potential is one of the main goals of GIEOAS. In our case, greenness stands for reflecting the energy consumption and the resulting GHG emissions of a running IT infrastructure. The foundation for this evaluation is an up-to-date model of the running IT infrastructure with resource usage details and inherent support for advanced analysis, as defined by the first goal G1 Resource-Aware Modeling.

A greenness evaluation can not be adequately done with one singular general metric, as it would abstract too many details. Therefore, the evaluation should offer fine-grained metrics - such as energy consumption per application and cluster - and enable evaluation scenarios like energy comparisons of specific technologies or detecting energy hotspots in distinct subparts of the IT infrastructure. Additionally, predefined metrics and evaluation scenarios should be extensible since foreseeing every possibility at design time is not feasible.

However, even fine-grained metrics and evaluation scenarios make it challenging to assess unambiguously the greenness of the IT infrastructure. For example, a single VM hosting multiple applications might exhibit high energy consumption and be flagged as an energy hotspot. Nonetheless, this design decision is still more energy-efficient than running several VMs, each hosting a single application, which would collectively consume more energy. Therefore, the evaluation should be able to detect and account for the usage of common energy-efficient design patterns. Moreover, it should warn if a pattern is violated, and be extensible for including new emerging patterns.

Finally, evaluating the IT infrastructure using a single, up-to-date model can be misleading due to fluctuating application loads over time. These load fluctuations lead to varying energy consumption, presenting a challenge in accurately assessing the greenness of the IT infrastructure.

In summary, the evaluation system of GIEOAS must solve the previously mentioned four challenges. Firstly, GIEOAS should incorporate several standard fine-granular metrics that accurately reflect the energy consumption. Secondly, it must provide and verify energy-efficient design patterns to avoid decision-making based solely on energy consumption. Thirdly, the evaluation system should be extensible to allow the addition of new emerging metrics and patterns. Finally, GIEOAS should be able to provide a reliable and accurate energy assessment despite fluctuating application loads and their associated fluctuations in energy consumption.

3.1 Current Green Metrics, Patterns and Approaches for Workload Predictions

Existing research offers several solutions that address isolated parts of the previously described green infrastructure evaluation problem, including metrics that assess energy consumption, patterns for energy-efficient design and methods for managing fluctuating resource demands. This section provides a short overview of these approaches and thereby lays the foundation for our vision that aims to solve the evaluation problem more holistically and extensible.

Kipp et al. [22] propose a set of Green Performance Indicators (GPIs) to measure the energy efficiency of IT service centers. These indicators assess energy consumption, efficiency, potential savings and other energy-related factors at different system levels, whilst also taking into account the application and its development and execution environment. For example, the energy impact metric Input/Output Operations Per Second (IOPS) per Watt measures how efficiently energy is being used to perform I/O operations by calculating the number of I/O transactions or throughput per unit of energy consumed. In line of this work, Brunnert et al. [7] introduces a new method to calculate the carbon emission of a software system. The emissions are calculated solely based on the software system's resource demands (e.g., CPU, memory, storage and network usage) and the known carbon emission values of the respective resource types, such as the emissions per CPU hour or gigabyte of memory consumed.

An early approach to determining the impact of design decisions on the energy consumption of a software system was stated by Seo et al. [29]. They developed a framework that allows the specification of an energy cost model for a system architecture style, such as a client-server, and can be used to assess its energy consumption. Their framework proved to predict the energy consumption of a system architecture style accurately.

More recent efforts have shifted toward identifying specific patterns across domains that can reduce the energy consumption of a software system. One such source is the Green Software Pattern catalog of the Green Software Foundation (GSF), which contains software patterns tailored for reducing the emissions of AI, cloud, and web systems [17].

The changing load of applications challenges the accurate evaluation of the energy consumption of IT infrastructure. To overcome this, Kistowski et al. [23], propose two machine learning models. The first model, the Workload Deployment Power Prediction Model, predicts the total power consumption of a system based on its current deployment topology and its current workload. The second, the Single Server Power Prediction Model, estimates the power consumption of a server reeling only on performance counters, and so enabling predictions without the need for external power measurement devices.

3.2   Vision: The Green Evaluation Framework

To evaluate the current IT infrastructure for energy efficiency and GHG emission and thereby fulfill the goal G2 Evaluation, we propose an evaluation framework that builds upon the aIPG generated from the Resource-aware Infrastructure Miner. This framework consists of the following four components: a custom DSL, an evaluator, a database containing KPIs and green patterns, and a simulation engine. Our evaluation process involves two main procedures: the core evaluation process and the scenario-driven evaluation process.

Core Evaluation. The core evaluation process focuses on assessing KPIs and green patterns on a given aIPG instance. To facilitate this, we design a DSL, the Green Evaluation Language (GEL), which is based on the concepts defined in the aIPG Schema and serves as the language used for the evaluation. GEL is used to implement KPIs and green patterns, and these implementations are stored in a dedicated database, the Green KPI and Pattern Database (GKPD). Initially, this database will include the GPIs from Kipp et al. [22], the emission approximation from Brunnert et al. [7], and the cloud and web patterns from the GSF [17]. Furthermore, we also plan to continuously search for new emerging metrics and green patterns to include in the database.

To perform this evaluation, a configuration file, which is written in GEL, and an aIPG instance are passed to the evaluator. The configuration file specifies a set of KPIs and green patterns, which are retrieved from the database. Based on the configuration file, the evaluator assesses the KPIs and green patterns on the aIPG instance and generates an evaluation result in the form of structured text and an extended aIPG instance. The structured text contains the calculated KPIs and the status of the analyzed patterns. The KPIs are mapped to their nodes, and nodes and edges are highlighted in the graph to indicate why a specific pattern is satisfied or violated. The extended aIPG instance visualizes these results differently by mapping the KPIs to their corresponding nodes and now highlighting nodes and edges in the graph to indicate why a specific pattern has been satisfied or violated.

The core evaluation satisfies the desired properties of providing standard KPIs, and green software architecture patterns, as well as a DSL to keep both extensible. Additionally, the proposed visualization of the extended aIPG instance partially addresses goal five explainability. The varying application loads are addressed in the following section Scenario-driven Evaluation.

Scenario-Driven Evaluation. To deal with fluctuations in application loads, such as changing communication between applications, or high GPU load during nightly ML trainings, we introduce a simulation component, a snapshot component, and an aIPG database. The snapshot component regularly requests from the Resource-aware Infrastructure Miner an updated version of the aIPG for a specified IT infrastructure, and stores it in the aIPG database. The simulation component is the orchestrator of the scenario-driven evaluation processes and enables the realization of the following three scenarios:

Evaluation Over Time: In this scenario, users can execute the core evaluation over a past time window and thus evaluate the infrastructure over a longer period of time under varying loads. The time interval can be defined in the configuration file via a query. The query can specify the time interval either explicitly (e.g., last week from 10 AM to 1 PM), or based on conditions (e.g., a one-hour interval when server A's utilization exceeded 75%)). To achieve this, the simulation component evaluates the time query and retrieves all matching aIPG snapshots from the database. The core evaluation is then performed on each snapshot, and the results are presented to the user in a UI that facilitates the detailed analysis of each snapshot.

Evaluation of Different Loads: In order to better understand the impact of varying loads on the infrastructure, users can change the loads of components in an aIPG and then perform the core evaluation process on the adapted aIPG. For example, a load change could involve increasing the number of requests for an application or the communication between two servers. To support this, the simulation component provides a UI where users can select and modify an aIPG. Depending on the adaptation, certain data points for calculating the KPIs may be missing. For instance, if the number of requests increases, the resulting resource consumption might be unknown. To address this, we plan to utilize the

Single Server Power Prediction Model by Kistowski et al. [23] to predict missing values. The simulation component checks the adapted aIPG for missing data points, predicts these values when necessary, and then passes the aIPG to the core evaluation process. Finally, it returns the results to the user.

Evaluation of a New Deployment Configuration.   To experiment with new deployment configurations without altering the running infrastructure, users can modify an aIPG, similar to the previous scenario. For example, a user might consolidate two servers into one. This scenario is realized in the same way as the Evaluation of Different Loads.

4   Green IT Infrastructure Optimization

In this section, we focus on the third objective of GIEAOS, the optimization of an IT infrastructure model to enhance the IT infrastructure's energy efficiency and reduce its greenhouse gas emissions. Therefore, the optimization process reconfigures the deployment. Reconfiguration may include actions such as moving an application from server A to server B to minimize communication overhead, replacing two VMs with a larger one, or shutting down unused servers or applications. However, the optimization process is limited by several constraints:

1. The available reconfiguration options for a given infrastructure model must be clearly identified. For example, does the infrastructure allow a larger VM to be dynamically provisioned?
2. Service Level Agreements (SLAs) must not be violated. For instance, consolidating VMs might place redundant components on the same host, increasing the risk of a single point of failure.
3. Timely execution of the optimization process to be able to react to changes in the infrastructure.

To accomplish this, we first explore related optimization approaches, such as general cloud optimizers and energy efficient schedulers in data centers. This is followed by a description of how GIEAOS optimizes an IT infrastructure model for energy consumption while addressing the three constraints mentioned above.

4.1   Related Optimization Approaches

The optimization of IT infrastructures has been a subject of research for some time. An early approach was presented by Simarro et al. [30], who proposed a cloud broker architecture to schedule VMs across different vendors to minimize total costs. The cloud broker consists of a scheduler, which determines the optimal deployment configuration, and a cloud manager, which executes the proposed reconfiguration. The scheduler aims to find the most cost-effective VM deployment configuration while minimizing potential performance degradation due to the reallocation of VMs. To calculate this optimal distribution, the problem is modeled using AMPL and solved with MINOS, a nonlinear programming solver.

Frey et al. [13] introduced a genetic algorithm, CDOXplorer, that generates an optimal cloud deployment option - a configuration of cloud resources (CDO), architecture, and scaling rules - for migrating enterprise software to a cloud environment. The cost efficiency, response time, and Service Level Agreement (SLA) compliance of the resulting deployment plan are considered during the optimization process. Therefore, the CDOXplorer takes an architectural model of the enterprise software, the current deployment model, a workload profile, a typical usage pattern based on historical data, and a cloud profile containing performance details about VMs as input. As final output, it provides a deployment model with reconfiguration rules that enable automatic adaption to changes through horizontal or vertical scaling. During the optimization process, the CDOXplorer utilizes a simulation component to evaluate potential CDOs for efficiency, response time, and SLA compliance.

Over time, the focus has shifted from pure cost optimization to energy-efficient optimization. In a recent survey, Hou et al. [19] conducted a literature review on energy-efficient task scheduling for data centers using Deep Reinforcement Learning (DRL). They analyzed 16 papers published between 2018 and 2022 that address this issue. For instance, Ding et al. [10] propose a two-step scheduling approach: first, a queuing model assigns tasks to servers within the data center; then, a Q-learning-based scheduler allocates the tasks from each server to its virtual machines, optimizing for energy efficiency and service quality. Overall, Hou et al. conclude that DLR-based approaches are promising for reducing the overall energy consumption of data centers.

4.2   Vision: Green Optimization Framework

The presented optimization approaches mainly optimize the scheduling of deployments within data centers or the migration to the cloud. In contrast to this, GIEOAS goal is to optimize the energy consumption of existing IT infrastructures, which could be realized on-premises, in the public could, or mixed. To achieve this, we introduce a green optimization framework. Notably, changing the IT infrastructure without human oversight might lead to unintended consequences, so the optimization framework only adapts an aIPG. This improved aIPG can be either used by a human to manually perform the proposed changes or be given to the alignment operator, which is introduced in the next section, to perform the changes as far as possible automatically. The details of the green optimization framework are described below.

Building on the literature review by Hou et al. [19], which demonstrates the successful application of DRL techniques to reduce data center energy consumption, our goal is to explore a Reinforcement Learning (RL) approach to optimize the aIPG. However, we intentionally leave the choice of the specific RL method, such as deep reinforcement learning, actor-critic, or alternative RL frameworks, open for further investigation and focus only on outlining the foundational elements (Agent, Environment, State, Actions, and Reward [21]) of the RL approach.

Agent: The RL agent represents the decision-making entity. It interacts with the IT architecture model to adjust configurations, such as combining two VMs or moving deployments between servers to reduce energy consumption.

Environment: The IT architecture modeled in the aIPG builds the environment by covering all elements that impact energy consumption, including servers, network configurations, and deployment settings.

State: The state is a refined, decision-oriented representation of the IT architecture and is also modeled using an aIPG. The aIPG is enriched with three key additions. Firstly, results from the green evaluation framework, such as KPIs and detected pattern violations. Secondly, the compliance with the SLA, including redundancy requirements. Thirdly, information on available resources, for example, VMs and servers that could be dynamically added to replace existing ones. To support SLA compliance and additional resource availability, we will provide a UI to add these elements to an aIPG. Moreover, an SLA checker will be implemented to evaluate SLA adherence.

Actions: Actions represent the possible adjustments the agent can make to improve the energy efficiency of the IT infrastructure. This could include changing deployment settings, optimizing resource allocations, or adjusting network configurations. Over time, the agent should learn which reconfiguration actions are valid by receiving the reward.

Reward: The reward function will be derived from the outputs of the green evaluation framework and the SLA checker. It primarily focuses on minimizing energy consumption and adhering to green patterns while ensuring SLA compliance. Actions that lead to increased energy consumption or SLA violations will be penalized with a negative reward. In contrast, positive rewards will be given for efficiency improvements achieved without breaching SLA requirements.

For the training process, we will develop a synthetic IT infrastructure generator that creates aIPGs based on parameters such as the number of applications, on-premises or cloud usage, and communication load. In order to evaluate the performance of the trained RL agent, we propose to carefully construct a benchmark dataset of real-world and synthetic IT infrastructures. We will then use mathematical optimization methods and heuristics to find energy-efficient solutions for these benchmarks and compare them against the solution from the RL agent for a detailed performance assessment.

The proposed optimization framework, visualized in Fig. 1, will improve the energy efficiency of an IT infrastructure model and respect the previously identified constraints. The constraints of valid reconfiguration options and adherence to SLAs are complied with due to the design of the reward function. Additionally, the trained RL agent can execute the optimization of an IT infrastructure model promptly. Overall, an optimized IT infrastructure model builds the foundation for aligning the real IT infrastructure with the model to realize energy saving.

5   IT Infrastructure Alignment

To finally reduce the energy consumption and the correlating GHG emissions of the real-world IT infrastructure, thereby addressing Goal 4 of GIEOAS, the optimized model generated by the previously introduced green optimization framework must be translated back into the real-world IT infrastructure. We define this translation process as IT infrastructure alignment since it systematically adjusts the current IT infrastructure to match the optimized IT infrastructure model. Therefore, the IT infrastructure alignment process must perform the following three steps.

First, the difference between the current state of the IT infrastructure and the proposed model must be identified.

Second, the differences between the current and the optimal state must be visualized appropriately. The visualization should, on the one hand, enable experts to validate these changes and, on the other hand, explain why these changes result in an energy reduction. This visualization thus contributes to goal 5 of GIEOAS, explainability, and ensures that the IT infrastructure alignment does not introduce negative impacts, such as mistakenly shutting down a relevant but temporarily unused server during the optimization process.

Third and finally, the optimized IT infrastructure model must be translated into IaC to align the current infrastructure with the optimized model. The generated IaC must ensure alignment with minimal disruptions, avoiding issues such as application downtime during server migration or data loss in stateful applications.

The following section presents related work for each of the three steps, beginning with techniques for efficiently identifying differences in large graphs, then graph visualization methods, and concluding with approaches for generating IaC.

5.1   Approaches for Graph Visualization and IaC Generation

Delugach et al. [9] proposed a technique called difference graphs to identify and visualize differences between two graphs. Their algorithm first determines the Least Common Generalization (LCG) of both graphs, which represents the shared structure. Then, it determines the operations required for transforming the first graph into the LCG and subsequently the operations to convey the LCG into the second graph or vice versa. Based on these transformation operations, they determine which parts of the graphs differ. For the visualization, color coding is used in the source graph to highlight common and different elements. Bach et al. [4] introduced GraphDiaries, a visualization tool to enhance the temporal exploration of dynamic, node-link-based graphs, such as networks. One core aspect of this visualization is to highlight changes between graphs more comprehensibly making the changes better identifiable and understandable for users, to improve users' ability to identify and understand these changes. Therefore, Bach et al. [4] employed staged animated transitions, which split the animation process into separate steps rather than executing all changes simultaneously. They defined three transition stages to handle the removal of elements, the transformation of the layout, and the addition of elements, each in a dedicated animation stage. Finally, they validated in a user study that this approach reduces task completion time and error rates, thus proving effective for exploring complex dynamic node-link-based graphs.

For the generation of IaC the two recent approaches of Wurster et al. [32] and Chiari et al. [8] are considered in more detail. The modeling details of Chiari et al. [8] have already been given in the Sect. 2.2, however details about IaC generation have been left out and are introduced in this section.

Wurster et al. [32] presented the Essential Deployment Metamodel (EDMM), a summary of the key components in 13 different declarative deployment technologies, with the goal of simplifying the comparison, selection, and migration of different deployment technologies. Additionally, they defined how the EDMM concepts can be semantically mapped to concrete deployment technologies such as Puppet or OpenStack Heat. Hence, EDMM can be used to create technology-agnostic model instances describing application topologies that can then be translated into specific IaC languages.

Chiari et al. [8] built upon the idea of EDMM to provide a technology-agnostic concept that can be translated into several IaC languages. In contrast to EDMM, they introduce a technology-agnostic cloud modeling language, DevOps Modeling Language (DOML), that can, next to application deployment and configuration, also deal with infrastructure provisioning. A DOML model is translated from DOML into IaC by the Infrastructure Code Generator (ICG). For this translation, the ICG first digests the provided DOML model, then creates an Intermediate Representation (IR) from it, and finally fills in predefined templates for each IaC target language, e.g., Terraform and Ansible, based on the IR.

5.2   Vision: Performing the IT Infrastructure Alignment

The IT Infrastructure Alignment framework transforms the current IT infrastructure into an energy-optimized state. To accomplish this transformation, the framework takes as input the current state of the IT infrastructure, which is determined by the resource-aware infrastructure miner and captured in an IPG and an aIPG, termed current IPG and current aIPG. Furthermore, the framework receives a second aIPG, referred to as optimized aIPG, representing the energy-optimized state generated by the green optimization framework. The alignment process of transforming the current state into the desired state is divided into two stages that enable the realization of the three alignment steps mentioned above.

Validation and Explainability: This stage aims to enable experts to validate the proposed energy-optimized model to ensure no unintended side effects are introduced. Additionally, explanations are provided to clarify how the changes contribute to energy reduction. In the first step, the differences between the current aIPG and the optimized aIPG are determined by adapting the difference graph approach of Delugach et al. [9] to property graphs. We plan to visualize these difference in addition with the evaluation results in a suitable UI, which will visualize the difference via the proposed staged animation transitions for graphs from Bach et al. [4]. Furthermore, filters and queries will be supported to enable a fine-grained difference analysis, such as determining what happened to specific applications, servers, or VMs. The staged animation transition combined with the filters, queries, and evaluation results should enable users to validate and understand the optimized model. Once no undesirable side effects have been detected, the optimized model can be released for the next stage.

IaC Generation and Alignment: With the validation completed, this stage focuses on transforming the optimized aIPG into executable IaC to align the current IT infrastructure with the optimized state. Therefore, the necessary alignment operations are determined by analyzing the previously determined differences between the current aIPG and the optimized aIPG. For example, the analysis might indicate that two VMs have been consolidated or an application moved to a different server. However, as aIPGs are technology-agnostic, they are unsuitable to determine the appropriate IaC language. To overcome this, we use the provided current IPG, which contains details about the technology, to select the correct IaC language for each affected component. After selecting the correct IaC language for each affected component, we translate them into IaC using templates similar to Chiari et al. [8]. The generated IaC can then be executed by users either automatically or manually.

After completing both stages, users may choose to initiate a new mining-evaluation-optimization-alignment cycle due to changed constraints or test the optimized IT infrastructure over an extended period to assess whether the anticipated energy reductions are achieved.

6   Innovation Potential

GIEOAS consists of four main components: Resource-aware Infrastructure Mining, Green Evaluation Framework, Green Optimization Framework, and the IT Infrastructure Alignment Framework, as visualized in Fig. 1. Each component contributes its part to the overall goal of reducing the energy consumption and GHG emissions of running IT infrastructures through its innovative approaches. The following section highlights the innovation aspects of each component and overall GIEOAS.

Resource-Aware IT Infrastructure Mining. The function of this component is to create an up-to-date model of a running IT infrastructure, capturing details about active components, their relationships, and resource usage. Therefore, we advance both infrastructure mining approaches and modeling concepts.

For infrastructure mining, we built on the approaches presented in [5], [18], and [15], significantly enhancing them by integrating resource usage information and utilizing IaC files as a data source.

In case of modeling concepts, we propose to disengage from MOF and static meta-model-based approaches, such as [8], and instead adopt a dynamic, graph-based approach. While [6] introduced graphs to capture IT components and relationships, we refine this model further to address the required dynamic and static capabilities. This refinement involves using property graphs as the underlying structure, with a dynamic model to capture an arbitrary IT infrastructure and a static model to which it can be transformed, providing a solid foundation for the evaluation, optimization and IaC generation processes.

Green Evaluation Framework. The Green Evaluation Framework determines KPI values and green pattern fulfillment for a given IT infrastructure model. Also, it allows the simulation of different scenarios to assess their impact on these KPIs and patterns. The calculation of the KPI values is similar to the Impact Framework of the GSF [12], which calculates environmental impacts, such as carbon emissions, of various software environments. However, our framework goes significantly beyond calculating KPIs by evaluating green design pattens, supporting the simulation of scenarios and providing extensibility for custom KPIs and patterns. Additionally, it will visualize the results appropriately in a UI to enhance understanding and decision-making.

Green Optimization Framework. Based on the IT infrastructure model and its evaluation results, the Green Optimization Framework restructures the model to reduce energy consumption without violating service level agreements. To achieve this, we plan to use a reinforcement learning approach that, to our knowledge, is the first attempt to optimize IT infrastructures holistically for energy consumption. Unlike existing methods focused solely on data centers, as seen in the survey by Hou et al. [19], our approach considers a complete IT infrastructure combined with its SLAs.

IT Infrastructure Alignment. The final step of GIEOAS is to align the current IT infrastructure with the model to realize energy reduction. This alignment is achieved by translating the optimized model into IaC. While Chiari et al. [8] generated IaC from MOF-based models, we transfer this approach to property graphs. Additionally, we focus on explainability and traceability to compare the IT infrastructure's current (as-is) and optimized (to-be) states to avoid errors and make the changes comprehensible. To support this, we utilize staged animation transitions for graphs in combination with filters and queries to highlight the differences between the graphs.

7   Conclusion

Our vision for the Green Infrastructure Evaluation, Optimization, and Alignment System (GIEOAS) aims to reduce the energy consumption and greenhouse gas emissions of the ICT sector by transforming existing IT infrastructures into energy-efficient systems. We defined five goals for GIEOAS: (G1) create an always up-to-date model of the active IT infrastructure; (G2) enable the assessment of an IT infrastructure model's greenness based on KPIs, patterns and simulations; (G3) optimize an IT infrastructure model for energy reduction while considering other objectives; (G4) align the current infrastructure with the optimized model through automatic IaC generation; and (G5) provide explainable visualizations of proposed adjustments throughout the entire process.

To achieve these goals, we proposed several innovative solutions. The Resource-aware Infrastructure Miner captures detailed information about the active components, their relationships, and resource usages from various data sources, including APIs, network traffic, and IaC files. These results are stored in a two-layered IT infrastructure model that combines the flexibility of a metamodel-free approach with the strictness of metamodel-based structures. Moreover, the green evaluation framework defines the Green Evaluation Language (GEL), which allows the definition of KPIs, patterns, and simulation scenarios. In the green optimization framework, these evaluation results are further enriched with service level agreements (SLAs) to ensure that the optimizer changes the IT infrastructure model without violating SLAs or causing sub-optimal performance. A reinforcement learning process executes the reconfiguration of the model. The produced optimized model is then compared to the current state of the IT infrastructure, and their differences are visualized with staged animation transition for better comprehensibility. After validation, suitable IaC is generated to transform the active IT infrastructure into the proposed optimized state.

In future work, we will implement these components, evaluate their effectiveness, and refine the concepts to support various infrastructures and evolving technologies.

Acknowledgments. Transparency note: The readability of this text has been improved with the support of AI technologies.

Disclosure of Interests. The authors have no conflicts of interest to declare that are relevant to the content of this article.

References

1. Andrae, A., Edler, T.: On global electricity usage of communication technology: trends to 2030. Challenges **6**(1), 117–157 (2015). https://doi.org/10.3390/challe6010117
2. Angles, R., et al.: PG-schema: schemas for property graphs. Proc. ACM Manage. Data **1**(2), 1–25 (2023). https://doi.org/10.1145/3589778
3. Artac, M., Borovssak, T., Di Nitto, E., Guerriero, M., Tamburri, D.A.: DevOps: introducing infrastructure-as-code. In: 2017 IEEE/ACM 39th International Conference on Software Engineering Companion (ICSE-C), pp. 497–498 (2017). https://doi.org/10.1109/ICSE-C.2017.162
4. Bach, B., Pietriga, E., Fekete, J.D.: GraphDiaries: animated transitions and temporal navigation for dynamic networks. IEEE Trans. Visual Comput. Graphics **20**(5), 740–754 (2013). https://doi.org/10.1109/tvcg.2013.254

5. Binz, T., Breitenbücher, U., Kopp, O., Leymann, F.: Automated discovery and maintenance of enterprise topology graphs. In: 2013 IEEE 6th International Conference on Service-Oriented Computing and Applications, pp. 126–134 (2013). https://doi.org/10.1109/soca.2013.29

6. Binz, T., Fehling, C., Leymann, F., Nowak, A., Schumm, D.: Formalizing the cloud through enterprise topology graphs. In: 2012 IEEE Fifth International Conference on Cloud Computing, pp. 742–749 (2012). https://doi.org/10.1109/cloud.2012.143

7. Brunnert, A.: Green software metrics. In: Companion of the 15th ACM/SPEC International Conference on Performance Engineering, pp. 287–288 (2024). https://doi.org/10.1145/3629527.3652883

8. Chiari, M., Xiang, B., Canzoneri, S., Nedeltcheva, G.N., Nitto, E.D., Blasi, L., Benedetto, D., Niculut, L., Škof, I.: DOML: a new modeling approach to Infrastructure-as-Code. Inf. Syst. **125**, 102422 (2024). https://doi.org/10.1016/j.is.2024.102422

9. Delugach, H., Moor, A.D.: Difference graphs. In: Contributions to ICCS 2005, pp. 41–53 (2005)

10. Ding, D., Fan, X., Zhao, Y., Kang, K., Yin, Q., Zeng, J.: Q-learning based dynamic task scheduling for energy-efficient cloud computing. Futur. Gener. Comput. Syst. **108**, 361–371 (2020). https://doi.org/10.1016/j.future.2020.02.018

11. Farwick, M., Agreiter, B., Breu, R., Ryll, S., Voges, K., Hanschke, I.: Automation processes for enterprise architecture management. 2011 IEEE 15th International Enterprise Distributed Object Computing Conference Workshops, pp. 340–349 (2011). https://doi.org/10.1109/edocw.2011.19

12. Foundation, G.S.: Impact framework (2024). https://if.greensoftware.foundation/. Accessed 10 Oct 2024

13. Frey, S., Fittkau, F., Hasselbring, W.: Search-based genetic optimization for deployment and reconfiguration of software in the cloud. In: 2013 35th International Conference on Software Engineering (ICSE), pp. 512–521 (2013). https://doi.org/10.1109/icse.2013.6606597

14. Hammadi, A., Mhamdi, L.: A survey on architectures and energy efficiency in data center networks. Comput. Commun. **40**, 1–21 (2014). https://doi.org/10.1016/j.comcom.2013.11.005

15. Harzenetter, L., Breitenbücher, U., Binz, T., Leymann, F.: An integrated management system for composed applications deployed by different deployment automation technologies. SN Comput. Sci. **4**(4), 370 (2023). https://doi.org/10.1007/s42979-023-01810-4

16. Hidas, P., Gartner: roadmap for your infrastructure - the gartner infrastructure maturity model.pdf. https://www.unicom-systems.com/Blog/Roadmap%20for%20Your%20Infrastructure%20-%20The%20Gartner%20Infrastructure%20Maturity%20Model.pdf. Accessed 17 Oct 2024

17. Holczmann, A., Hsu, S., Johnson, B., Hussain, A.: Green software patterns. https://patterns.greensoftware.foundation/. Accessed 02 Oct 2024

18. Holm, H., Buschle, M., Lagerström, R., Ekstedt, M.: Automatic data collection for enterprise architecture models. Softw. Syst. Model. **13**(2), 825–841 (2012). https://doi.org/10.1007/s10270-012-0252-1

19. Hou, H., Jawaddi, S., Ismail, A.: Energy efficient task scheduling based on deep reinforcement learning in cloud environment: a specialized review. Futur. Gener. Comput. Syst. **151**, 214–231 (2024). https://doi.org/10.1016/j.future.2023.10.002

20. Hüttermann, M.: Building blocks of DevOps. In: DevOps for Developers, pp. 33–47. Apress, Berkeley (2012). https://doi.org/10.1007/978-1-4302-4570-4_3

21. Kaelbling, L.P., Littman, M.L., Moore, A.W.: Reinforcement learning: a survey. J. Artif. Intell. Res. **4**, 237–285 (1996). https://doi.org/10.1613/jair.301

22. Kipp, A., Jiang, T., Fugini, M., Salomie, I.: Layered green performance indicators. Futur. Gener. Comput. Syst. **28**(2), 478–489 (2012). https://doi.org/10.1016/j.future.2011.05.005

23. Kistowski, J.v., Deffner, M., Kounev, S.: Run-time prediction of power consumption for component deployments. In: 2018 IEEE International Conference on Autonomic Computing (ICAC), pp. 151–156 (2018). https://doi.org/10.1109/icac.2018.00025

24. Machiraju, V., Dekhil, M., Wurster, K., Garg, P., Griss, M., Holland, J.: Towards generic application auto-discovery. In: NOMS 2000. 2000 IEEE/IFIP Network Operations and Management Symposium 'The Networked Planet: Management Beyond 2000' (Cat. No.00CB37074), pp. 75–87 (2000). https://doi.org/10.1109/noms.2000.830376

25. Malmodin, J., Lövehagen, N., Bergmark, P., Lundén, D.: ICT sector electricity consumption and greenhouse gas emissions - 2020 outcome. Telecommun. Policy **48**(3), 102701 (2024). https://doi.org/10.1016/j.telpol.2023.102701

26. Menzel, M., Klems, M., Lê, H.A., Tai, S.: A configuration crawler for virtual appliances in compute clouds. In: 2013 IEEE International Conference on Cloud Engineering (IC2E), pp. 201–209 (2013). https://doi.org/10.1109/ic2e.2013.12

27. Nie, M., Chen, D., Wang, D.: Reinforcement learning on graphs: a survey. IEEE Trans. Emerg. Top. Comput. Intell. **7**(4), 1065–1082 (2023). https://doi.org/10.1109/tetci.2022.3222545

28. Object Management Group: Meta Object Facility (2016). Version 2.5.1

29. Seo, C., Edwards, G., Popescu, D., Malek, S., Medvidovic, N.: A framework for estimating the energy consumption induced by a distributed system's architectural style. In: Proceedings of the 8th international workshop on Specification and verification of component-based systems, pp. 27–34 (2009). https://doi.org/10.1145/1596486.1596493

30. Simarro, J.L.L., Moreno-Vozmediano, R., Montero, R.S., Llorente, I.M.: Dynamic placement of virtual machines for cost optimization in multi-cloud environments. In: 2011 International Conference on High Performance Computing & Simulation, pp. 1–7 (2011). https://doi.org/10.1109/hpcsim.2011.5999800

31. Wu, Z., Pan, S., Chen, F., Long, G., Zhang, C., Yu, P.S.: A comprehensive survey on graph neural networks. IEEE Trans. Neural Netw. Learn. Syst. **32**(1), 4–24 (2021). https://doi.org/10.1109/tnnls.2020.2978386

32. Wurster, M., et al.: The essential deployment metamodel: a systematic review of deployment automation technologies. SICS Softw.-Intensive Cyber-Phys. Syst. **35**(1–2), 63–75 (2019). https://doi.org/10.1007/s00450-019-00412-x

33. Zaman, T.: A few weeks post Twitter-acquisition in 2022 we found 700 V100 GPUs (pcie, lol) in the datacenter (2024). https://x.com/tim_zaman/status/1815495006469365889?ref_src=twsrc%5Etfw. Accessed 17 Oct 2024

Organic Computing for Adaptive and Resilient Electricity Grid Management

Mischa Ahrens[2] and Hartmut Schmeck[1,2]([✉])

[1] Karlsruhe Institute of Technology (KIT), Kaiserstraße 12,
76131 Karlsruhe, Germany
hartmut.schmeck@kit.edu
[2] FZI Research Center for Information Technology, Haid-und-Neu-Straße 10-14,
76131 Karlsruhe, Germany
{ahrens,schmeck}@fzi.de
https://www.kit.edu, https://www.fzi.de

Abstract. The motivation behind the research initiative "Organic Computing" has been the need for system architectures supporting the self-organized response of decentralized technical application systems to changing requirements and disturbances in their operational environment, while respecting functional objectives as defined by their users. Such a property is strongly resembling the notion of system resilience. In this paper, we report some results on the achievement of resilience in energy systems by an adaptive system for regional management of electricity grids. This system mitigates physical disturbances of the grid in different ways, depending on the availability of information on the current status of the relevant entities, such as the distribution system operator, grid equipment, and (active) buildings. Our approach is based on an extension of the Organic Smart Home, a software framework for energy system management, simulation, and optimization, which has been strongly influenced by the observer/controller architecture, which emerged as a core generic architectural concept for organic computing systems. Depending on the current availability of communication infrastructure, the adaptive grid management switches between centralized, distributed, and completely decentralized derivation of control decisions in response to undesired deviations of voltages at grid connection points, high transformer temperatures, or high line currents. Based on power profiles from real-life studies and simulation models, our evaluation shows that even the completely decentralized strategy is still capable of guaranteeing the desired resilience.

Keywords: Organic computing · Energy management · Electricity grid · Resilience · Communication infrastructure

G. Ernst et al. (Eds.): Wolfgang Reif Festschrift, LNCS 15765, pp. 242–261, 2025.
https://doi.org/10.1007/978-3-031-92196-4_12

1   Introduction and Related Work

Around 20 years ago the Computer Engineering section of the German Informatics Society successfully initiated the priority research program "Organic Computing" (funded by the German Research Association (DFG)). Besides an emphasis on the effects and potential of (controlled) self-organization, an essential aspect was the requirement to investigate the research topics with respect to specific technical application areas. Hence, the research projects considered robotics [9,13], highway traffic [8], control of city traffic and traffic lights [23], office buildings [29], and industrial production lines [20], to name a few.

The main rationale driving this research program was the insight that it is not a question whether self-organizing systems will emerge in future technologies but how we can design adequate system architectures and methodology in order to cope with inevitably arising artifacts and devices in our vicinity with the ability to autonomously communicate and adapt their behavior. While the research groups of Christian Müller-Schloer, Hartmut Schmeck, and Theo Ungerer focused on the investigation of fundamental effects of self-organization and on the design of system architectures and algorithms, the group of Wolfgang Reif developed software design patterns for Organic Computing taking as an example industrial production systems.

It turned out that one of the major challenges with respect to the dissemination of the research results into real-world applications was the question to what extent these self-organizing organic systems would be capable to provide dependable functionality and performance. In particular, dependability addresses the ability to cope with disturbances due to failures or misconduct of components in the environment, typically summarized under terms like fault tolerance, resilience, or trustworthiness. The latter term has been the core topic of the "OC Trust" initiative of W. Reif, T. Ungerer, and C. Müller-Schloer [27], which investigated self-organizing systems under uncertainty about the behavior of their components. Among others, decentralized energy management was chosen as one of their technical application areas [5].

Independently, the research group of H. Schmeck moved its attention from traffic to energy systems, transferring the concepts of Organic Computing into the development of decentralized energy management systems. This was driven by the necessity to detect the potential flexibility of energy consumption and generation schedules and to optimize the exploitation of this flexibility for supporting the stability of an electricity grid that is increasingly influenced by the volatile and highly decentralized power generation from renewable sources. The necessary integration of decentralized active power generators and consumers into the grid management and control relies on the availability of communication between these actors.

Motivated by this requirement, in this paper we present some results of an investigation of a concept for an adaptive electricity grid management, which is supposed to provide resilience even in the case of failures of the communication system. The central tool for our investigation is based on the *Organic Smart Home - OSH*, a software framework for decentralized energy manage-

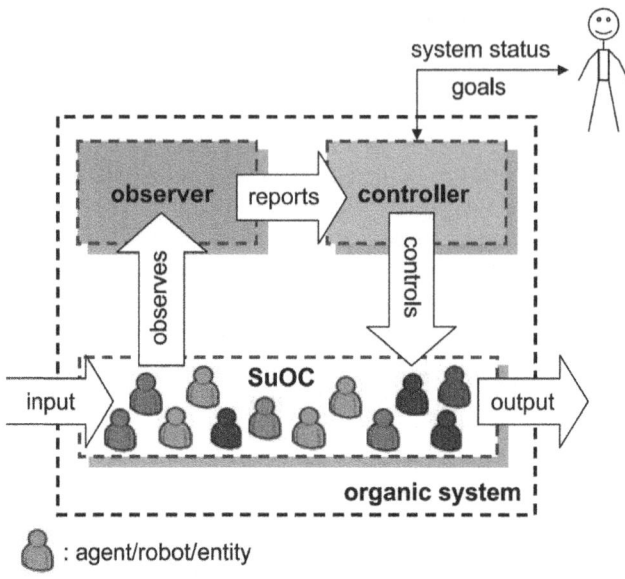

Fig. 1. Observer/Controller Architecture (see [6])

ment, simulation, and optimization. This framework is strongly based on the observer/controller (O/C) architecture (see [6, 24, 28] and Fig. 1), which has been developed as a suitable architectural design pattern for Organic Computing systems. In Sect. 2, after a brief presentation of the OSH framework, we introduce several strategies for a resilient electricity grid management, which differ with respect to the availability of communication between the active components. Some results of an evaluation of these strategies with respect to their capability to provide resilience against disturbances of the electrical and disruptions of the communication system are provided and discussed in Sect. 3. This is followed in Sect. 4 by a brief discussion of some aspects relevant to the practical implementation of the presented adaptive and resilient grid management system. The paper concludes in Sect. 5 with a brief summary and an outlook to further research.

2   Methodology

In this section, we give an overview of the developed adaptive electricity grid management system and its functionality. This system is based on the building energy management system (BEMS) and simulation framework OSH, which we present in Sect. 2.1. This is followed by an introduction of the different grid management strategies used by the system to respond to different types of communication disruption in Sect. 2.2. Section 2.3 then deals with the types of electrical disturbances that can be mitigated with the developed system. The counter measures to such disturbances utilized by the system are subsequently described in Sect. 2.4.

Some of the methodology described in this section has already been published with respect to an earlier development status in [3] and [1]. Therefore, some of the wording used in this section is taken from these publications. Due to the limited page budget, we do not include mathematical descriptions or parameterizations of the algorithms and methods used in this paper. Substantially more detailed descriptions will be published in [2]. The specific parameter values for the electrical disturbance criteria in Sect. 2.3 as well as for the algorithms and methods introduced in Sect. 2.4 were determined in parameter studies, which will be published in [2] as well. The determined parameter values are already used for the evaluation in Sect. 3.

2.1 Organic Smart Home

The open source BEMS and simulation framework used to implement and evaluate the system presented in this paper is the OSH, which has been developed at the *Karlsruhe Institute of Technology* and the *FZI Research Center for Information Technology*. To enable self-organization for the systems managed by the OSH, it is based on a hierarchical implementation of the O/C architecture [28], a structural view is given in Fig. 2. Each device has its own local O/C-unit which is responsible for low level observation and control as well as the communication of relevant data to and the implementation of commands from higher levels. The local O/C-units are observed and controlled by a global O/C-unit, which handles the energy management of the building. It aggregates the data communicated from lower levels and retrieves external signals, such as variable electricity tariffs and user inputs, to make optimized control decisions using an evolutionary algorithm. The resulting control sequences are subsequently communicated to and implemented by the local O/C-units.

The first OSH-version was developed in [4]. Based on extensions of the OSH architecture introduced in [16], the OSH was enhanced in [17] and [15] to become a multi-modal BEMS. An energy-related degree of freedom was added to form a multi-commodity simulation and optimization allowing to switch energy carriers for hybrid home appliances when advantageous. Furthermore, simulation models for new device types, such as interruptible and hybrid appliances, were implemented, and previous models, such as for a combined heat and power plant (CHPP), substantially revised. The possibility of simulating and optimizing battery energy storage systems (BESSs) was added in [19] and extended in [18]. Electric vehicle (EV) charging and simulation was introduced in [10].

The OSH was further developed in [14] to include co-simulation capabilities. Among other things, these enable the simultaneous simulation of multiple buildings and their interactions with a simulated low-voltage grid, including transformers. These new options were used to implement and evaluate a regional energy management system coordinating the use of the flexibility of simulated or real buildings in a grid-supportive manner [14]. This OSH-version formed the basis for the system presented in this paper.

The OSH-version 4.0 (released in 2016) is available under GPL 3.0 license at [22]. This version already contains the appliance, photovoltaic (PV), H0, heating, and domestic hot water power profiles used in Sect. 3. Figure 2 shows the OSH architecture with respect to the development status in [15].

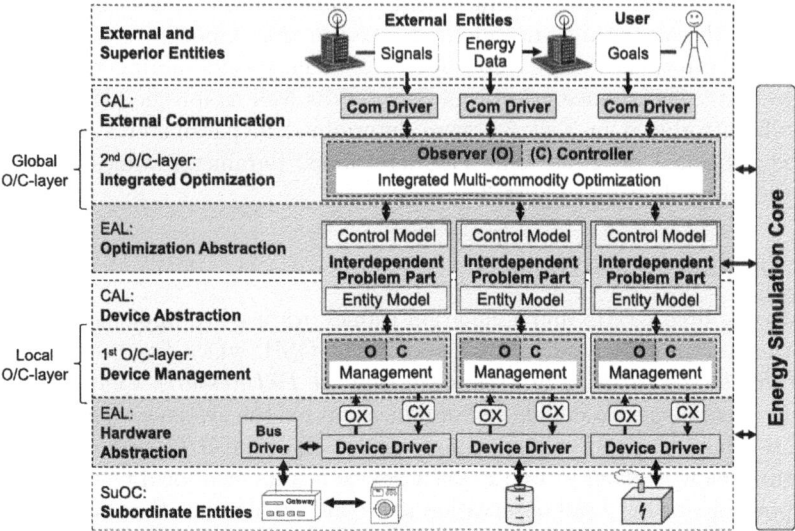

Fig. 2. Overview of the Organic Smart Home (OSH) with respect to the development status in [15], CAL: communication abstraction layer, Com: communication, EAL: entity abstraction layer, OX: observation exchange object, CX: control exchange object, SuOC: system under observation and control, reproduced from [15] (original image partially based on [17])

2.2   Grid Management Strategies

Electricity grid management usually requires functional communication infrastructure between the relevant entities to exchange grid and device status data as well as commands. This makes grid management systems susceptible to communication disruption, which can occur, for example, due to equipment failure or cyber attacks. To ensure resilient grid operation in the event of communication disruption, we propose the adaptive use of different grid management strategies depending on the current availability of the utilized communication infrastructure. These strategies are described in the following and visualized in Fig. 3. After the strategies are introduced, the functionality of their adaptive application is described in more detail. The considered segment of the distribution system consists of a low-voltage distribution transformer and all buildings connected to grid lines below this transformer.

Centralized Strategy. This strategy is only used when the distribution system operator (DSO) is able to regularly receive data from buildings connected to and the transformer supplying a low-voltage grid. Furthermore, the DSO has to be able to send commands to the buildings, when an electrical disturbance occurs that can potentially lead to electricity outages. The grid is operated by the DSO's O/C-unit based on the data communicated by the connected buildings as well as the transformer and additional data calculated in regularly performed power-flow studies using part of the received data as inputs. The data communicated by the buildings comprises voltages, active and reactive powers at their grid connection points (GCPs), measured by smart meters, as well as information on currently implemented and potentially implementable counter measures to electrical disturbances. The data communicated by the transformer comprises its load factor, hot-spot temperature, and primary voltage. The data calculated in the power-flow studies are the currents on all line segments and all GCP voltages. The latter can be compared to the respective measured voltages, to assess the accuracy of the power-flow studies. If the DSO detects an electrical disturbance, it derives counter measures and sends corresponding commands to the global O/C-units of the buildings. The considered electrical disturbances are detailed in Sect. 2.3. The utilized counter measures are described in Sect. 2.4.

Distributed Strategy. This strategy is used, if the DSO does not communicate commands anymore, but communication between the buildings and the transformer is still active. For example, this could be the case, if the DSO is subject to a cyber attack, which prevents it from communicating commands. The distributed strategy can also be used exclusively in combination with the decentralized strategy. The use of the centralized strategy is not required, even if no communication disruptions occur. In the distributed strategy, each building regularly sends the same data it would send to the DSO in the centralized strategy to all other buildings. The transformer communicates the data it would send to the DSO to all buildings as well. Based on the data received from other buildings and the transformer as well as its own measured data, each building regularly performs power-flow studies to gain an understanding of the current grid status. In doing so, each building has the same information as all other buildings and knows whether and how it should react to electrical disturbances. This means that the observation of the grid and the determination of counter measures are shifted from the DSO to the buildings. Each building uses the same algorithms as the DSO to autonomously determine and implement counter measures.

Decentralized Strategy. This strategy is used by buildings that can not communicate with other buildings or the transformer, for example, due to communication network outages. It is also used by buildings for which the grid status can not be observed with sufficient accuracy, which is explained later in this section. Each building using the decentralized strategy has to detect and remedy electrical disturbances on its own and without full knowledge of the current grid status. In this case, the only data available to a building are the active and reactive power as well as the voltage at its GCP. Although the building can only detect local, voltage-based electrical disturbances, there are synergies with other electrical disturbances. If the local GCP voltage gets too high or low, this can indicate that other GCP voltages on the same grid line are as well and that line and transformer congestion is imminent. Therefore, counter measures triggered by local voltages, such as reducing active power feed-in or consumption, can prevent the congestion of lines or the transformer as well. The physical counter measures to electrical disturbances in the decentralized strategy are the same as for the other strategies. However, the corresponding algorithms and electrical disturbance criteria are modified so that counter measures can be implemented solely based on the local GCP voltage, the building's position in the grid, and previously implemented counter measures (see also Sects. 2.3 and 2.4).

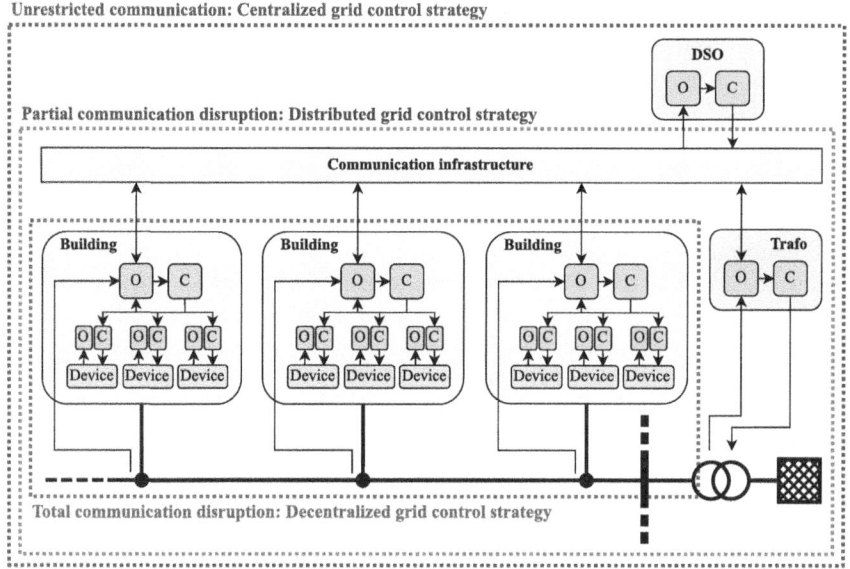

Fig. 3. Visualization of the centralized, distributed, and decentralized grid management strategies, O: Observer, C: Controller, DSO: distribution system operator, Trafo: transformer, image based on [3]

Adaptive Grid Management. To be able to adapt their grid management strategy in response to communication disruptions, the buildings have to detect these autonomously. They achieve this by continuously monitoring the grid status at their respective GCPs and communicating relevant data not only to the DSO, but also to all other buildings. The transformer sends the data it communicates to the DSO to the buildings as well. Based on locally measured data as well as the data received from other buildings and the transformer, each building regularly performs its own power-flow studies and is consequently aware of the current status of the entire low-voltage grid segment it is connected to. If a building detects an electrical disturbance but does not receive a corresponding command from the DSO, it switches from the centralized to the distributed strategy. Since this method for detecting disruptions of the communication to the DSO entails regular communication among the buildings and the transformer, a larger disruption that also disrupts the communication among these entities is easily detectable. If a building does not receive data from other buildings or the transformer anymore, it starts acting according to the decentralized strategy.

The centralized strategy can still be partially used, if the transformer, the DSO, and some buildings can still communicate with each other, while other buildings have stopped communicating. If the DSO can partially observe the current grid status with sufficient accuracy, using the current data provided by communicating buildings and the last communicated data provided by non-communicating buildings, it can still determine commands for communicating buildings. To assess whether the accuracy of the power-flow studies is still sufficiently high to determine commands for a particular communicating building, the DSO compares the measured GCP voltage of each communicating building to the corresponding voltage determined in the power-flow studies. If a particular deviation between these voltages becomes too large, the accuracy of the power-flow studies is considered too low for the corresponding building. At what time this point is reached for a specific building, depends not only on the percentage of buildings across the grid that do not communicate current data anymore, but also on the respective positions of the communicating and non-communicating buildings in the grid as well as on the grid topology. If a communicating building is situated on a grid line with many non-communicating buildings, the power-flow study accuracy can become insufficient earlier than for a building situated on a line with no or few non-communicating buildings. If the power-flow studies are determined to be insufficiently accurate for a building, the DSO does not send commands to this building. A building that does not receive commands from the DSO would normally switch to the distributed strategy. However, all buildings regularly monitor the deviation between their measured GCP voltages and the respective voltages calculated in their own power-flow studies as well. Each communicating building that detects an excessive deviation for its own GCP switches directly to the decentralized strategy. All buildings that still receive commands from the DSO keep acting according to the centralized strategy.

Similar to the centralized strategy, the distributed strategy can be partially used, if the DSO and some buildings have stopped communicating, but the transformer and other buildings can still communicate with each other. In such a situation, all buildings that can communicate and assess their own measured GCP voltage to be sufficiently close to the respective voltage determined in their own power-flow studies use the distributed strategy. All buildings that can not communicate or detect an excessive deviation between their measured and calculated GCP voltages switch to the decentralized strategy.

When the communication to the other buildings and the transformer is restored after a disruption and the deviation between the measured and calculated GCP voltages is acceptable, a building switches back to the distributed or centralized strategy, depending on whether it receives commands from the DSO again as well. When a building uses the distributed strategy but receives one or more commands from the DSO, it switches back to the centralized strategy.

The described decision processes are summarized in Fig. 4. It is important to note that the decisions to switch between the different strategies are performed locally by the buildings based on the data received locally.

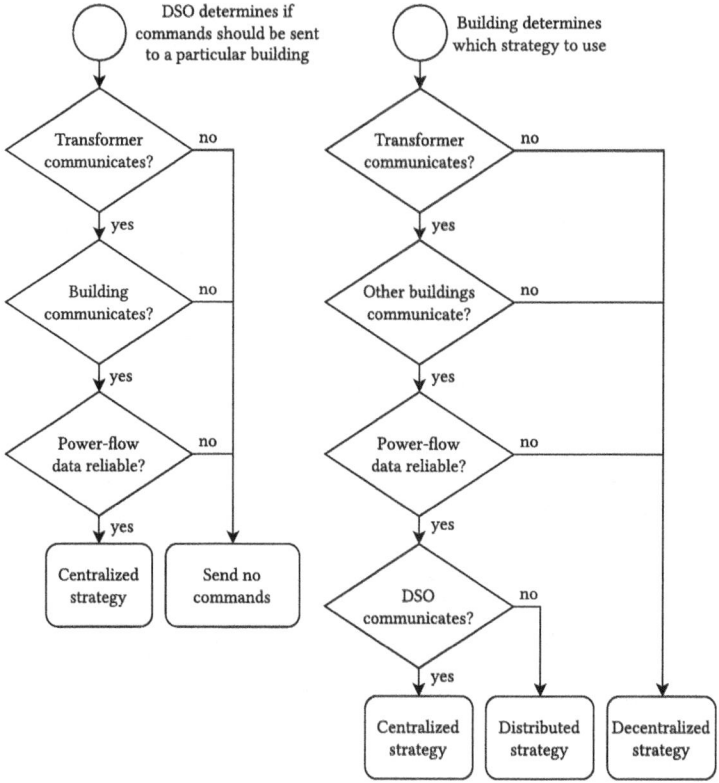

Fig. 4. Decision processes for determining the right strategy during communication disruptions, from the perspectives of the DSO (left) and a building (right)

2.3   Electrical Disturbance Criteria

We consider three types of electrical disturbances that are relevant in low-voltage grids. Remedying these disturbances can substantially increase the resilience of a given grid [12]. It is important to note that by electrical disturbances, we do not refer to electricity outages, but to conditions that will likely lead to electricity outages, if no appropriate counter measures are implemented. For more information on the counter measures to electrical disturbances, see Sect. 2.4.

Voltage-Based Disturbance Criteria. The first disturbance type is *critical voltages*. In low-voltage grids, voltages have to stay within a range of $\pm 10\,\%$ around the nominal voltage [11]. To prevent voltage range deviations due to sudden spikes in power consumption or feed-in, we propose more narrow voltage ranges to trigger the activation of grid management measures. The centralized and distributed strategies use fixed boundaries. In the decentralized strategy, which functions without comprehensive knowledge on the current grid status, the boundaries are varied depending on a building's position on the grid line. Buildings positioned closer to the transformer use more narrow voltage ranges than buildings positioned further away from the transformer. This is done because of the voltage drop or increase that occurs along lines, which would otherwise cause strongly deviating reactions by differently positioned buildings that only know their own GCP voltage. The other strategies do not need this voltage range variation, as here all buildings on a line know and always react to the most critical voltage on the line. Criteria for the implementation of measures against voltage jumps are used as well. Since a voltage jump can indicate a further voltage rise at the following measurement time, which could already violate the permissible voltage range, it can be useful to implement counter measures, even if the voltage boundaries are not yet violated. In the decentralized strategy, the minimum voltage jump and resulting voltage to trigger counter measures are varied depending on the building's position on the line as well.

Line-Based Disturbance Criteria. The second electrical disturbance type is the imminent *congestion of grid lines* due to excessive power transmission. When the current on a line reaches a certain percentage of its nominal current, the use of active power flexibility and reactive power compensation is triggered. Only when the current reaches a predefined higher percentage, PV systems are curtailed so that currents stay below this percentage. The two current boundaries are only applicable for the centralized and distributed strategies, since their application requires functioning communication infrastructure and sufficiently accurate power-flow studies. In the decentralized strategy, a building can only measure the power and voltage at its GCP. Therefore, active power flexibility use and reactive power compensation are triggered based on the locally measured GCP voltage and the building's position on the line. PV curtailment is implemented based on static curtailment settings learned while using the centralized and distributed strategies.

Transformer-Based Disturbance Criteria. The third type of electrical disturbance considered is imminent *transformer overheating* due to congestion. The measures against transformer overheating are triggered according to two different load factor boundaries, as well as three different hot-spot temperature boundaries. Reaching the higher load factor boundary triggers the use of active power flexibility and reactive power compensation, while reaching the lower boundary coming from a higher level stops the use of these measures. Reaching the medium-temperature boundary triggers PV curtailment if the temperature is currently rising. Reaching the high-temperature boundary entails PV curtailment, regardless of whether the temperature is currently rising. When the lower temperature boundary is reached while the temperature is falling, PV curtailment is gradually scaled back. We determined the specific values of the temperature boundaries in compliance with the temperature limits given in [7]. As for line-based disturbances, the aforementioned criteria can only be applied in the centralized or distributed strategies, which require the communication of transformer data to the DSO and buildings. For the decentralized strategy, the criterion for triggering the use of reactive power compensation and active power flexibility is the same as for line-based disturbances. Static PV curtailment settings, learned during the previous use of the centralized or distributed strategies, are used to prevent transformer overheating as well.

2.4   Counter Measures to Electrical Disturbances

This section describes the utilized counter measures to the electrical disturbances introduced in Sect. 2.3.

Reactive Power Compensation. In the context of line or transformer congestion, reactive power compensation can be useful. Here, a PV inverter's capability to generate or consume reactive power is used to compensate for the reactive power consumption or generation of a building. This decreases or eliminates the reactive power that has to be transferred by the grid. This can be especially useful in situations where utilization of grid equipment is already high.

Reactive-Power-Based Voltage Maintenance.   In addition to reactive power compensation, reactive power generation or consumption by PV inverters can also be used to increase or decrease grid voltages, making it an effective tool for voltage maintenance [26]. If voltages are critical according to the voltage boundaries and voltage jump criteria mentioned in Sect. 2.3, we always prefer the use of reactive power for voltage maintenance over reactive power compensation, since voltage range deviations generally tend to result in damaged equipment [25] faster than congestion [7,30]. We utilize reactive-power-based voltage maintenance algorithms that are designed to limit the additional load on the grid introduced by the transmission of reactive power. However, to prevent oscillation effects, the algorithms reduce the reactive power consumption or generation only in very small increments when voltages are back inside the undisturbed range and have passed a dead band in which the reactive power target for the PV inverter is kept constant. The algorithm to determine reactive power targets in the centralized and distributed strategies differs from the one used in the decentralized strategy. While the former considers all GCP voltages on a grid line, the latter only considers a building's local GCP voltage and position on the line.

Adaptive Photovoltaic Generation Curtailment. PV inverters can curtail the active power generation of attached PV systems. This functionality can be used to implement algorithms that adjust the generated active power depending on the grid status and the available active power flexibility. It is important to note that we differentiate between curtailment, where load is reduced without substitution, and utilizing flexibility, where load is temporally shifted or temporarily reduced but later increased. We do not use curtailment for other devices than PV systems, as all other high-power devices present in residential buildings, such as heat pumps (HPs) and EVs, provide some operational flexibility that can be leveraged without curtailment, using some type of load shifting or shaping. Since different types of electrical disturbance require adjusted algorithms, we use different algorithms for PV curtailment, depending on whether the occurring disturbance is line- or transformer-based. These algorithms are utilized in the centralized strategy by the DSO and in the distributed strategy by each building. The most substantial curtailment settings determined by these algorithms during the use of the centralized and distributed strategies are saved to be statically applied during the use of the decentralized strategy, if needed. Hence, the curtailment can be more substantial than necessary in the decentralized strategy. This is necessary as the actual current grid status is not known in this case. While the algorithm to prevent transformer overheating dynamically adjusts the curtailment based on the current transformer hot-spot temperature, the algorithm to prevent line congestion learns individual curtailment settings for each line, which are constantly applied to each connected building. This distinction is made to account for the thermal inertia of transformers, which allows to exceed a transformer's nominal power for a certain amount of time until a critical temperature is reached. The algorithms do not directly target the active power generated by a PV system, but the apparent power at the building's GCP, since this is the actual power that stresses lines and the transformer. For example, if a BESS or an EV is currently charged using part or all of the active power generated by the PV system, the latter does not have to be curtailed as much or at all to reach a particular apparent power target at the building's GCP.

Active Power Flexibility. The exploitation of a building's energy flexibility to adjust its active power consumption or feed-in can be used to reduce the load on grid lines and transformers [14]. An example of a flexible device is a BESS that can be charged or discharged at different speeds. If PV curtailment is used, active power flexibility is not directly needed to remedy electrical disturbances. However, since the curtailment algorithms utilized in this paper apply to the GCP of the building, the use of active power flexibility can reduce the actual curtailment of the PV system. This is desired, as the curtailment-induced loss of generable energy reduces the usefulness of PV systems.

The grid-supportive use of active power flexibility is achieved by optimizing the energy flows of a building with respect to an active power target. This target can be set by the DSO in the centralized or by the buildings themselves in the distributed and decentralized strategies. In this paper, the target is always set to 0 W, when the use of active power flexibility is triggered. This is done for comprehensively exploiting the flexibility of a building, which most effectively reduces the need for PV curtailment. The implementation of an active power target requires a building's BEMS to allow for optimized scheduling based on such targets. In the OSH the already implemented evolutionary algorithm and device models are used in conjunction with an objective function implemented in [21]. This objective function minimizes the difference between the expected active power profile and the active power target for the duration of the optimization horizon.

3   Evaluation and Discussion

In the chosen simulated evaluation scenario (specified in Table 1), different disruptions of communication occur at different times of the day. July 1 is chosen as the basis for this scenario, as it is a day with a very high amount of PV generated active energy in the utilized PV generation profile and consequently a very high potential for electrical disturbances.

The default grid management strategy for this scenario is the centralized strategy. As shown in Fig. 5, the first disruption occurs at 08:00, when the DSO stops to communicate (indicated by a thick vertical red line). At 09:00, the first building stops to communicate. After that, every three minutes an additional building stops to communicate until 10:57 (indicated by a sequence of thin vertical red lines). At 11:00, the communication between the buildings is restored, but the transformer stops to communicate (also indicated by a thick vertical red line). At 12:00, the communication is fully restored (indicated by a thick vertical green line). Figure 5 shows the resulting maximum GCP voltage, transformer hot-spot temperature, maximum normalized line current, and aggregated PV active power profiles for two different communication disruption mitigation configurations. The first ("Static grid management") exclusively uses the centralized strategy and does not utilize strategy adaption in response to communication disruptions. The second ("Adaptive grid management") uses the strategy adaption detailed in Sect. 2.2 in response to communication disruptions. Both configurations use all counter measures to electrical disturbances described in Sect. 2.4. The profiles for a control configuration with no counter measures to electrical disturbances at all ("No grid management") are given as a reference. The horizontal dotted red lines indicate the respective permissible maximum and minimum values for voltage, temperature, and line current.

As can be seen in Fig. 5, the communication disruptions lead to violations of the maximum admissible transformer hot-spot temperature, if no communication disruption mitigation is implemented. This is caused by the insufficiently low PV curtailment settings communicated by the DSO before it stops communicating. However, since the settings to prevent violations of nominal line currents are already sufficiently low before the first occurring communication disruption, all line currents stay below the respective nominal line currents. The maximum GCP voltage stays inside the admissible voltage range as well, as the

Table 1. Grid and building configurations for the performed simulations

Parameter	Value(s)
Transformer nominal power	400 kVA
Voltage on primary transf. side	Contains a random component as in [14]
Grid topology	Radial
Feeders	5
Distributing cabinets	5
Grid line type	NAYY 4x150 SE, buried cable
Avg. distance between GCPs	53 m
Grid nodes	62
GCPs (buildings)	55
Inhabitants per building	1–5, average: 2.927
Base load profile	H0, adjusted for weekday, Saturday, Sunday, season, number of inhabitants
Simulated appliances (additional to base load, based on real-life load profiles)	Dishwasher, washing machine, dryer, induction hob, electric oven (probability based starting times [15])
Flexible appliances	Dishwasher, washing machine, dryer
PV system peak power	6–30 kW$_p$ (Inverter: 6.316 kVA–33.333 kVA)
PV system power profile	Measured at FZI House of Living Labs in 2013, scaled to respective peak power
BESS capacity	6–10 kWh
BESS max. active charge/discharge power	4.2–7 kW ($cos(\varphi) = 1$), power exchange with grid permitted
CHPP (if no HP is used)	1 kW$_{el}$ ($cos(\varphi) = 0.9$), 4 kW$_{th}$, 5.682 kW$_{gas}$
Air source HP (if no CHPP is used)	6.7 kW$_{el}$ ($cos(\varphi) = 0.95$), coefficient of performance (COP): 1.2–5.1 (depends on air and tank temperature)
Hot water storage tank	750 L, 60/80 °C (CHPP), 40/60 °C (HP)
EV battery capacity	85 kWh
EV max. active charge power	22 kW ($cos(\varphi) = 0.99$), no discharging
Evolutionary algorithm stopping criteria	1000000 generations reached or fitness delta $\leq 5 \cdot 10^{-5}$ for 20 generations
Evol. algo. population size	100 solution candidates
Random seeds	5 per building, different for each building
Building simulation resolution	1 s
PV profile resolution	1 min
Power-flow studies resolution	1 min
Grid management resolution	1 min
Optimization algo. resolution	5 min

DSO already communicates reactive power targets before 08:00 in this scenario. However, as these targets are not updated after the communication disruption to the DSO, the highest maximum GCP voltage almost reaches the maximum admissible voltage. Only after the communication is fully restored, the DSO heavily adjusts the PV curtailment settings of all buildings, such that the transformer hot-spot temperature decreases back below its admissible maximum.

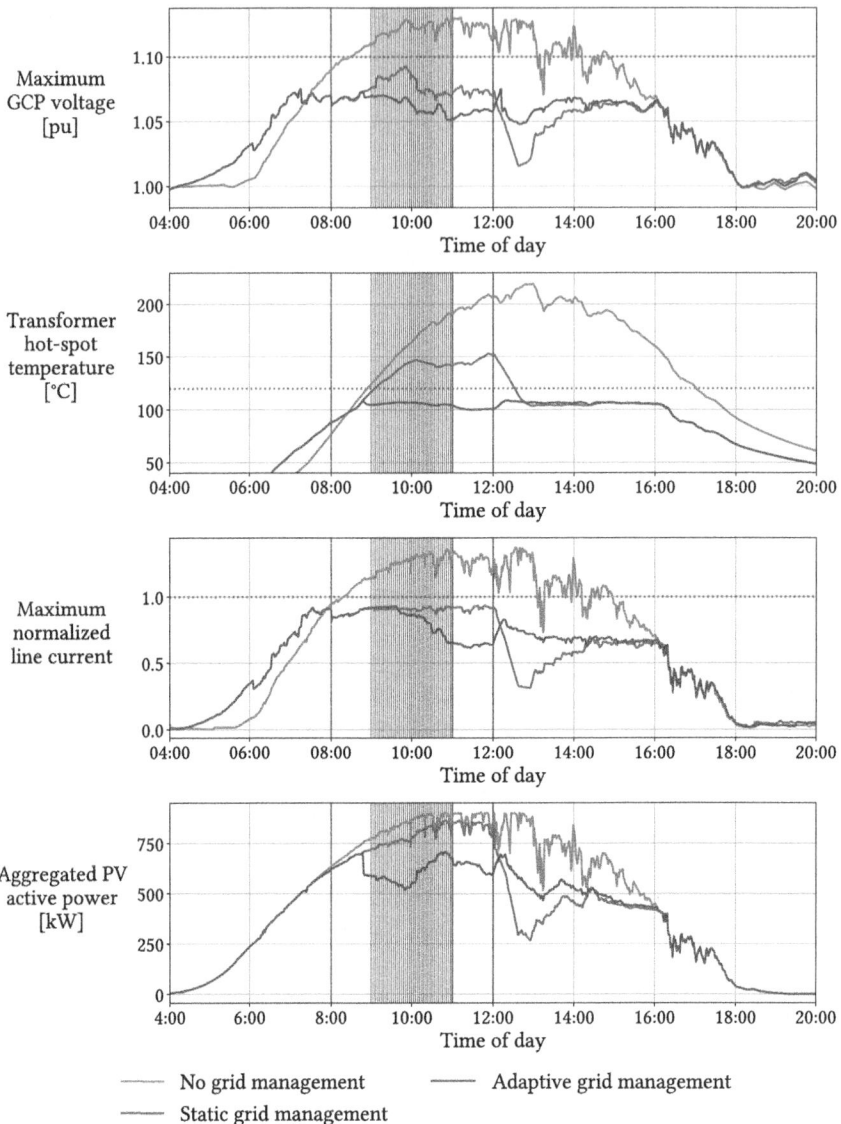

Fig. 5. Profiles of the maximum GCP voltage in the grid, the transformer hot-spot temperature, the maximum normalized line current in the grid, and the active power generated by PV systems for July 1 (more details are given in the text)

If the communication disruption mitigation measures presented in Sect. 2.2 are implemented, all grid status indicators stay inside their respective admissible ranges. When the communication to the DSO starts to be disrupted, this is detected by the buildings, which consequently switch to the distributed strategy and implement measures against electrical disturbances on their own. When the buildings cease to communicate one by one, all buildings that can not communicate anymore switch to the decentralized strategy. Buildings that still receive data from other buildings keep using the distributed strategy, as long as the voltages measured at their GCPs do not deviate too far from the voltages calculated in their power-flow studies. If a building detects an excessive deviation, it switches to the decentralized strategy, even if it can still communicate. These behaviors can not be reliably observed in the presented profiles, since buildings that still use the distributed strategy adjust their reactive power and PV curtailment settings to compensate for differing settings implemented by buildings using the decentralized strategy. When the transformer ceases to communicate, all buildings switch to the decentralized strategy. This can be observed in the transformer hot-spot temperature profile, which drops to a lower level, as the decentralized strategy uses the previously learned PV curtailment setting to prevent transformer overheating, which is lower than the setting needed for this particular time period. When the communication is fully restored, the buildings switch back to the centralized strategy and the DSO relaxes the curtailment settings again, which causes the transformer hot-spot temperature to rise again. As a consequence, the maximum GCP voltage and line current increase as well.

It is important to note that the counter measures against electrical disturbances presented in this paper are able to fully prevent all electricity outages, which would otherwise occur in this scenario, as long as occurring communication disruptions are mitigated as well.

4   Prerequisites, Scalability, Economic Aspects, and Potential Misconduct

To utilize the presented electricity grid management system, the current status of the managed distribution grid segment has to be observable to a certain extent. Its application at the temporal resolution and power-flow study accuracy used for the evaluation in Sect. 3 would require a suitable smart meter infrastructure, enabling the measurement and communication of the active and reactive power as well as the voltage at the GCP of each building connected to the managed grid segment at least once every minute. For less problematic grid segments than the one evaluated in this paper, lower temporal resolutions may be feasible as well. To reduce the number of needed smart meters, different methods of state estimation could be beneficial, but fall outside the scope of this paper. The communication of messages and measurements in the centralized and distributed strategies needs a suitable infrastructure, such as a cellular, satellite, copper, or fiber optic cable network. However, due to the adaptive design of the presented system, this network does not have to be fail-safe to effectively counteract electrical disturbances. In some cases, power-line communication may also be feasible, making a dedicated communication network non-essential. Another prerequisite is the implementation of a suitable communication protocol, for example, using a hosted message broker or peer-to-peer communication.

The limiting factor in scaling the application of the presented system to larger distribution grid segments than the one evaluated in Sect. 3 is the requirement that the power-flow studies have to be computed in time on the low-power hardware usually utilized to run a BEMS. The computation time for each power-flow study increases exponentially with the number of considered GCPs. However, for the 55-GCP grid evaluated in this paper, the solution time for each power-flow study is in the low single digit millisecond range on a single core processor. Evaluations for a 130-GCP grid segment, which will be published in [2], show that computation times are still several orders of magnitude below the grid management resolution of one minute. We therefore assume that the system would be able to scale to substantially larger numbers of GCPs as well. The solution times of the evolutionary algorithm utilized by the OSH do not impact scalability, since each building only optimizes its own energy flows. The current design of the system only considers low-voltage grids. However, its principles and methods could be transferred to higher grid levels as well, resulting in a multi-layer adaptive system.

To assess the economic feasibility of implementing a system like the one presented, one has to consider the alternative. If grids with a high amount of decentralized power consumption and generation, where peak powers can exceed the grid capacity, are not intelligently managed, they have to be physically expanded. This would entail the replacement of grid equipment, such as transformers and lines, with more capable and expensive equipment as well as expensive construction measures. Since the presented system only requires low-power computers, a communication network, and a suitable smart meter infrastructure, we strongly suspect this to be less expensive than grid expansion measures. In most inhabited regions of the world, a communication network is already present. Furthermore, the utilization of BEMS-hardware can already be economically advantageous by enabling the participation in energy markets and benefiting from variable electricity tariffs alone, especially if a building consumes or generates power to an extent that would negatively impact the adjacent distribution grid.

Since a system that actively involves its participants in distributed or decentralized decision making can be susceptible to participant misconduct, we also consider counter measures to intentional misconduct and sensor failures in the design of the presented system. A detailed description and evaluation of these counter measures will be published in [2].

5   Summary and Outlook

Resilience against disturbances is one of the key requirements for the electricity grid, as it is the most crucial of our critical infrastructures. In this paper we have shown how core architectural concepts of Organic Computing can be utilized for the design of an adaptive grid management, which is switching between centralized, distributed, and decentralized strategies for mitigating electrical disturbances depending on the current reliability of the communication system. After a brief description of the underlying Organic Smart Home software framework, we sketched typically occurring electrical disturbances as well as appropriate physical responses to mitigate their undesired effects and we introduced the three different strategies and their integration into an adaptive grid management. The practical relevance of this system design has been evaluated with respect to typical load profiles. In this paper we only presented a few positive results of a more extensive investigation, which will be described comprehensively in the upcoming dissertation of the first author [2]. Some preliminary details of this investigation have also been published in [3] and [1].

Acknowledgments. We gratefully acknowledge the support of the Federal Ministry of Education and Research (BMBF), which funded our research on grid resilience within the project OCTIKT (Grant No. 01IS18064A). We acknowledge support by the state of Baden-Württemberg through bwHPC.

Disclosure of Interests. The authors have no competing interests to declare that are relevant to the content of this article.

References

1. Ahrens, M.: Increasing power grid resilience with a multi-agent system of smart buildings. In: Tomforde, S., Krupitzer, C. (eds.) Organic Computing - Doctoral Dissertation Colloquium 2021, pp. 15–31. kassel university press (2022). https://doi.org/10.17170/kobra-202202215780
2. Ahrens, M.: Preliminary title: adaptive and resilient electricity grid management with smart buildings. Ph.D. thesis, Karlsruhe Institute of Technology (KIT) (2025). (Unpublished thesis)
3. Ahrens, M., Kern, F., Schmeck, H.: Strategies for an adaptive control system to improve power grid resilience with smart buildings. Energies **14**(15) (2021). https://doi.org/10.3390/en14154472
4. Allerding, F.: Organic smart home - energiemanagement für intelligente Gebäude (English: Organic smart home - energy management for intelligent buildings). Ph.D. thesis, Karlsruhe Institute of Technology (KIT) (2013). https://doi.org/10.5445/KSP/1000038928
5. Anders, G., Siefert, F., Steghöfer, J.P., Reif, W.: Trust-based scenarios - predicting future agent behavior in open self-organizing systems. In: Elmenreich, W., Dressler, F., Loreto, V. (eds.) Self-Organizing Systems, pp. 90–102. Springer, Heidelberg (2014)
6. Branke, J., et al.: Organic computing - addressing complexity by controlled self-organization. In: Second International Symposium on Leveraging Applications of Formal Methods, Verification and Validation (isola 2006), pp. 185–191 (2006). https://doi.org/10.1109/ISoLA.2006.19
7. DIN IEC 60076-7 (VDE 0532-76-7): 2008-02, Power transformers - Part 7: Loading guide for oil-immersed power transformers (DIN IEC 60076-7:2005). Standard, DIN Deutsches Institut für Normung, International Electrotechnical Commission (IEC) (2008)
8. Fekete, S., Schmidt, C., Wegener, A., Fischer, S.: Hovering data clouds for recognizing traffic jams. In: IEEE Proceedings of the Second International Symposium on Leveraging Applications of Formal Methods, Verification and Validation

(isola 2006), pp. 198–203. IEEE (2007). https://doi.org/10.1109/ISoLA.2006.30. 2nd International Symposium on Leveraging Applications of Formal Methods, Verification and Validation, ISoLA 2006; Conference date: 15-11-2006 Through 19-11-2006

9. Hoffmann, A., Nafz, F., Schierl, A., Seebach, H., Reif, W.: Developing self-organizing robotic cells using organic computing principles. In: Meng, Y., Jin, Y. (eds.) Bio-Inspired Self-Organizing Robotic Systems. Studies in Computational Intelligence, vol. 355, pp. 253–273. Springer, Heidelberg (2011). https://doi.org/10.1007/978-3-642-20760-0_11

10. Hummel, G.: Integration von Elektrofahrzeugen in ein Gebäudeenergiemanagementsystem (English: Integration of electric vehicles into a building energy management system). Bachelor thesis, Karlsruhe Institute of Technology (KIT) (2016)

11. IEC 60038:2009, IEC standard voltages. International standard, International Electrotechnical Commission (2009)

12. Ipakchi, A., Albuyeh, F.: Grid of the future. IEEE Power Energ. Mag. 7(2), 52–62 (2009). https://doi.org/10.1109/MPE.2008.931384

13. Jakimovski, B., Meyer, B., Maehle, E.: Swarm intelligence for self-reconfiguring walking robot. In: 2008 IEEE Swarm Intelligence Symposium, pp. 1–8 (2008). https://doi.org/10.1109/SIS.2008.4668286

14. Kochanneck, S.: Systemdienstleistungserbringung durch intelligente Gebäude (English: Provision of system services by intelligent buildings). Ph.D. thesis, Karlsruher Institut für Technologie (KIT) (2019). https://doi.org/10.5445/IR/1000089171. 37.06.01; LK 01

15. Mauser, I.: Multi-modal building energy management. Ph.D. thesis, Karlsruhe Institute of Technology (KIT) (2017). https://doi.org/10.5445/IR/1000070625. 37.06.01; LK 01

16. Mauser, I., Hirsch, C., Kochanneck, S., Schmeck, H.: Organic architecture for energy management and smart grids. In: 2015 IEEE International Conference on Autonomic Computing, pp. 101–108 (2015). https://doi.org/10.1109/ICAC.2015.10

17. Mauser, I., Müller, J., Allerding, F., Schmeck, H.: Adaptive building energy management with multiple commodities and flexible evolutionary optimization. Renew. Energy 87, 911–921 (2016). https://doi.org/10.1016/j.renene.2015.09.003. optimization Methods in Renewable Energy Systems Design

18. Müller, J., Ahrens, M., Mauser, I., Schmeck, H.: Achieving optimized decisions on battery operating strategies in smart buildings. In: Sim, K., Kaufmann, P. (eds.) EvoApplications 2018. LNCS, vol. 10784, pp. 205–221. Springer, Cham (2018). https://doi.org/10.1007/978-3-319-77538-8_15

19. Müller, J., März, M., Mauser, I., Schmeck, H.: Optimization of operation and control strategies for battery energy storage systems by evolutionary algorithms. In: Squillero, G., Burelli, P. (eds.) EvoApplications 2016. LNCS, vol. 9597, pp. 507–522. Springer, Cham (2016). https://doi.org/10.1007/978-3-319-31204-0_33

20. Nafz, F., Ortmeier, F., Seebach, H., Steghöfer, J.-P., Reif, W.: A universal self-organization mechanism for role-based organic computing systems. In: González Nieto, J., Reif, W., Wang, G., Indulska, J. (eds.) ATC 2009. LNCS, vol. 5586, pp. 17–31. Springer, Heidelberg (2009). https://doi.org/10.1007/978-3-642-02704-8_3

21. Ochs, P.: Dezentrale Spannungshaltung in Stromverteilnetzen durch intelligente Gebäude (English: Decentralized voltage maintenance in electricity distribution grids by intelligent buildings). Bachelor thesis, Karlsruhe Institute of Technology (KIT) (2020)

22. OSH website. http://organicsmarthome.fzi.de. Accessed 28 Oct 2024
23. Prothmann, H., Tomforde, S., Branke, J., Hähner, J., Müller-Schloer, C., Schmeck, H.: Organic traffic control. In: Müller-Schloer, C., Schmeck, H., Ungerer, T. (eds) Organic Computing—A Paradigm Shift for Complex Systems. Autonomic Systems, vol. 1, pp. 431–446. Springer, Basel (2011). https://doi.org/10.1007/978-3-0348-0130-0_28
24. Richter, U.M.: Controlled self-organisation using learning classifier systems. KIT Scientific Publishing, Karlsruhe (2009). https://doi.org/10.5445/KSP/1000013138
25. Seljeseth, H., Rump, T., Haugen, K.: Overvoltage immunity of electrical appliances laboratory test results from 60 appliances. In: 21st International Conference on Electricity Distribution (2011). Paper No. 946
26. SMA Solar Technology AG: SMA shifts the phase - Why reactive power is important - and no problem with SMA technology. https://www.sma.de/en/partners/knowledgebase/sma-shifts-the-phase.html. Accessed 28 Oct 2024
27. Steghöfer, J.P., Reif, W.: OC-trust: towards trustworthy organic computing systems. In: Müller-Schloer, C., et al. (eds.) Organic Computing—A Paradigm Shift for Complex Systems, pp. 593–595. Springer, Basel (2011). https://doi.org/10.1007/978-3-0348-0130-0_43
28. Tomforde, S., Prothmann, H., Branke, J., Hähner, J., Mnif, M., Müller-Schloer, C., Richter, U., Schmeck, H.: Observation and control of organic systems. In: Müller-Schloer, C., Schmeck, H., Ungerer, T. (eds) Organic Computing—A Paradigm Shift for Complex Systems. Autonomic Systems, vol. 1, pp. 325–338. Springer, Basel (2011). https://doi.org/10.1007/978-3-0348-0130-0_21
29. Trumler, W., Petzold, J., Bagci, F., Ungerer, T.: AMUN: an autonomic middleware for the smart doorplate project. Pers. Ubiquitous Comput. 10(1), 7–11 (2006). https://doi.org/10.1007/S00779-005-0029-4
30. Yang, Y., Harley, R.G., Divan, D., Habetler, T.G.: Thermal modeling and real time overload capacity prediction of overhead power lines. In: 2009 IEEE International Symposium on Diagnostics for Electric Machines, Power Electronics and Drives, pp. 1–7 (2009). https://doi.org/10.1109/DEMPED.2009.5292772

Specifying and Implementing Interface Moore Machines by a Logic of Actions

Manfred Broy[⊠] [iD]

School of Computation, Information and Technology, Technical University Munich, Munich,
Germany
broy@in.tum.de

Abstract. Interface Moore machines are used as the operational model for interface specifications of systems. A logic of actions (LA) is defined to specify interface Moore machines which implement interface specifications. Specific notations
and methods for specifying, implementing, composing, and verifying interface
Moore machines are introduced. A concurrent composition operator is defined for
Moore machines. It is shown how to refine interface specifications into Moore
machine specifications and further on into Moore machine programs and how
to relate, derive and prove implementations, invariants, and functional interface
specifications for Moore machines. The interface behavior of the concurrent composition of Moore machines is identical to the concurrent composition of their
interface behavior.

Keywords: System Specification and Refinement · Concurrent Composition ·
Moore Machines

1   Introduction

In [5] a specification technique for interface specifications of timed interactive systems in
introduced by which the interface behavior of systems is described. It supports modular
concurrent composition and this way the design of architectures. There it is shown how
such specifications relate closely to Moore machines (see [13]) and that their specific
properties are justified by properties of Moore machines. In the following, we introduce
a specification technique for Moore machines, inspired by Lamport's Temporal Logic
of Actions (TLA, see [11]), by a logic of actions (LA). However, we avoid temporal
logic as used in TLA to express liveness properties. We include liveness properties (see
[2]) explicitly into the specification of Moore machines. Moreover, we use a concept of
strict encapsulation and information hiding leading – in contrast to TLA – to a model
of functional behavior of Moore machines by interface abstraction. Both for Moore
machines and interface specification, concurrent composition is defined, such that the
interface abstraction of the concurrent composition of Moore machines is identical to
the concurrent composition of the interface abstractions of the Moore machines. This
results in a modular approach to the development of concurrent systems.

© The Author(s), under exclusive license to Springer Nature Switzerland AG 2025
G. Ernst et al. (Eds.): Wolfgang Reif Festschrift, LNCS 15765, pp. 262–281, 2025.
https://doi.org/10.1007/978-3-031-92196-4_13

The key idea of this approach to concurrent, distributed, interactive systems is to provide a modular model with explicit concurrency, real time properties, encapsulation, as well as a specification technique with a refinement and verification calculus which is sound and complete, supporting concurrent composition, with adequate abstraction by interface specification and a solid operational model leading to a practical implementation technique in terms of Moore machines.

1.1 Related Work

This paper uses the stream approach of [10]; it is an extension of and based on [5]. There, an extensive theory is presented for the specification of the functionality, the interface behavior, of nondeterministic systems:

- A system interface behavior is a relation between the involved input and output streams. Interface predicates and interface assertions represent such relations.
- The streams considered are timed by an incorporated timescale.
- A key concept for input and output in data flow is causality. It describes the relationship between input and output by indicating which input is causal for which output. This is expressed by a simple formula refining relations on timed streams.
- Strong causality leads to specifications as implemented by Moore machines the functional behavior of which follows the principle of causality.
- Interface Moore machines show fully realizable interface behaviors (see [5]).

In [4] a sound and relative complete calculus for the refinement and verification of interface specifications and their concurrent composition is presented.

1.2 Structure of the Content

In [5] it is shown that interface Moore machines define and calculate fully realizable interface specifications and that every fully realizable interface specification can be described by an interface Moore machine. We briefly repeat the basic concepts of interface specification from [5]. In the following, we introduce interface Moore machines for the implementation of interface specifications and a specification formalism for interface Moore machines and define concurrent composition of Moore machines in this formalism. Specifications of interactive systems are refined into specifications of Moore machines and further into implementations by Moore machines. We show how interface specifications can be derived from interface Moore machines and are proved correct. A number of small examples is given to illustrate the approach.

2 Basic Definitions

In the following, we introduce the basic structures and concepts of our approach. In this and the following section, we introduce a model for concurrent, distributed, interactive systems. By $\mathbb{N}$ we denote the natural numbers including 0, $\mathbb{N}_+ = \mathbb{N}\backslash\{0\}$.

2.1 Sequences, Timed and Untimed Streams and Histories

Let set M be the universe of messages. Let M* denote the set of finite sequences over M which formally can be represented by functions (here $[1{:}n] \subseteq \mathbb{N}_+$ denotes finite intervals of natural numbers for $n \in \mathbb{N}$)

$$M^* = \cup_{n \in \mathbb{N}}([1:n] \to M)$$

By M^ω we denote the set of infinite sequences over M represented by functions:

$$M^\omega = (\mathbb{N}_+ \to M)$$

We consider the set $(M^*)^\omega$ of infinite *timed streams* over a message set M:

$$\left(M^*\right)^\omega = \left(\mathbb{N}_+ \to M^*\right)$$

We write Tstr M for the type of timed streams over data type M. A stream $x \in (M^*)^\omega$ is called *timed*, since we understand $\mathbb{N}_+$ to represent time in terms of an infinite sequence of time intervals. Each time interval is numbered by a number $t \in \mathbb{N}_+$. Then x(t) with t $\in \mathbb{N}_+$ denotes the sequence of messages communicated via the timed stream x in time interval t. The set $M^{*|\omega} = M^* \cup M^\omega$ denotes the set of untimed streams. The set of finite streams is denoted by M*. The set of finite timed streams is denoted by $(M^*)^*$.

For timed streams $x \in (M^*)^\omega$ and $t \in \mathbb{N}$ we denote a *time cut* of length t as follows

$$x \downarrow t \in \left([1:t] \to M^*\right) \quad \text{where for all } n \in \mathbb{N}: \quad (x \downarrow t)(n) = x(n) \Leftarrow 1 \le n \le t$$

$x \downarrow t$ denotes the first t sequences of x, representing the messages transmitted in the first t time intervals. We use the same notation for channel histories (see Sect. 2.1).

By s^z we denote the concatenation of finite sequences (also of finite timed streams) s with finite or infinite sequences (also finite or infinite timed streams resp.) z. For a timed stream $x \in (M^*)^\omega$ we denote by $\overline{x} = x(1)^{\wedge}x(2)^{\wedge}\ldots$ The untimed stream which is the result of the concatenating of all the sequences in x.

By $\sqsubseteq$ the *prefix* order between sequences and also between streams is denoted

$$x \sqsubseteq y \Leftrightarrow \exists z : x^{\wedge}z = y$$

2.2 Syntactic Interfaces and Interface Behavior

Given a set of typed channel names (a family of channels)

$$X = \{x_1 : \text{Tstr } T_1, \ldots, x_n : \text{Tstr } T_n\}$$

$\overrightarrow{X}$ denotes the set of channel histories given by families of timed streams, one timed stream for each of the channels:

$$\overrightarrow{X} = \left(X \to \left(M^*\right)^\omega\right)$$

$\overrightarrow{X}_{fin}$ denotes histories of finite untimed streams. We assume that for each history $x \in \overrightarrow{X}$ the timed stream $x(c_k)$, $1 \le k \le m$, for channel $c_k \in X$ carries messages of type T_k. By

$$\overline{X} = \left(X \rightarrow \left(M^{*|\omega} \right) \right)$$

we denote histories of untimed streams and by $\overline{X}_{fin}$ histories of finite untimed streams.

Given two sets of typed channels

$$X = \{x_1 : \text{Tstr } T_1, \ldots, x_n : \text{Tstr } T_n\} \quad Y = \{y_1 : \text{Tstr } P_1, \ldots, y_m : \text{Tstr } P_m\}$$

the syntactic interface of a system S is denoted by $(X \blacktriangleright Y)$. A system with syntactic interface $(X \blacktriangleright Y)$ takes a history $x \in \overrightarrow{X}$ as input and generates a history $y \in \overrightarrow{Y}$ as output. The behavior and the implementation of a system with syntactic interface $(X \blacktriangleright Y)$ can be described and implemented by an interface Moore machine (see Sect. 4).

2.3 Specifying Interface Behavior

A system with the syntactic interface $(X \blacktriangleright Y)$ is specified by an *interface assertion* which is a logical formula with the channels names in the sets X and Y as free variables for timed streams.

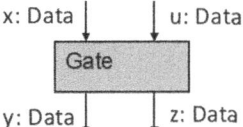

x: Data | u: Data

y: Data | z: Data

Fig. 1. System or service Gate as a data flow node

An example of an interface specification for the system shown in Fig. 1 reads as follows (note that we write x: Data instead of x: Tstr Data to keep the notation short):

IFP Gate = (x: Data, u: Data $\blacktriangleright$ y: Data, z: Data):
$$\forall d \in \text{Data: } d\#x + d\#u = d\#y + d\#z \land (\#x + \#u = \infty \Rightarrow (\#y = \infty \land \#z = \infty))$$

Here $d\#x$ denotes the number of copies of message d in stream $x \in (M^*)^\omega$, we also write $S\#x$ for $S \in M$ to denote the number of copies of elements in set $S \subseteq M$ and $\#x$ for $M\#x$. The specification defines a syntactic interface, in this case (x: Data, u: Data $\blacktriangleright$ y: Data, z: Data).

An interface predicate, in this case Gate, is specified by an interface assertion

$$\forall d \in \text{Data: } d\#x + d\#u = d\#y + d\#z \land (\#x + \#u = \infty \Rightarrow (\#y = \infty \land \#z = \infty))$$

which is a logical expression with the channel names used as identifiers for timed streams. We also write Gate(x, u, y, z) for the interface assertion.

An interface predicate Q for the syntactic interface $(X \blacktriangleright Y)$ is a predicate

$$Q : \overrightarrow{X} \times \overrightarrow{Y} \rightarrow \mathbb{B}$$

The proposition $Q(x, y)$ denotes the result of this predicate applied to the input history $x \in \overrightarrow{X}$ and the output history $y \in \overrightarrow{Y}$. Given a syntactic interface $(X \blacktriangleright Y)$, an interface assertion, which is an expression of higher order predicate logic with free identifiers from X and Y for timed streams, defines an interface predicate. We write $Q::(X \blacktriangleright Y)$ to express that Q is an interface predicate for the syntactic interface $(X \blacktriangleright Y)$.

An interface predicate $Q::(X \blacktriangleright Y)$ is defined by the equation

$$Q = (X \blacktriangleright Y): A$$

where A is an interface assertion for the syntactic interface $(X \blacktriangleright Y)$.

Given two sets of typed channels

$$X = \{x_1 : \text{Tstr } T_1, \ldots, x_n : \text{Tstr } T_n\} \quad Y = \{y_1 : \text{Tstr } P_1, \ldots, y_m : \text{Tstr } P_m\}$$

then (mis-using the order in which the identifiers are listed in the sets X and Y)

$$Q(x_1, \ldots, x_n, y_1, \ldots, y_m)$$

defines the interface assertion for predicate Q.

An interface predicate $Q'::(X \blacktriangleright Y)$ is called a *refinement* of interface predicate $Q::(X \blacktriangleright Y)$ if $Q' \Rightarrow Q$.

Note the difference between this approach based on logic and a semantic model in contrast to the approach of SysML which provides just graphical syntax (see [6]).

2.4 *Focus* – Nondeterministic Real Time Data Flow Systems: The Model

Let the syntactic interface $(X \blacktriangleright Y)$ be given. A function on histories

$$f : \overrightarrow{X} \rightarrow \overrightarrow{Y}$$

is called *strongly causal*, if

$$\forall t \in \mathbb{N}, x, z \in \overrightarrow{X} : x \downarrow t = z \downarrow t \Rightarrow f(x) \downarrow t + 1 = f(z) \downarrow t + 1$$

Then we write $SC[f]$. An interface specification $Q::(X \blacktriangleright Y)$ is called *strongly causal* if

$$\forall t \in \mathbb{N}, x, z \in \overrightarrow{X} : x \downarrow t = z \downarrow t \Rightarrow \{y \downarrow t + 1 : Q(x, y)\} = \{y \downarrow t + 1 : Q(z, y)\}$$

As shown in [5], Chapter 5, there exists a weakest interface predicate, denoted by $Q^{\circledcirc}::(X \blacktriangleright Y)$ for every interface specification $Q::(X \blacktriangleright Y)$ in the set of strongly causal refinements of Q, which is strongly causal (in [5], Chapter 5, an explicit definition for $Q^{\circledcirc}$ is given). The weakest strongly causal refinement Gate$^{\circledcirc}$ of Gate reads as follows:

IFP Gate$^{\circledcirc}$ = (x: Data, u: Data ▶ y: Data, z: Data):
$$\forall\, d \in \text{Data}: d\#x + d\#u = d\#y + d\#z \wedge (\#x + \#u = \infty \Rightarrow (\#y = \infty \wedge \#z = \infty))$$
$$\wedge \, \forall\, t \in \mathbb{N}_+: d\#x \downarrow t + d\#u \downarrow t \geq d\#y \downarrow (t+1) + d\#z \downarrow (t+1)$$

The additional assertion

$$\forall t \in \mathbb{N}+ : d\#x \downarrow t + d\#u \downarrow t \geq d\#y \downarrow (t+1) + d\#z \downarrow (t+1)$$

in Gate$^{©}$ specifies that at each time point t the number of copies of message d in the output streams y and z until time point $t+1$ are not larger than the number of copies of d in input streams x and u till time point t. This indicates that input is causal for output.

An interface specification Q::(X▶Y) is called *realizable*, if there exists a strongly causal function f such that

$$\forall x \in \overrightarrow{X} : Q(x, f(x))$$

The function f is called a *realization* of Q. As shown in [5], there exist Moore machines which implement specification Q if and only if Q is realizable.

System specification Q::(X▶Y) is called *fully realizable*, if it is realizable and if for all $x \in \overrightarrow{X}, y \in \overrightarrow{Y}$ for which Q(x, y) holds there exists a realization f of Q where $y = f(x)$. Formally, interface assertion Q for the syntactic interface (X▶Y) is *fully realizable*, if

$$Q(x, y) = (\exists f : \overrightarrow{X} \to \overrightarrow{Y} : y = f(x) \wedge Q[f])$$

and Q is realizable. The proposition Q[f] holds if f is a realization for Q:

$$Q[f] = (SC[f] \wedge \forall x \in \overrightarrow{X} : Q(x, f(x)))$$

According to [5], the specification Q is fully realizable if and only if there exists a Moore machine with a functional behavior (see Sect. 3) identical to Q.

Systems are implemented (realized) by generalized Moore machines which define the operational model. Interface Moore machines compute interface behaviors, which are fully realizable. Every fully realizable interface assertion Q corresponds to an interface Moore machine which computes Q (see [13]). Every fully realizable predicate Q is a correct and complete logical description of the functional behavior of the most general, the most nondeterministic Moore machines implementing Q.

We define specification Q$^{®}$ for a given specifying predicate Q as follows

$$Q^{®}(x, y) = \exists f : \overrightarrow{X} \to \overrightarrow{Y} : y = f(x) \wedge Q[f]$$

Interface predicate Q$^{®}$ is false if Q is unrealizable and otherwise fully realizable being the weakest refinement of Q in the set of fully realizable refinements of Q (see [5]). In many cases we have $Q^{®} = Q^{©}$. This holds, for instance, for our running example Gate.

Interface predicates Q can be refined by refining them to strongly causal ones or even to fully realizable ones provided they are realizable. Fully realizable interface predicates imply all interface properties that hold for all implementations of Q by Moore machines. Actually, the concurrent composition of the interface behavior of two Moore machines yields a fully realizable interface behavior identical to the functional behavior of the Moore machine being the result of the concurrent composition of the two Moore machines. Therefore, the calculus, which comprises concurrent composition (see next section) and the rule to refine interface predicates into fully realizable ones, is sound and relative complete (see [5]). The calculus and the theory allow us to derive fully realizable interface predicates from given interface predicates by refinement – provided implementations for the interface predicate exist.

2.5 Concurrent Composition with Interaction via Feedback Loops

Concurrent composition means that systems are composed by putting them side by side introducing feedback loops for mutual communication via their channels that fit together. We describe the concurrent composition only for two systems. The generalization to a finite number of systems and thus to large architectures is obvious.

Systems are composed as shown in Fig. 2 where z_1 and z_2 are feedback channels.

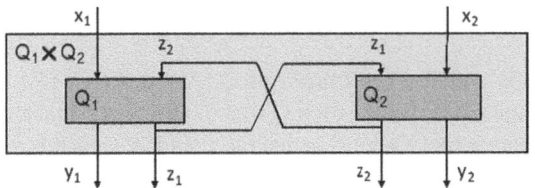

Fig. 2. Composition of systems S1 and S2

Systems with syntactic interfaces $(X_k \blacktriangleright Y_k)$ for $k = 1, 2$ are called *composable*, if the channel types of their sets of channels are consistent and their sets of output channels are disjoint: $Y_1 \cap Y_2 = \emptyset$.

We define the composition of composable systems interface specifications with feedback which corresponds to the composition of Moore machines with feedback (see Sect. 5). Concurrent composition of two systems with behaviors described by interface assertions is specified by logical conjunction of the interface assertions. Feedback channels may stay visible or become hidden by existential quantification.

Consider composable interface predicates $Q_k = (X_k \blacktriangleright Y_k): A_k$ (where the A_k are the specifying interface assertions) for $k = 1, 2$. Define

$$X = (X_1 \cup X_2) \backslash Z \quad \textit{input channels for the composed system}$$
$$Y = Y_1 \cup Y_2 \qquad \textit{output channels for the composed system}$$
$$Z = Y \cap (X_1 \cup X_2) \textit{ feedback channels } - \textit{ internal channels}$$

Z denotes the set of feedback channels in the composition (see Fig. 2). For the specifications Q_k let

$$Q_k^{\circledR} = (X_k \blacktriangleright Y_k): B_k$$

denote their refinement to full realizability (or to false) for $k = 1, 2$, as described in Sect. 2.4 (B_k is the interface assertion for $Q_k^{\circledR}$). Concurrent composition is defined by

$$(Q_1 \times Q_2) = (X \blacktriangleright Y): (B_1 \wedge B_2)$$

Replacing specification Q_k by $Q_k^{\circledR}$ in the definition of concurrent composition is justified by the fact that $Q_k^{\circledR}$ denotes the strongest specification which is fulfilled by all interface Moore machines that implement Q_k. Thus $Q_k^{\circledR}$ is implemented by the same set of Moore machines as Q_k but expresses explicitly additional properties that hold for all its implementations. From $Q_k^{\circledR}$ we can deduce all the functional properties that hold for all Moore machines implementing Q_k. As shown in [5], strongly causal functions have unique fixpoints. This is the reason why concurrent composition can be defined without referring to specific partial orders for defining least fixpoints.

We get for the given interface assertions A_k (since $B_k \Rightarrow A_k$)

$$(Q_1 \times Q_2) \Rightarrow (X \blacktriangleright Y): (A_1 \wedge A_2)$$

The specifying assertion $A_1 \wedge A_2$ holds for the composite system.

We use our running example to demonstrate the concept of concurrent composition.

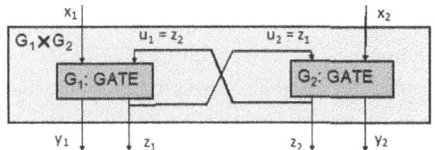

Fig. 3. Composition of two gates

It is specified by the concurrent composition of the interface specification of two gates as shown in Fig. 3:

IFP TG = (x_1: Data, x_2: Data $\blacktriangleright$ y_1: Data, z_1: Data, y_2: Data, z_2: Data):
$$\text{Gate}^{\circledR}(x_1, z_2, y_1, z_1) \wedge \text{Gate}^{\circledR}(x_2, z_1, y_2, z_2)$$

Unfolding $\text{Gate}^{\circledR}(x_1, z_2, y_1, z_1)$ and $\text{Gate}^{\circledR}(x_2, z_1, y_2, z_2)$ delivers the interface predicate (recall $\text{Gate}^{\circledR} = \text{Gate}^{\copyright}$)

IFP TG = (x_1: Data, x_2: Data $\blacktriangleright$ y_1: Data, z_1: Data, y_2: Data, z_2: Data):
$$\forall\, d \in \text{Data:} \quad d\#x_1 + d\#z_2 = d\#y_1 + d\#z_1 \wedge (\#x_1 + \#z_2 = \infty \Rightarrow (\#y_1 = \infty \wedge \#z_1 = \infty))$$
$$\wedge\, d\#x_2 + d\#z_1 = d\#y_2 + d\#z_2 \wedge (\#x_2 + \#z_1 = \infty \Rightarrow (\#y_2 = \infty \wedge \#z_2 = \infty))$$
$$\wedge\, \forall\, t \in \mathbb{N}_+: d\#x_1\!\downarrow t + d\#z_2\!\downarrow t \geq d\#y_1\!\downarrow(t+1) + d\#z_1\!\downarrow(t+1)$$
$$\wedge\, d\#x_2\!\downarrow t + d\#z_1\!\downarrow t \geq d\#y_2\!\downarrow(t+1) + d\#z_2\!\downarrow(t+1)$$

A simple property which we prove from assertion TG(x_1, x_2, y_1, z_1, y_2, z_2) is as follows

$$\forall d \in \text{Data: } d\#x_1 + d\#x_2 = 0 \Rightarrow d\#y_1 + d\#z_1 + d\#y_2 + d\#z_2 = 0$$

It can be proved by straightforward induction on the timing of the streams. This property cannot be proved from Gate(x_1, z_2, y_1, z_1) $\wedge$ Gate(x_2, z_1, y_2, z_2) since z_1 and z_2 are not causal. This demonstrates that assertion $\text{Gate}^{\circledR}(x_1, z_2, y_1, z_1) \wedge \text{Gate}^{\circledR}(x_2, z_1, y_2, z_2)$ is logically stronger than assertion Gate(x_1, z_2, y_1, z_1) $\wedge$ Gate(x_2, z_1, y_2, z_2). Note that TG is fully realizable since $\text{Gate}^{\circledR}$ is fully realizable (see [5]). As proved in [5], concurrent composition of fully realizable predicates yields fully realizable predicates.

The presented theory is the basis for a proof calculus that is sound and relatively complete [4]. It works as follows: Given interface predicates Q specifying the properties forming the nuclei of systems, we refine the predicate Q into $Q^{\circledR}$ by full realizability. Actually, in nearly all practical cases, it is sufficient to refine Q by strong causality (see [5]) if the result $Q^{\circledR}$ is fully realizable anyway (as explained this applies in the case of the interface specification Gate).

3 Interface Moore Machines

In this section, a formal model for interface Moore machines and their concurrent composition is introduced.

3.1 Defining Interface Moore Machines

An interface Moore machine is a Moore machine (see [13]) with input and output as described by a syntactic interface (X▶Y). For syntactic interface (X▶Y), a generalized nondeterministic (total) interface Moore machine with state space Σ is a pair (Δ, Λ) where.

- $\Lambda \subseteq \Sigma$ is a *nonempty set of initial states*
- Δ is a *total state transition function*

$$\Delta : \left(\Sigma \times \overline{X}_{\text{fin}} \right) \to \wp\left(\Sigma \times \overline{Y}_{\text{fin}} \right) \setminus \{\emptyset\}$$

where for $a \in \overline{X}_{\text{fin}}, b \in \overline{Y}_{\text{fin}}, \sigma, \sigma' \in \Sigma$

$$\left(\sigma', b \right) \in \Delta(\sigma, a)$$

and the output b does not depend on the input a but only on the state σ. Formally

$$\forall a, a' \in \overline{X}_{\text{fin}}, b \in \overline{Y}_{\text{fin}} : \left(\exists \sigma' \in \Sigma : \left(\sigma', b \right) \in \Delta(\sigma, a) \right) \Leftrightarrow \left(\exists \sigma' \in \Sigma : \left(\sigma', b \right) \in \Delta(\sigma, a') \right)$$

We write $(\Delta, \Lambda)::(X▶Y)$ to indicate that (Δ, Λ) is an interface Moore machine for syntactic interface (X▶Y). Given a Moore machine (Δ, Λ) for syntactic interface (X▶Y) with state space Σ a computation is given by.

- an infinite stream of states $\{\sigma_i \in \Sigma: i \in \mathbb{N}_+\}$
- an input history of $x \in \overrightarrow{X}$
- an output history $y \in \overrightarrow{Y}$

such that $\sigma_1 \in \Lambda$ (σ_1 is an initial state) and

$$\forall i \in \mathbb{N}_+ : (\sigma_{i+1}, y(i)) \in \Delta(\sigma_i, x(i))$$

This way a Moore machine defines an interface predicate:

$$[\![\Delta, \Lambda]\!] : \overrightarrow{X} \times \overrightarrow{Y} \to \mathbb{B}$$

where for $x \in \overrightarrow{X}, y \in \overrightarrow{Y}$

$$[\![\Delta, \Lambda]\!](x, y) = \exists \sigma \in \overline{\Sigma} : \forall i \in \mathbb{N}_+ : (\sigma(i + 1), y(i)) \in \Delta(\sigma(i), x(i))$$

which is the result of interface abstraction: the information hiding of state space.

A Moore machine M' is called a *functional refinement* of Moore machine M if

$$[\![M']\!] \Rightarrow [\![M]\!]$$

As shown in [5], for every Moore machine M its interface abstraction $[\![M]\!]$ is fully realizable and thus also strongly causal. Therefore, if interface Moore machine M fulfills Q then interface Moore machine M fulfills $Q^{\copyright}$ and $Q^{\circledR}$ (see [5] and Sect. 6).

3.2 Concurrent Composition of Interface Moore Machines

As defined in Sect. 2.5 two syntactic interfaces $(X_k \blacktriangleright Y_k)$ for $k = 1, 2$, are called *composable* if $Y_1 \cap Y_2 = \emptyset$ and the channels in $X_1 \cup X_2$ and $Y_1 \cup Y_2$ have consistent types. Channels in $(X_1 \cup X_2) \cap (Y_1 \cup Y_2)$ are (called) feedback channels. Moore machines as well as interface predicates are called *composable*, if their syntactic interfaces are composable – composability is a syntactic notion.

Moore machines $(\Delta_k, \Lambda_k)::(X_k \blacktriangleright Y_k)$, $k = 1, 2$, with composable syntactic interfaces, are composed concurrently to a composite Moore machine with feedback

$$(\Delta, \Lambda)::(X \blacktriangleright Y)) \quad \text{where} \quad X = (X_1 \cup X_2) \backslash Y, Y = Y_1 \cup Y_2,$$

defined by

$$(\Delta, \Lambda) = ((\Delta_1, \Lambda_1) \times (\Delta_2, \Lambda_2))$$
$$\Sigma = (\Sigma_1 \times \Sigma_2)$$
$$\Lambda = \{(\sigma_1, \sigma_2) : \sigma_1 \in \Lambda_1 \wedge \sigma_2 \in \Lambda_2\}$$
$$\Delta : \left(\Sigma \times \overline{X}_{\text{fin}}\right) \rightarrow \wp\left(\Sigma \times \overline{Y}_{\text{fin}}\right) \backslash \{\emptyset\}$$

$$\Delta((\sigma_1, \sigma_2), x)$$
$$= \left\{((\tau_1, \tau_2), y) : (\tau_1, y|Y_1) \in \Delta_1(\sigma_1, x \oplus y|X_1) \wedge (\tau_2, y|Y_2) \in \Delta_2(\sigma_2, x \oplus y|X_2)\right\}$$

Here we write $y|Y_1$ for the restriction of mapping y to the set Y_1 and we write $x_1 \oplus x_2$ where $x_1 \in \overrightarrow{X}_1$ and $x_2 \in \overrightarrow{X}_2$ (where X_1 and X_2 are disjoint sets of channels) for the history in $\overrightarrow{X}$ (where $X = X_1 \cup X_2$) defined by

$$c \in X_1 \Rightarrow (x_1 \oplus x_2)(c) = x_1(c) \quad c \in X_2 \Rightarrow (x_1 \oplus x_2)(c) = x_2(c)$$

Note the definition of concurrent composition is consistent for syntactically composable Moore machines; for them there exists always the Moore machine as defined by concurrent composition. However, this does not apply for Mealy machines (see [12]), in general! There does not exist such a straightforward definition of concurrent composition with feedback for general Mealy machines. Using the defining equations for concurrent composition for Moore machines naively for general Mealy machines would lead to an inconsistent definition.

4   Logic of Actions for Interface Moore Machines

We specify a Moore machine for the syntactic interface $(X \blacktriangleright Y)$ in a more specification and programming oriented style by defining its state space, its set of initial states, and its state transition function by a kind of programming notation. The state space is defined by a set of typed program variables

$$V = \{v_1 : W_1, \ldots, v_k : W_k\}$$

Here W_i is the data types of variable v_i. By $\widehat{V}$ we denote the set of valuations for each of the programming variables by elements of the specified type. The state space Σ is defined by

$$\Sigma = \widehat{V} \times \overline{Y}_{fin}$$

Here we choose to include the output $y \in \overline{Y}_{fin}$ as part of the state since, according to the properties of Moore machines, the output y can directly be computed from the state. A state transition then reads as follows

$$((v, y), y^\circ) \in \Delta((v^\circ, y^\circ), x)$$

where (v°, y°) denotes the current state, x denotes the current input and y° denotes the current output while (v, y) denotes the next state and thus y denotes the next output.

Let init: $\widehat{V} \times \overline{Y}_{fin} \to \mathbb{B}$ be a predicate on the state space. We specify the set of initial states by the initial state assertion init(v, y) as follows

$$\left\{ (v, y) \in \widehat{V} \times \overline{Y}_{fin} : init(v, y) \right\}$$

Here we assume that this set is not empty. We specify the state transition relation by an assertion with free variables from the sets V°, Y°, X, V, and Y. The sets V° and Y° rename V and Y. They contain for every variable $z \in V$ (or $z \in Y$) the variable $z^\circ \in V^\circ$ (or $z^\circ \in Y^\circ$). The variables in V° and Y° denote the state in which the transition is activated (the "old" state and the "old" output), the identifiers in X denote the current input, and the variables in V and Y denote the post-state (the new state and the next output). In the assertion the variables from X, Y, and Y° stand for finite sequences of the specified type, which are the sequences of messages communicated in the respective time interval. Let sta be the state transition predicate

$$sta: \widehat{V^\circ} \times \overline{Y^\circ}_{fin} \times \overline{X}_{fin} \times \widehat{V} \times \overline{Y}_{fin} \to \mathbb{B}$$

which specifies the state transition relation. We define the state transition function Δ by the state transition assertion sta$(v^\circ, y^\circ, x, v, y)$ as follows

$$((v, y), y^\circ) \in \Delta((v^\circ, y^\circ), x)$$

$$\Leftrightarrow$$

$$\left(sta(v^\circ, y^\circ, x, v, y) \vee \left((v, y) = (v^\circ, y^\circ) \wedge \neg \exists v' \in \widehat{V}, y' \in \overline{Y}_{fin} : sta(v^\circ, y^\circ, x, v', y') \right) \right)$$

where sta($v°$, $y°$, x, v, y) is called the *state transition assertion*. It specifies an action (see Sect. 7). This interpretation expresses that if a state transition is not enabled, the state remains unchanged. This guarantees that the state transition function is total. If

$$\forall v° \in \widehat{V}, y° \in \overline{Y}_{fin}, x \in \overline{X}_{fin} \; : \exists v \in \widehat{V}, y \in \overline{Y}_{fin} : sta\big(v°, y°, x, v, y\big)\big)$$

holds, then the definition above simplifies to

$$\big((v, y), y°\big) \in \Delta\big((v°, y°), x\big) \Leftrightarrow sta\big(v°, y°, x, v, y\big)$$

Similar to TLA, we describe a Moore machine MM by a syntactic scheme of actions:

MoMa MM = (X ▶ Y): **State** V **Init** Init **Transition** Trans

Here Init is an assertion specifying the initial state and assertion Trans specifies the state transition relation. Similar to interface predicates we describe the initial state assertion and the state transition assertion by assertions that contain the state variables and the channels as free variables as well as the variables and the output channels in $V°$ and $Y°$. The channel variables denote sequences of the respective type.

As a first simple example we specify a Moore machine simply forwarding its input

MoMa Forward = (x: M ▶ y: M): **State** void **Init** y = $\langle\rangle$ **Transition** y = x

Note that in this case "void" indicated that we do not need to introduce local state variables but only use the output channel y as a state attribute.

Forwarding messages with some unspecified time delay is specified by

MoMa Delayed_Forward = (x: M ▶ y: M):
    State   z : Seq M
    Init   z = $\langle\rangle \wedge$ y = $\langle\rangle$
    Transition   y$\hat{\;}$z = z°$\hat{\;}$x $\wedge$ (#z°$\hat{\;}$x > 0 $\Rightarrow$ #y > 0)

Note that y denotes the output in the next step. Here we use a brute force progress property that makes sure that every input eventually occurs as output.

Of course, we could use here the idea of TLA to define liveness properties (see [2]) by some kind of fairness assumptions formally expressed by temporal logic. We do not introduce temporal logic here and represent liveness conditions directly by the state and the state transition assertion (as an alternative to temporal logic).

We continue with a number of quite simple examples. To use our technique for more complicated ones, see the Moore machine specifications at the end of this section.

A Moore machine implementing a buffer is specified as follows

MoMa Forgetful_Buffer = (x: Data, r: {req} ▶ y: Data):
    State   b : Seq Data
    Init   b = $\langle\rangle \wedge$ y = $\langle\rangle$
    Transition   y$\hat{\;}$b = b°$\hat{\;}$x $\wedge$ #y = min{#r, #b°$\hat{\;}$x}

This form of a buffer returns received messages as required by the request channel r but ignores and forgets requests for which not enough messages are currently available. A buffer remembering requests which cannot be served immediately reads as follows

MoMa Mindful_Buffer = (x: Data, r: {req} ▶ y: Data):

State	s : Seq Data, n : Nat
Init	$s = \langle \rangle \wedge y = \langle \rangle \wedge n = 0$
Transition	$y\hat{}s = s°\hat{}x \wedge \#y = \min\{n°+\#r, \#s°\hat{}x\} \wedge n = n°+\#r-\#y$

A store which is more flexible and more nondeterministic than the buffer is a data pool:

MoMa Data_Pool = (x: Data, r: {req} ▶ y: Data):

State	s : Seq Data, n : Nat
Init	$s = \langle \rangle \wedge y = \langle \rangle \wedge n = 0$
Transition	$\forall d \in$ Data: $d\#(y\hat{}s) = d\#(s°\hat{}x)$
	$\wedge \#y = \min\{n°+\#r, \#s°\hat{}x\} \wedge n = n°+\#r-\#y$

The data pool is not fair. It may happen that some data is never produced as output. Only if the data pool is made empty from time to time this is definitively avoided. We may introduce an explicit fairness condition by a prophecy variable (see fair merge below) for specifying that every input is eventually returned as output provided enough requests arrive.

Finally, we specify the Moore machine MM_Gate:

MoMa MM_Gate = (x: Data, u: Data ▶ y: Data, z: Data):

State	n : Int
Init	$n = 0 \wedge y = \langle \rangle \wedge u = \langle \rangle$
Transition	$\forall d \in$ Data: $d\#y+d\#z = d\#x+d\#u \wedge \#y-\#z = n-n° \wedge -1 \leq n \leq 1$

Actually, the interface behavior of Moore machine MM_Gate is a refinement of the interface specification Gate (for a proof see Sect. 6).

Next we define the merging of two streams and begin with simple fair merge

MoMa Simple_Fair_Merge = (x: Data, z: Data ▶ y: Data):

State	void
Init	$y = \langle \rangle$
Transition	$y \in \text{mix}(x, z)$

The function mix delivers the set of all merges of two finite sequences. It is defined for sequences r and s and elements d and e by

$$\text{mix}(\langle \rangle, s) = \text{mix}(s, \langle \rangle) = \{s\}$$
$$\text{mix}(\langle d \rangle\hat{}r, \langle e \rangle\hat{}s) = \{\langle d \rangle\hat{}v : v \in \text{mix}(r, \langle e \rangle\hat{}s)\} \cup \{\langle e \rangle\hat{}v : v \in \text{mix}(\langle d \rangle\hat{}r, s)\}$$

Simple_Fair_Merge is a simple implementation of fair merge where all received input is produced as output in the very next step.

We get a more general fair merge by choosing the state space $\hat{V} = \vec{Y} \times \vec{Y}$:

MoMa Fair_Merge = (x: Data, z: Data ▶ y: Data):

State	p, q : Tstr Data
Init	#p = 0 ∧ #q = 0 ∧ y = ⟨⟩
Transition	y = mix(p(1), q(1))

$$\wedge\ \overline{p} = \overline{rt(p°)}\widehat{\ }x \wedge rt(p°) \precsim p \wedge \overline{q} = \overline{rt(q°)}\widehat{\ }z \wedge rt(q°) \precsim q$$

The used auxiliary functions and logical relations are defined as follows (let r be a sequence or an element and z be finite or infinite sequence, for instance a timed or an untimed stream):

$$ft\big(\langle r\rangle^{\wedge}z\big) = r \quad rt\big(\langle \mathbf{r}\rangle^{\wedge}z\big) = z$$

The relation $x \precsim y$ holds for timed streams x and y if all the messages in x occur in the beginning of y at the same time in the same order. It is formally defined as follows

$$x \precsim y = \exists t \in \mathbb{N} : x \downarrow t = y \downarrow t \wedge x(t+1) \sqsubseteq y(t+1) \wedge \forall n \in \mathbb{N}_+ : (n > t+1 \Rightarrow x(n) = \langle\rangle)$$

The interface behavior of Moore machine FairMerge is a general fair merge as specified by the interface predicate

IFP FairMerge = (x: Data, z: Data ▶ y: Data): ∃ x', z' ∈ (Data*)ᵒ:
$$x' \in delay(x) \wedge z' \in delay(z) \wedge \forall\ d \in Data, t \in \mathbb{N}_+: y(t) = mix(x'(t), z'(t))$$

where for timed streams x the set delay(x) which contains all time shifts of x is specified as follows

$$delay(x) = \Big\{ z \in \big(Data\ ^* \big)^{\omega} : \forall t \in \mathbb{N} : \overline{z \downarrow t+1} \sqsubseteq \overline{x \downarrow t} \wedge \overline{z} = \overline{x} \Big\}$$

This illustrates the power of the specification framework due to the flexibility of higher order predicate logic.

5 Concurrent Composition of Interface Moore Machines

Given two composable interface Moore machines described by the logic of actions. We require that the sets V_1 and V_2 of their state variables are disjoint (no shared memory),

MoMa MM_k = $(X_k \blacktriangleright Y_k)$: **State** V_k **Init** $Init_k$ **Transition** $Trans_k$

We define their concurrent composition leading to a composite Moore machine described in the logic of actions. We define the composed machine ConComp as follows:

MoMa ConComp = $(X \blacktriangleright Y)$: **State** V_1, V_2 **Init** $Init_1 \wedge Init_2$ **Transition** $Trans_1 \wedge Trans_2$

The fact that the behavior of Moore machines is specified by state transition assertions makes it easy to define concurrent composition.

We derive the specification of the Moore machine MMTG formed by the concurrent composition of Moore machines MM_Gate[x_1, $z_2 \blacktriangleright y_1$, z_1] and MM_Gate[x_2, $z_1 \blacktriangleright y_2$, z_2] according to Fig. 3. Here we write MM_Gate[x_1, $z_2 \blacktriangleright y_1$, z_1] for the Moore machine with syntactic interface (x_1, $z_2 \blacktriangleright y_1$, z_1) and interface assertion MM_Gate(x_1, z_2, y_1, z_1). Along the lines shown in Sect. 6 it is proved that MMTG realizes TG:

MoMa MMTG = (x_1: Data, x_2: Data $\blacktriangleright$ y_1: Data, z_1: Data, y_2: Data, z_2: Data):

State	n_1 : Int, n_2 : Int
Init	$n_1 = 0 \wedge n_2 = 0 \wedge \#y_1 = 0 \wedge \#y_2 = 0 \wedge \#z_1 = 0 \wedge \#z_2 = 0$
Transition	$\#y_1 + \#z_1 = \#x_1 + \#z_2 \wedge \#y_1 - \#z_1 = n_1 - \overset{\circ}{n_1} \wedge -1 \leq n_1 \leq 1$
	$\wedge \; \#y_2 + \#z_2 = \#x_2 + \#z_1 \wedge \#y_2 - \#z_2 = n_2 - \overset{\circ}{n_2} \wedge -1 \leq n_2 \leq 1$

Note that the state transition assertion is a logical formula from which we may deduce further formulas as shown in the following section.

6 Proving Invariants and Interface Properties of Interface Moore Machines

Given a Moore machine specification, we derive and prove properties about their functional behavior in terms of interface predicates and assertions. We illustrate this for the example MM_Gate without giving all details or working out the general proof theory which is rather straightforward and can be understood from the example MM_Gate:

MoMa MM_Gate = (x: Data, u: Data $\blacktriangleright$ y: Data, z: Data):

State	n : Nat
Init	$n = 0 \wedge y = \langle\rangle \wedge z = \langle\rangle$
Transition	$\forall \; d \in$ Data: $d\#y + d\#z = d\#x + d\#u \wedge \#y - \#z = n - n^\circ \wedge -1 \leq n \leq 1$

We show rules that allow us to prove invariants about the states. An invariant holds, if it can be proved from the initial assertion and provided it holds for old state variables and the old output it can be derived with the help of the state transition relation for the state variables and the output after the transition. For instance, we immediately prove this way that $-1 \leq n \leq 1$ is an invariant for the Moore machine MM_Gate.

More interesting properties are proved by deriving interface predicates from the specifications of interface Moore machines. In the initial state assertion we replace variables c for the state attributes and the output channels by c(1). In the state transition assertion we replace for every output channel c and for every state attribute c the identifier c° by c(t), for every input channel x the identifier x by x(t) and for every output channel c and every state attribute c the identifier c by c(t + 1).

We get from the initial state assertion of the Moore machine MM_Gate for the timed streams $n \in \mathbb{N}^\omega$, y, z $\in$ (Data*)$^\omega$:

$$n(1) = 0 \wedge \#y(1) = \langle\rangle \wedge \#z(1) = \langle\rangle$$

and from the state transition assertion the assertion for the streams x, u, y, z that holds for all times $t \in \mathbb{N}_+$ and all data $d \in Data$.

$$d\#y(t+1) + d\#z(t+1) = d\#x(t) + d\#u(t) \wedge \#y(t+1) - \#z(t+1)$$
$$= n(t+1) - n(t) \wedge -1 \le n(t+1) \le 1$$

From this assertions we prove by induction on $t \in \mathbb{N}_+$ for the timed streams x, z, y:

$$\forall t \in \mathbb{N}_+ : -2 \le \#y \downarrow t - \#z \downarrow t \le 2 \wedge d\#y + d\#z = d\#x + d\#u$$

The proof is quite straightforward and left to the reader. From the derived assertion we prove the liveness condition:

$$\#x + \#u = \infty \Rightarrow (\#y = \infty \wedge \#z = \infty)$$

This assertion is shown by first proving the assertion

$$\forall t \in \mathbb{N}+ : \#x \downarrow t + 1 + \#u \downarrow t + 1 = \#y \downarrow t + z \downarrow t$$

by induction from the derived assertion above. Moreover, if $\#x + \#u = \infty$ then by the assertion above we conclude $\#y + \#z = \infty$ and since

$$-2 \le \#y \downarrow t - \#z \downarrow t \le 2$$

we conclude

$$\#y = \infty \wedge \#z = \infty$$

This concludes the proof that Moore machine MM_Gate fulfills the liveness condition of Gate. Thus for the interface predicate $[\![MM\_Gate]\!]$ we conclude

$$[\![MM\_Gate]\!] \Rightarrow Gate$$

Note that we can apply the theory worked out in [5]; every interface behavior of a Moore machine is fully realizable. Therefore, if we prove for a Moore machine an interface predicate Q then the Moore machine also fulfills the predicates $Q^{\copyright}$ and $Q^{\circledR}$ since these are the weakest predicates that are refinements of Q and that are strongly causal and fully realizable respectively. Thus, from $[\![MM\_Gate]\!] \Rightarrow Gate$ we immediately conclude (since $Gate^{\circledR}$ holds for all state machines that implement Gate).

$$[\![MM\_Gate]\!] \Rightarrow Gate^{\circledR}$$

This shows how the theory simplifies the verification of proof obligations. To prove a specifying predicate $Q^{\circledR}$ for a Moore machine it is sufficient to prove Q.

For the concurrent composition of Moore machines we apply this theory. Since

$$[\![MM\_Gate]\!](x_1, z_2, y_1, z_1) \Rightarrow Gate^{\circledR}(x_1, z_2, y_1, z_1)$$
$$[\![MM\_Gate]\!](x_2, z_1, y_2, z_2) \Rightarrow Gate^{\circledR}(x_2, z_1, y_2, z_2)$$

we conclude, due to the fact that the concurrent composition of Moore machines which are refinements of interface predicates is a refinement of the concurrent composition of the interface predicates

$$MMTG \Rightarrow TG$$

This shows how we prove from specifications of Moore machines properties of their interface behavior and further on from the concurrent composition of these interface behaviors the properties of the interface behaviors of the composed Moore machines. This leads to a modular development of systems with encapsulation of the states of Moore machines, information hiding of the states by interface abstraction of Moore machines and the derivation of the interface properties of architectures as the result of the concurrent composition of the interface abstractions of the Moore machines.

7 Implementing Interface Moore Machines

We implement interface Moore machines by statements for the state transition assertions. We demonstrate this by our running example Gate. We use n, x, u, y, z as programming variables in the following statement called PRG formulated in Dijkstra's guarded command language (see [8])

$$\{ y, z := \langle\rangle, \langle\rangle;$$

$$\textbf{do } n \leq 0 \wedge \#x > 0 \ \rightarrow x, y, n := rt(x), y^\wedge\langle ft(x)\rangle, \ n+1$$
$$[] \quad n \geq 0 \wedge \#x > 0 \ \rightarrow x, z, n := rt(x), z^\wedge\langle ft(x)\rangle, \ n-1$$
$$[] \quad n \leq 0 \wedge \#x > 0 \ \rightarrow x, y, n := rt(x), y^\wedge\langle ft(x)\rangle, \ n+1$$
$$[] \quad n \geq 0 \wedge \#x > 0 \ \rightarrow x, z, n := rt(x), z^\wedge\langle ft(x)\rangle, \ n-1 \qquad\qquad \text{PRG}$$
$$[] \quad n \leq 0 \wedge \#u > 0 \ \rightarrow u, y, n := rt(u), y^\wedge\langle ft(u)\rangle, \ n+1$$
$$[] \quad n \geq 0 \wedge \#u > 0 \ \rightarrow u, z, n := rt(u), z^\wedge\langle ft(u)\rangle, \ n-1$$
$$[] \quad n \leq 0 \wedge \#u > 0 \ \rightarrow u, y, n := rt(u), y^\wedge\langle ft(u)\rangle, \ n+1$$
$$[] \quad n \geq 0 \wedge \#u > 0 \ \rightarrow u, z, n := rt(u), z^\wedge\langle ft(u)\rangle, \ n-1 \ \textbf{od} \ \}$$

We show that this statement implements the state transition assertion (here x° and u° are used to denote the values of x and u before the transition starts)

$$\forall d \in \text{Data: } d\#y + d\#z = d\#x^\circ + d\#u^\circ \wedge \#y - \#z = n - n^\circ \wedge -1 \leq n \leq 1$$

To prove that PRG implements the state transition assertion we show the following post condition (where n, x, u, y, z denote the values of these program variables after executing PRG) for PRG:

$$\forall d \in \text{Data: } d\#y + d\#z = d\#x^\circ + d\#u^\circ \wedge \#y - \#z = n - n^\circ \wedge -1 \leq n \leq 1$$

The proof is carried out by proving that the following assertion called INV

$$\{\forall d \in \text{Data: } d\#y + d\#z + d\#x + d\#u = d\#x^\circ + d\#u^\circ \wedge \#y - \#z = n - n^\circ \wedge -1 \leq n \leq 1\} \quad \text{INV}$$

is an invariant for the loop in PRG. The proof is rather straightforward and thus is left to the reader.

Finally note that INV holds as a precondition for the loop. Assertion $-1 \leq n \leq 1$ holds always before the state transition is executed. Thus, assertion INV holds as a precondition for the loop since precondition $y = \langle \rangle \wedge z = \langle \rangle \wedge -1 \leq n \leq 1$ holds.

Since the loop always terminates in a state where $\#x + \#u = 0$ holds we derive from INV the expected postcondition

$$\left\{ \forall d \in \ Data: d\#y + d\#z = d\#x^\circ + d\#u^\circ \wedge \#y - \#z = n - n^\circ \wedge -1 \leq n \leq 1 \right\}$$

for PRG. This proves that Moore machine MM_Gate_Prg based on program PRG

MoMaProg MM_Gate_Prg = (x: Data, u: Data ▶ y: Data, z: Data):
State	n : Nat
Init	n, y, z := 0, $\langle \rangle$, $\langle \rangle$
Action	PRG

correctly implements the state transition specification of Moore machine MM_Gate. As shown in Sect. 6, MM_Gate is correct for Gate; thus we conclude that the program MM_Gate_Prg is correct for Gate.

The proof can also be carried out in Hehner's logic of predicative programming (see [9]) taking the state transition assertion as specification of the state transition action, in our example the program PRG. Note that the program based implementation MM_Gate_Prg of a Moore machine directly corresponds to a "runnable" being a schedulable thread at the level of operating systems.

8 Discussion and Related Work

What is presented lifts the Focus approach to the architectural level. The examples in the paper are quite small. For applying the method to cyber-physical systems see [4]. There the fixed time steps of the streams may have to be extended for real-time systems. For a comparison with data-flow-based synchronous languages, like Lustre, that seem to offer a similar support for concurrent composition see the extended discussion in [6].

Compared to TLA, the present does not introduce interleaving of statements nor fairness assumptions expressed by temporal logic for our operational semantics in terms of Moore machines. This way we avoid a number of complications which might make it more difficult for practical engineers who are to use the approach. This way we avoid temporal logic completely. Engineers who use the presented approach have to understand only the idea of streams, in particular, the idea of timed streams. There is no implicit assumption about fairness, which as soon as one has to argue about the correctness of programs, would have to be made explicit – say by fairness assumptions expressed, for instance, by temporal logic.

An interesting question is – in this context – how to represent liveness conditions. In the presented approach we can formulate liveness conditions in all variations and have not to stick to fairness only. For interface predicates they are just part of the interface assertion sometimes referring explicitly to infinite streams and general to infinity. At the level of Moore machines liveness can be delt with by prophecy variables. This enables a lot of flexibility.

A second key concept is the notion of refinement and the notion of abstraction. In some sense, they are dual to each other. Let us just talk about refinement: as shown, to specify the interface behavior of a system, which is at the same time the ultimate require-ment specification of all functional requirements including real time properties, we may concentrate to begin with only essential properties and ignore issues like causality and realizability. We speak of a *nucleus* of the specification. Even this nucleus can be worked out by a sequence of steps of refinement introducing more and more relevant properties explicitly.

Moreover, we may structure the specification according to a number of sub-services which are offered by the system. We did not demonstrate this in this paper, the foundations for this idea can be found in [1].

Having an interface specification for the system, we have two possibilities: either we use the system in a bottom-up mode by a concurrent composition with other systems to produce a more powerful composite system, or we can use the system specification in a top-down approach by decomposing it into a number of subsystems which offer the required functionality by their concurrent cooperation. An example demonstrating how to do that is given in [3].

In any case, a key question is how to go on from an interface specification to an implementation. This is basically carried out in two steps: One step of refinement goes from the given interface specification to a fully realizable specification and then we go on to a Moore machine representation. Each of these steps have to be effective refinement steps. Of course, it is possible to go on from an interface specification which is not fully realizable, but just realizable to a state machine implementation. Note that for a non-deterministic interface specification there exists a rich set of implementations, as demonstrated in [5].

One question is why we use Moore machines. The answer is simple: Moore machines show a strongly causal, fully realizable functional behavior. For them concurrent com-position can be defined in a straightforward way coherent to concurrent composition of interface specifications. Finally, we can work out Moore machine specifications, and going on, we can give a program to implement it along the line shown in Sect. 7. A key property of the approach is its *modularity*. For each subsystem implemented by a Moore machine we have to prove its correctness for its interface specification given by interface predicates while at the level of distributed architecture we only deal with these interface predicates.

9   Conclusion

What we have presented here is in some sense the final and in these details missing step in our approach to go from high level interface specifications over to architectures, which form our systems and their interface specifications, and the interface specifications of the subsystems (see [7]). This way, we are able to prove that the implementation of the composed system is a refinement of the original interface specification of the composite system, and then in the next steps we can go on and work out an implementation for the individual subsystems which involves requirement steps finally ending up with Moore machines, their specification and their implementation which again can be proved correct while at the level of distributed systems and their architectures we work with interface predicates.

This shows that we have ended up with a general formal logical and programming framework based on a complementing set of system models for systems engineering. Note that in this framework we can also include the specification of physical systems by Moore machines, as demonstrated in [3].

The particular system model we have used models systems which are typical, for instance, such as cyber-physical systems as found in modern cars, production systems, in airplanes, and in communication networks, including the internet. By the system model, we completely avoid for methodological reasons any shared storage and any form of interleaving semantics, and also temporal logic.

Acknowledgement. It is a pleasure to acknowledge useful discussions with Harald Ruess as well as helpful remarks about draft versions by him and useful hints by the referees.

References

1. Broy, M.: Multifunctional software systems: structured modeling and specification of functional requirements. Sci. Comput. Program. **75**, 1193–1214 (2010)
2. Broy, M.: Refining the safety-liveness classification of temporal properties according to realizability. In: Bartocci, E., Falcone, Y., Leucker, M. (eds.) Formal Methods in Outer Space: Essays Dedicated to Klaus Havelund on the Occasion of His 65th Birthday, pp. 10–31. Springer, Cham (2021). https://doi.org/10.1007/978-3-030-87348-6_2
3. Broy, M.: Software system documentation: coherent description of software system properties. In: Margaria, T., Steffen, B. (eds.) Leveraging Applications of Formal Methods, Verification and Validation. Software Engineering, ISoLA 2022. LNCS, vol. 13702, pp. 10–27. Springer, Cham (2022). https://doi.org/10.1007/978-3-031-19756-7_2
4. Broy, M.: A calculus for the specification, design, and verification of distributed concurrent systems. Formal Aspects Comput. **36**(3), 1–54 (2024)
5. Broy, M.: Specification and verification of concurrent systems by causality and realizability. Theoret. Comput. Sci. **974**, 5 (2023)
6. Broy, M., Rumpe, B.: Development use cases for semantics-driven modeling languages. Commun. ACM **66**(05), 62–71 (2023)
7. Broy, M.: Time, causality, and realizability: engineering interactive, distributed software systems. J. Syst. Softw. **210** (2024)
8. Dijkstra, E.W.: A Discipline of Programming. Prentice-Hall (1976)
9. Hehner, E.C.R.: A Practical Theory of Programming. Springer, New York (1993)
10. Kahn, G., MacQueen, D.B.: Coroutines and Networks of Parallel Processes. In: IFIP Congress, pp. 993–998 (1977)
11. Lamport, L.: Specifying Systems: The TLA+ Language and Tools for Hardware and Software Engineers. Addison-Wesley (2002)
12. Mealy, G.H.: A method for synthesizing sequential circuits. Bell Syst. Tech. J. 1045–1079 (1955)
13. Moore, E.F.: Gedanken-experiments on sequential machines. Autom. Stud. Ann. Math. Stud. **34**, 129–153 (1956)

Artificial Intelligence in Software Documentation: Embracing the Documentation as Code Paradigm

Florian Nafz(✉) ⬗, Magdalena Krajinovic, and Martin Ley

Munich University of Applied Sciences HM, 80335 Munich, Germany
nafz@hm.edu
https://hm.edu

Abstract. A frequently neglected area of software engineering is documentation. In this article, we take a closer look at how the trend towards faster release cycles can be reconciled with software documentation by applying documentation as code and artificial intelligence.

The article focuses on the automation of documentation close to code and shows where generative AI can be useful here. The potential and challenges arising from the combination of documentation as code and artificial intelligence are discussed and demonstrated with practical examples using GitHub Copilot. In addition to documentation in parallel to development, legacy code documentation is also considered in this article. It brings additional challenges like bad quality or missing design documents. The correctness and completeness of the documentation will also be addressed and possible improvements to current approaches will be discussed.

Keywords: Software engineering · Software documentation · Artificial intelligence · Natural language processing · Documentation as code · Knowledge graphs

1 Introduction

Today, software is developed iteratively and incrementally, and delivered regularly at short intervals. Continuous integration and delivery (CI/CD) [16] as well as automated testing are standard in software development and there is a wide range of tools to support these activities. DevOps [28], with its notion of fully automation and its possibility of multiple releases per day, shortens the release cycles even further. This requires that the documentation must keep pace with this. As a result, documentation lacks quality, is sometimes omitted

---

Dedication: This article is dedicated to Wolfgang Reif on the occasion of his 65th birthday, in recognition of his scientific leadership throughout his career. This article looks at an intersection of his research areas – artificial intelligence, software engineering and correctness/completeness of systems.

G. Ernst et al. (Eds.): Wolfgang Reif Festschrift, LNCS 15765, pp. 282–302, 2025.
https://doi.org/10.1007/978-3-031-92196-4_14

or delivered late. This challenge can be met by a high degree of automation and optimization of the documentation process.

From this perspective, Artificial Intelligence (AI) can be seen as a logical extension of automation. It can automate documentation steps that were previously done manually or semi-automated.

In addition to the development of new software, there is a large number of legacy systems today, some of them poorly documented or not documented at all. Often these systems have "evolved historically" and have therefore a non-modular structure with little cohesion. A so called "big ball of mud" [18]. To refactor or renew these systems, developers need to spend a lot of time understanding the code and designing a migration path. In essence, the following principal questions emerge from this:

1. *How and to what extent can documentation be automated?* This includes the development of suitable tools, but also a seamless integration into the development processes. In order to automate, it is necessary to bring the documentation closer to the code. Documentation as Code (DaC) tries to close this gap by using developer tools for documentation. This means that documentation can directly benefit from the advantages of development pipelines and automation.

2. *Where can AI be used effectively and what are the consequences?* The use of AI must have advantages over manual documentation or other non-AI tools. At least, the generated documentation must fulfill the same quality requirements. With the breakthrough of large language models (LLMs) the idea of automated documentation seems feasible. AI generated documentation is tempting as it could save a lot of time and effort and is sometimes a chore. On the other hand, the need for proper documentation remains.

3. *How to obtain and extract the required information?* The aspect of information retrieval is very important to provide necessary information. This information must be context-specific and vary according to target groups. In addition, there is usually no single source of truth or at least information is spread through different data sources.

There are many works that deal with artificial intelligence in software engineering [36]. These mainly focus on activities in the individual phases. Cross-cutting activities such as documentation are rarely or only partially considered.

Software documentation is effective for stakeholders only if the relevant information is provided - i.e. the required information in a given situation is delivered to the specific target group in the right format. Typically, the management aspect is supported by specialized tools such as Component Content Management Systems (CCMS). The information itself is delivered either on paper, as online-help or via so-called Content Delivery Portals (CDP). As the portfolio of different software products emerge and more and more complex software configurations arise, this process becomes even more challenging. This typically results in fragmented workflows, redundant efforts, extended release cycles, and even a potential mismatch in software product and documentation. The complexity even increases if information from other source systems such as a Enterprise Resource Planning System (ERP) or Customer Relationship Management System (CRM) need to be included, leading to data inconsistency and a heterogeneous information system landscape.

In this article, we want to address these high level questions and give an overview over current approaches, the challenges in this field and discuss ideas how to tackle those. In a way, DaC is a prerequisite for the use of AI here. Therefore, we first look at automation with DaC and its integration into the development process. In the next step, we examine the use of generative AI to further enhance automation using DaC.

As the types of documentation are manyfold, we focus on documentation that concerns the current state of the software and that is close to the code, like code comments and architecture (including UML diagrams). We briefly discuss more high level and non-technical documentation, as the challenges increase here.

The article is structured as follows: First of all, we will describe the term Documentation as Code and provide a short classification before we elaborate on the challenges and requirements for the use of AI in documentation. After that, we will take a closer look at code-related documentation. Specifically, we will answer the question of whether an AI is suitable for code-related documentation. In particular, whether an AI can document in a semantically abstract way and not just in a declarative way. We will then look at management and delivery of documentation, which is not bound to code related documentation. Finally, we will discuss some aspects like correctness and completeness of AI-based documentation and potential approaches to solve this.

2   Documentation as Code

This section takes a closer look at the automation of documentation and its integration into the respective steps of the development process.

2.1   Requirements on Code Documentation

Because the documents are for various stakeholders, each type of documentation has different requirements and needs. There are many legal or normative requirements for documentation, especially in the safety-critical area.

Theunissen et al. state the main challenge is that documentation is poor [46]. There are projects that produce little or no documentation. But there are also projects that produce far too much documentation. As Beck et al. write in [10] the objective is to document adequately. This means, among other things, documenting in a way that is concise, relevant to the intended audience, and free of redundancy. An important requirement for every documentation is to be up-to-date, correct and complete. In addition, the readability (clarity) and accessibility of the documentation is also very important to practitioners [6]. More information about best practices for good documentation and technical writing can be found in [32,34], for example.

Software becomes more and more complex. At the same time software release cycles get shorter. This leads to further requirements for documentation or more concise the documentation process, as this reduces the time available to adapt the documentation. Code should be tested before release; the same goes for the documentation. This should be adapted before release. In testing, automation is key to reduce effort and time. The same applies to documentation.

2.2 Prerequisites for Automation

In order to automate, the documentation must be as close as possible to the code. This is naturally the case for code comments, but the gap widens as soon as architectural documentation is involved or even documention more far away from source code like end user documentation.

The Documentation as Code paradigm (DaC) tries to close this gap and follow the DevOps idea of "everything as code". It refers to the idea to write documentation with the same tools as in software development and applies its techniques to technical documentation. Some more detailed information can be found in [20] or [31]. Two core aspects of DaC are the following:

– *Markup Languages:* Using markup languages for documentation. This allows to create and maintain documentation like code – with the same tools and even in the same code repository.
– *Source Code Management:* Using a source code management (SCM) tool like Git allows for establishing the same workflows as for code development. According to [29] this is one of the key enabler of DaC as it allows to store and release software and documention together.

These two aspects are essential to enable automation and to adopt DevOps principles like continuous integration and delivery. By using development tools and their integration into the CI/CD pipelines, documentation can adapt and benefit from the related practices, like review processes, build and deployment servers or automated tests. The objective is to achieve the same automation for documentation as currently available for coding in software development.

2.3 Integration Into the Development Pipeline

Seamless integration into the development pipeline - in terms of both tools and processes - is crucial. Figure 1 shows how a DaC integration into the development process could look like. In the following, we will take a brief look at each of those workflow steps.

Write Docs. The documentation is ideally written in a simple markup language (e.g. Markdown, reStructuredText, AsciiDoc). This allows to use the same tools the devlopers use for coding and maintaining the code. For example an Integrated Development Environment (IDE) or a simple text editor. There is no need to switch tools for writing documentation, especially for documentation written by the developers themselves.

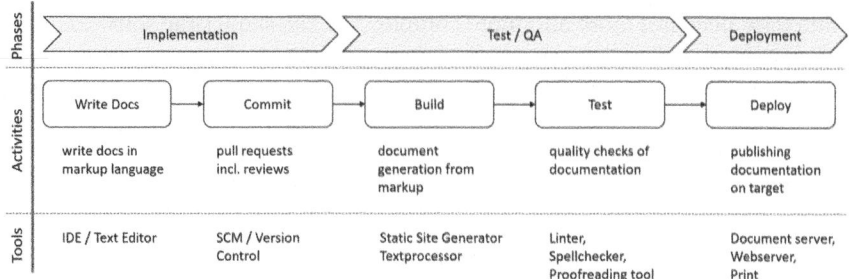

Fig. 1. Documentation as Code: Integration into the different software development phases.

On the other side, more complex documents need a modular approach based on information units rather than document type. The exclusive use of simple markup languages is not sufficient for this. Here Component-Content-Management-Systems (CCMS) or documentation tool suites offer very good support here and are preferred by technical writers. They allow for a modular and small-scale management of the document structure, including advanced link handling, detailed search and many other features. But this means a trade-off between using markup and developer tools and specialized tools for more comprehensive documentation. CCMS address this by offering suitable APIs for integration into to the CI/CD workflow and by automatic conversions, from Markdown to standards such as DITA XML [43], for example.

Commit. Using a source code management (SCM) tool like Git allows for establishing the same workflows as for code development (pull request, merge, rollback, branching). Platforms such as GitHub and GitLab facilitate collaboration between developers and writers by displaying the changes made to each file and providing a location to discuss and explain those changes.
The use of pull requests and peer reviews improves documentation quality and ensures knowledge transfer. Further version control increases traceability by allowing much better matching of requirements, documentation and code changes.

Build. Text-based formats allow for further processing into other target formats, such as HTML or PDF. Static Site Generators (e.g. Sphinx[1] or Jekyll[2]) or text processors (e.g. AsciiDoctor[3]) can be used to automate the generation process and integrate it into the build step.

Test. By integrating documentation checks into the continuous integration process, documentation errors can be detected and fixed early, which increases the

---

[1] https://www.sphinx-doc.org/en/master/index.html.
[2] https://jekyllrb.com/.
[3] https://asciidoctor.org/.

overall consistency and quality of the software. For example, this can be done by linting the documentation (e.g. with Textlint [2] or Vale [3]). Similar to code linting, a natural language linter checks documentation for syntax errors, consistent terminology or violations of style guidelines and can directly be integrated into the CI/CD pipeline. This requires that the documentation is available in a text-based format.

Deployment. Depending on the type of documentation there are several publishing options for the deployment – from a document server to a static website. Many of the build tools also already have support for publishing on different targets and in different formats.

2.4   Potential and Challenges

The idea of Documentation as Code is promising. There are numerous tools and automation options for creating documentation out of code comments that have been in use for years. There are some implementations for architecture documentation (architecture as code) like the docToolchain[4] or individual approaches like described in [29]. For other types in the field new practices and tools need to emerge.

Today's CI/CD pipelines are very flexible and have many interfaces and options for implementing hooks. For source code creation, several tools are usually integrated into the process for automation. This allows for an individual setup meeting the project requirements. DaC tries to follow a similar path and integrating various tools, based on documentation in plain text. The appropriate tools are then integrated to meet the requirements of the respective documentation leveraging their APIs. The idea is to having more a meshup of tools then a monolithic tool offside the process as there is usually no one fits all solution.

Besides the technical and workflow aspects of DaC, there are also organizational challenges. A large part of a system's documentation (user documentation, safety and warning information) is often written by a group of people with specialized skills (e.g. technical writers) organized in a separate team or department. Therefore, the development process itself must be considered. DaC requires technical knowledge about the used tools, which makes tool selection important. Even if CCMS are not strictly speaking part of DaC, it can make sense to integrate them.

3   AI Generated Documentation

In the previous section, we looked at the integration and automation of documentation within the development process. This is an important prerequisite for the next step – automated creation of documentation with generative AI.

---

[4] https://doctoolchain.org/.

In the field of documentation, artificial intelligence (AI) has so far mainly been used for machine translation or, more recently, for code commenting. Common tools are DeepL[5] for document translation or Github Copilot[6] for generating code or code comments.

However, there are many other types of documentation at different levels of abstraction and for a variety of target groups. From code comments through architecture and technical documentation to documentation addressing the end user. The question is, where can AI provide support here and how much?

In the following, some aspects of using AI are discussed more closely and some lessons learned from experiments with code comment generation and architectural documentation are described.

3.1   Code Comments

Inline code documentation is mainly written by the developer, and there are many tools that allow documentation to be produced in a variety of output formats. Besides many classic post processors, that parse annotated source code to produce documentation, modern code documentation platforms allow for even more. They include interactive search and integrate directly into the CI/CD pipeline.

Code comments should be clear, concise in language, and ideally have a consistent style throughout the code base. This consistency allows other developers to quickly grasp and understand the purpose and logic of the code segments. We conducted an experiment within a software development company to evaluate the quality and effectiveness of AI-generated comments. We focus on the subsequent documentation (by code comments) of a whole code base not on the parallel use of a copilot during coding. The code base was a self-written C# math library consisting of 316 scripts and a total of a little bit more than 200k lines of code. As tool for generation GitHub Copilot was used as it is currently one of the most prominent ones and also used for code generation as it is saves time and reduces effort [48]. The \doc command was used to generate the comments. After the commenting process the generated documentation was assessed based on predefined quality criteria such as clarity, accuracy, and adherence to documentation standards.

The evaluation revealed that 93% of the comments were considered high-quality, providing accurate explanations of functions and methods, and correctly describing the implementation of complex algorithms. These comments maintained consistency in terminology and style, similar to manually created documentation. In general, Copilot tended to make either too many comments (every line) or just a summary for a whole function. The documentation of the function and some of the "important" code lines would have been ideal. With additional prompts to the \doc command, it is possible to adapt documentation behavior. However, it is difficult to define what are important lines of code and

---

[5] https://www.deepl.com/.

[6] https://copilot.github.com/.

what are not, which would be optimal for the documentation of an existing code base. 7% of the comments contained inaccuracies. These errors were primarily due to the AI's misinterpretation of the code's functionality or the omission of critical implementation details necessary for a comprehensive understanding of the code.

The experiment showed that automatic code commenting works well in most cases. Incorporating a hybrid approach - where AI-generated comments are reviewed and refined by human developers - is recommended to ensure the documentation's accuracy and relevance ("human in the loop". This is also the basic idea of current copilots.

On the other hand, it is not ideal for subsequent commenting of an existing code base (e.g. legacy code) where quality assurance is more time-consuming, as the checking developer does not write the code at the same time or someone else has written it. For the documentation of legacy code other approaches are needed.

Since the developer may be working with a copilot anyway – in terms of coding – the use of a copilot for code documentation is useful. The quality assurance of the generated comments can be done directly after generation and is no extra step in the development process.

3.2 Impact of Code Quality

Studies such as those by Buse and Weimer [12] showed that code readability is influencing the quality of documentation. Clean code is defined by its adherence to clear and consistent naming conventions, modular design, and logical structure. It facilitates a better understanding of the code's functionality [33]. Ideally, this should reduce the amount of documentation required, as the code is to some extent self-explanatory. So, does the code quality impact the AI-generated documentation or is it just helpful for humans?

We analyzed this in another experiment. We took 50 small implementations of different mathematical and physical functions, with each function in a clean code and an obfuscated version. An example function implementing Snells Law[7] is shown in Listings 1.1 and 1.2).

Listing 1.1. clean code version with semantic variable names

```
static double snellsLaw(double refractionIdx1, double
    refractionIdx2, double angleIncidence) {
  double angleIncidenceRadians = angleIncidence * (Math.PI
      /180);
  double angleRefractionRadians = Math.Asin((refractionIdx1 /
      refractionIdx2) * Math.Sin(angleIncidenceRadians));
  double angleRefraction = angleRefractionRadians * (180 /
      Math.PI);
  return angleRefraction;
}
```

---

[7] Refraction of light when passing different media.

For the obfuscated version all semantic variables and method names were replaced by meaningless identifiers.

Listing 1.2. obfuscated version with meaningless variable names

```
static double func(double n1, double n2, double aI) {
    double aIR = aI * (Math.PI / 180);
    double aRR = Math.Asin((n1 / n2) * Math.Sin(aIR));
    double aR = aRR * (180 / Math.PI);
    return aR;
}
```

For both versions code comments were generated using Copilot's \doc command. The results were evaluated for quality by manual review. Special attention was given to whether the comments generated were semantic summaries or purely functional descriptions. For example, Listings 1.1 or 1.2 can be summarized in two ways:

- Functional description: *"This method gets three input values and performs a calculation using trigonometric functions"*
- Semantic summary: *"This method gets three double values. The refraction index of the first and second medium and the angle of incidence. It calculates the angle of refraction using Snell's Law"*

The first, functional description offers relatively low added value for development, as a developer can also see this directly from the code. The second, semantic summary, offers significant added value, as it contains more specialized information and describes what the formula calculates. Therefor, the goal is to generate semantic summaries.

Table 1. Quality of the commented functions

	# samples	correct description	semantic summary
clean code	50	50	50
obfuscated	50	50	37

The results of the experiment are shown in Table 1. All functions in both versions were described correctly and without errors. While every function in clean code was also commented semantically, 37 (74%) of the obfuscated functions were commented well.

Copilot often recognized the formulas or common constants used in the formulas, even in the obfuscated versions and was able to assign and describe them correctly. The summary for the shown implementation of Snell's Law, for example. Listing 1.3 shows the output.

Listing 1.3. Documentation with Github Copilot of SnellsLaw method

```
/// <summary>
/// Calculates the angle of refraction based on Snell's Law.
/// </summary>
/// <param name="n1">The refractive index of the first medium
   .</param>
/// <param name="n2">The refractive index of the second
   medium.</param>
/// <param name="aI">The angle of incidence in degrees.</
   param>
/// <returns>The angle of refraction in degrees.</returns>
static double func(double n1, double n2, double aI) {
[...]
```

Copilot was able to semantically describe all of the obfuscated versions with special constants or more complex formulas. The assumption is that the underlying model already knows many of these generally known formulas and their implementation. The experiment in Sect. 3.1 shows that Copilot occasionally makes mistakes in very specific implementations, if they do not contain semantic naming.

This indicates that the quality of the underlying code directly impacts the stability and predictability of the AI's output. Future work will be to perform a similar test with application-specific business functions to confirm this.

This would be a first step towards dealing with legacy code. The AI needs to be able to semantically summarize and describe the functionality of modules or larger building blocks. Ideally, even if they don't have semantic variable names.

3.3   Architecture Documentation

Another form of code near documentation is the documentation of the software architecture. Architecture documentation consists not only of text, but much more importantly of diagrams (e.g. BPMN [39] or UML [40]) and both, structural views and behavioral views are required.

An important goal of architectural documentation is Architecture Consistency (AC). This means that the code matches the intended architecture. Ideally, the architectural design precedes the implementation and leads to the documentation. AC aims to align the implemented system with its intended architecture [8].

Adersberger and Philippsen define in [5] two topics that need to be addressed in order to check consistency. As prerequisite for tool-based or automatic checks a mapping of architecture elements to code is needed (architecture-to-code traceability). This enables to check for inconsistencies. Either there is a dependency on the source code side but not in the architecture model (divergence) or vice versa (absence). Today, there are many tools that can create architecture models, do consistency checks from source code or perform static analysis.

Again, the DaC goal is to use development tools as much as possible. Tools such as plantUML[8] or Structurizr[9] provide a text-based language for diagrams and are therefore suitable. The use of these tools can support the AC as code and documentation are close together. It also allows the use of LLMs as the diagrams are text-based.

There are already some interesting works on the use of AI in software architecture. These are mainly concerned with architecture in the design phase. For example, how AI can support the iterative development of the architecture based on requirements [15]. Today Large Language Models (LLMs) can generate syntactically correct UML Diagrams. At the same time, studies show that LLMs have weaknesses in generating semantically correct diagrams and, for instance, generating appropriate domain models [14]. Further LLMs also tend to have a high variability in the generated output [9]. These properties make them difficult to use in software design and still require manual review.

3.4   Software Archaeology

For system documentation, especially for legacy systems with missing design documentation, the task is not so much to create a design from natural language requirements, but to extract it from an existing code base. This process is called software archaeology [25]. The retrieved documentation can be used, for example, to identify an integration path for refactoring or reimplementation. But also as an support for the developers, regarding a possible further development of the existing system.

So far software archaeology is a time consuming task with a lot of manual work, as original developers are often no longer available at this point, which means that the know how in their heads is also lost. Therefore, the ideal case would be that the architecture documentation can be automatically generated from the source code. Figure 2 shows an abstract view on a simplified retrieval process.

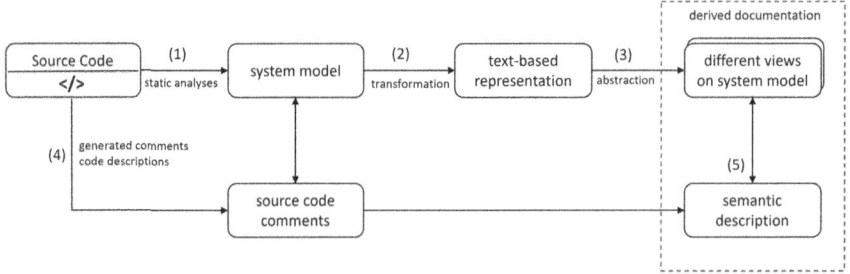

Fig. 2. Software Archaeology: A simplified documentation retrieval process

---

[8] https://plantuml.com/de/.
[9] https://structurizr.com/.

(1) Detailed Technical Model. For architecture consistency the generated documentation must match the code in any point of the generation process. As mentioned above, this step can be carried out properly with common tools available today. The result is a fine-grained system model that usually contains all the technical details. In terms of class diagrams, these are details such as interfaces, abstract classes, or helper classes. The advantage of this 1-to-1 mapping is that the generated model is consistent with the code.

(2) Transformation to Text-Based Representation. The generated model typically has a visual representation and depending on the tool used, the underlying representation may be proprietary or non-text based. In order to be able to process the model further with an LLM, it may be necessary to transform it into a text-based representation the LLM can work with.

(3) Abstraction of the System Model. Architecture work requires different views of the system, depending on the context. Therefore, diagrams are needed that abstract some of the implementation details. This requires that the code structure can be suitable abstracted to show the "important" aspects. The notion of 'important' can vary depending on the user. While developers may want a detailed technical view, also high-level views are needed to look at different aspects of the system. For example, to get an overview of the structure of the system and its components. Based on this, a target architecture can be developed and a migration path with necessary refactorings can be derived. If the system already has a clean structure in the form of components or services, it is easier to extract this from the system model created in step (1). Legacy systems usually don't have this, and are often historically grown or monolithic from the ground up and therefore it is more difficult to recognize a structure and identify components for a possible target architecture

LLMs could be used to identify those structures and create context-sensitive custom views at this point. This means that LLMs must be able to abstract an existing UML diagram appropriately. To maintain AC, dependencies must be included at the abstracted level. Dependencies may also be hidden if several classes are combined into one, for example.

(4) Source Code Comments. Missing documentation of the source code is documented in this step. As described in Sect. 3.1, this could be done automatically by a copilot. However, there are still open issues regarding the quality of the generated documentation. Legacy code in particular tends not to adhere to the clean code rules. The documentation generated in this way should therefore continue to be used with the awareness that it could be incorrect.

(5) Describing UML Diagrams. Besides a (visual) model of the system, some textual description is needed for a complete documentation. Analogous to the code summaries in Sect. 3.2 an open question is: Can a LLM semantically describe an UML diagram? We made a series of similar tests like the code comments in Sect. 3.1 by creating several small UML class diagrams in plantUML. These

diagrams were then given to GitHub Copilot[10] and ChatGPT[11] with the task of summarizing and describing them. GitHub Copilot tended to only describe the structure (number of variables, methods), while ChatGPT described the functionality as well. ChatGPT was hallucinating in some cases and started describing the functionality of methods in the class diagram (without having any implementation). Ideally, the description should be summarizing rather than structural. So, the results were mixed but promising, although there is still potential. The code comments generated in Step (4) could be used as additional information for the LLM here.

The goal of future research is to provide the generated technical UML model (in plantUML) and then be able to make natural language queries to the LLM. Although the allowed context lengths of LLMs are getting longer, they are still limited and can be too small to query the entire model, comments, and possibly even the code.

3.5   Combining AI and DaC

DaC enables the use of AI as it focuses on automation and can ideally be used for automating some of the manual steps. More concise Natural Language Processing techniques can be applied in the "Write Docs" and the "Test" step (see Figure 1).

So-called copilots assist developers in writing source code. They can both generate program code and document existing code (see 3.1). There are also a growing number of tools for in software architecture using AI methods. Same holds for tools doing static analysis. Another part where LLMs are already often used is machine translation [44].

The used AI methods must meet at least the same quality requirements as currently used methods without AI. Documentation is required to obey with writing and content standards. This can be achieved by running the same tests as for non-AI generation or by manually checking the results again each time after generation (human in the loop). From this perspective, AI is another option for the (partial) automation of activities.

With regard to DaC and integration into an agile development process, however, it is important not to introduce additional manual steps. Ideally, manual steps should be done in parallel with coding or directly by the developer. Similar to the approach of testing as early as possible (shift-left testing) [17], documentation should not wait until the end (shift-left documentation). Shifting manual activities to the left plus automation allows smooth integration into the development process. For documentation types that are further away from the source code this is more difficult.

4   Management and Delivery of Documentation

After having addressed the issues of automation and generation of software documentation in general, we now would like to focus on two additional aspects:

---

[10] uses gpt-3.5-turbo.
[11] used model: gpt-4-turbo.

how technical documentation or - to be more precise - documentation modules can be managed and delivered. This is particularly important for large systems, especially complex embedded systems, where both hardware and software documentation exist and must be consistent.

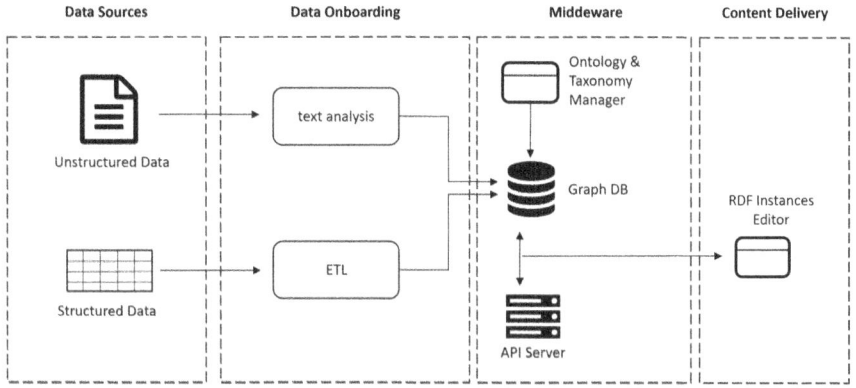

Fig. 3. A knowledge graph architecture (see also [30])

4.1 Knowledge Graphs

Laadhar et al. [30] suggest that knowledge graphs (KG) in general provide a robust solution when it comes to bridging information silos and harmonizing heterogeneous data sources. KGs allow the effective representation of complex information and can serve as a centralized "knowledge hub" that includes key components such as ontologies, thesauri, and instance data [42]. We suggest to extend this approach to cover software documentation as well, thereby addressing the challenges mentioned in the introduction. According to our conception, a KG is a middleware, covering (at least) these aspects (see Fig. 3):

- An ontology, defined as the "formal, explicit specification of a shared conceptualization of a phenomenon in the world" [4]
- A thesaurus (taxonomy), i.e. an ordered compilation of concepts and their natural language terms/labels, ideally represented in SKOS [23]
- Instance data, i.e. the actual data in a graph database that instantiate the classes of the ontology and are enriched with metadata from the thesaurus

A KG architecture can then be extended by three additional layers:

- Data sources, typically consisting of structured and instructed data from various source systems such as CCMS, CRM or ERP
- Data onboarding, allowing for the preparation and integration of data for use within the KG
- Serving layer, i.e. an interface and access point (such as a CDP) or a software-specific online help

4.2   An Ontology Network for Software Documentation

In [30] Ladhaar et al. propose a network of different ontologies, such as "Product Hierarchy Ontology", "Market Segmentation Ontology" or "Versioning and Change Tracking Ontology". They each address specific requirements, here in particular representing domain knowledge about physical products (in machine building) and for the purpose of building Product Information Management Systems (PIM). This modular approach is suitable for the integration of (at least two) additional ontologies required for software documentation. First of all, we recommend integrating iiRDS, the intelligent information request and delivery standard [21]. iiRDS is an RDF-based standard designed to exchange technical information across different platforms. It provides an ontology for representing different kind of information (documents, so-called topics, and fragments) and further classifying that information by a controlled vocabulary into corresponding topic-types or subject-types.

Thereby, iiRDS plays a crucial role in the overall approach described here. It is a widely acknowledged and supported standard for classifying and exchanging technical information. It allows us to uniquely describe our (automatically generated) documentation resources and to integration them into the KG.

In addition, the knowledge model requires an ontology for representing the software itself. We call this ontology the "Software Product Hierarchy Ontology". Analogous to the "Product Hierarchy Ontology", this ontology models the portfolio of different software applications, software variants, and software versions, etc. In addition, the "Software Product Hierarchy Ontology" shall be accompanied by a "Software Component Hierarchy Ontology". This ontology defines what functions, micro services or features a particular software product itself consists of, thus representing the modular structure of software.

4.3   Potentials and Benefits

Based on the three ontologies proposed, our approach allows us to link documentation to software features and consequently to software products. This in turn provides the possibility to automatically identify documentation gaps. If software products in turn are integrated into machines, it is even possible to determine which product contains which features (and therefore requires which parts of the documentation). This is made possible by creating a relation between a class of the "Software Product Hierarchy Ontology" and the already available "Product Hierarchy Ontology", thereby extending the digital twin [1,22] by information about the software (and its documentation). Finally, a KG-based solution could be used to automatically update software and its documentation.

5   Discussion

This section discusses some more aspects of automated documentation using artificial intelligence.

5.1   Documentation for Safety Critical Systems

There are general standards for software documentation such as ISO/IEC/IEEE 15289 or the ISO/IEC/IEEE 2651x series of standards [26,27]. Especially in highly regulated fields there are additional requirements on the documentation [24]. For safety critical system LLMs can currently not be used directly, as they create sometimes incorrect and incomplete results [47]. Therefore, procedures or possibilities are needed to ensure the correctness and completeness of the documentation. There is already a wide range of tools and options for system safety, from formal verification to fault tree analysis (FTA), failure mode and effects analysis (FMEA) or Deductive Cause Consequence Analysis (DCCA) [41] through to corresponding test approaches [11,13]. So far, correct documentation is currently ensured by establishing suitable processes plus manual quality assurance. Automation, especially with AI requires new methods to ensure correctness and completeness. Another important requirement is traceability and explainability of the AI generated documentation. Ideally it is possible to find out why the system wrote some documentation and based on what information.

There are some approaches that try to use Knowledge Graphs to avoid hallucinations [7,45]. Knowledge graphs are a method of storing and managing domain-specific knowledge in a structured way. They allow for symbolic reasoning on their data and dependencies. The downside is that they also have to be created somehow and may be incomplete themselves. Nevertheless, these first approaches indicate that they have the potential to ensure correct results when combined with LLMs.

5.2   Generating Useful Documentation

Documentation needs domain specific knowledge. Starting with the use of the domain's terminology through to the appropriate description of technical or business processes. Basic LLMs also need special knowledge about the applications context and have to be fine tuned. As not every project has enough data for a good fine tuning, other methods must be found.

Enriching the model with domain-specific knowledge is essential, especially for documentation, which should always be product-specific to add value. This knowledge could be provided by the knowledge graphs mentioned above, but also by existing documents from design or earlier versions of the software. Providing this specific knowledge can be time consuming and depends heavily on what information is available. In the case of the legacy systems discussed, this is difficult as there is usually no suitable original data available.

Further AI can be used for quality assurance. For instance, to detect semantic duplicates, to check terminology or for grammar- and spellchecking. These are all tasks that are already being solved very well by AI today.

5.3   Reproducibility of the Documentation

Another significant observation was the consistency of AI-generated comments. Multiple runs on the same code segments produced identical results, demonstrating a high degree of reproducibility and stability in the outputs. However, slight structural changes to the code led to changes in the documentation.

In [35] the robustness of code generation based on JavaDoc method description was investigated. The results showed that semantically identical but paraphrased descriptions led to different implementations. One property of LLMs is that slight changes in the input can produce significant changes in the output.

For documentation in general, reproducibility but also stability against small changes is an important property. For code comments this is not such an issue as they are mostly generated once and their scope is limited. But for more high level documentation which is regenerated for new versions, for example, it is important that the changes are local and only affect the changed functionalities.

Future work will include finding good integration and workflows for updating documentation, or adding documentation chunks for new features to the overall documentation. On the one hand, the documentation must be consistent in terms of terminology and language. On the other hand, it must also meet the other quality requirements mentioned earlier.

5.4   Human in the Core

AI is not a monolithic black box that will solve all tasks fully automated. AI tools (or AI agents) will support several processing steps. The next generation of workflows and processes will be more an orchestration of components/tools that are specialized on the individual task coordinated by human specialists.

The term "human in the loop" is overloaded and has several meanings depending on the context. Humans are seen as a part of the system [38]. In technical documentation they are often integrated as a verification step in quality assurance to ensure correctness of automatically generated documents. In the field of machine learning, humans are used to enhance knowledge in the training phase [37] or humans-in-the-loop guide the machine learning algorithm [19]. They all have in common, that the interaction between humans and machines is a clearly defined step in the process.

This approach is particularly useful for documentation in cases where the quality of the results cannot be guaranteed or there are special requirements for correctness. Research and tool developers try to improve these approaches to the point where manual review becomes obsolete.

What remains is the orchestration part. Similar to the setup of a DevOps pipeline, suitable tools must be selected for the DaC approach and integrated into the development process. Depending on the individual requirements and tasks, steps can be automated or semi-automated. The "human(s) in the core" must therefore know how the respective AI tools work. They must take care of both the upfront setup and the maintenance/adaptation of the tool chain. They must also be able to evaluate the results in terms of their quality. Quality control can be carried out on a case-by-case basis, for example, or in the form of a defined manual proofreading step ("human in the loop"). With the latter, particular attention must be paid to efficient integration so that there is no idle time or waiting time for others in the development process.

6   Summary and Outlook

At first glance, software documentation may seem simple – or even unnecessary. It is not that easy to produce good documentation. Especially if you want to look at all dimensions and aim for a fully integrated and highly automated approach. In this article, we looked at the automation of technical documentation with the use of AI, with the focus on documentation close to code. Docs-as-Code is an essential approach that brings documentation closer to the software and supports its use in AI tools such as copilots.

There are two scenarios to consider: First, design documents are available and documentation is ideally created in parallel with software development. The AI-generated results can be reviewed directly and quickly (human in the loop). Second, documentation is only created after development is finished, and there may be little or no design documentation or code comments.

In the first case, integration is easier and AI correctness can be ensured by in parallel. Initially automated documentation can be used for individual and supporting steps. In the second case, full automation is much more difficult, because the effort to check correctness and completeness is high and there is no or few initial data for the AI model as context.

To maximize the benefits of AI-driven documentation tools, it is crucial to maintain a clean and structured code base. This can be achieved by adopting clean code practices, conducting regular code reviews, and using tools that enforce coding standards and evaluate code quality. By improving the quality of the underlying code, developers can leverage AI documentation tools more effectively, reducing the need for extensive manual interventions and enhancing the overall quality of the documentation. Experiments with different quality of source code confirm this. Further domain specific input data is also important. As these allow the model to give specific descriptions based on contextual knowledge in addition to general answers. This can further reduce hallucination. It is therefore advisable to evaluate these aspects first and to start close to the source code, before turning to high-level or non-technical documentation.

The tests carried out and the literature show that there are still some aspects that need to be improved. Correctness and completeness of the answers are necessary, especially in a regulatory environment. In addition to these two aspects, there are also linguistic requirements that need to be met for good documentation, such as terminology, lack of redundancy or simple language. This applies in particular to technical documentation further away from the code, which we have not yet considered here, but is one of our next steps.

We are currently looking at the components for the legacy code pipeline described here. So far we have only looked at individual steps in the planned pipeline, the next step would be find suitable integration patterns for the different scenarios.

When managing and delivering documentation – both code-related and all other types of documentation – it is still a major challenge to combine the individual approaches consistently with each other. For code-related documentation, there are already some tools for implementing DaC and automation to support short release cycles. There are already a number of approaches for other types of documentation (see Sect. 4). It remains to find solutions how these can best be combined.

Another point we look at is how knowledge graphs can be used to improve the results. The aim is to enrich an LLM with a system model, code comments and, if necessary, knowledge in the form of knowledge graphs, which can then be queried accordingly to generate customized documentation.

Disclosure of Interests. The authors have no competing interests to declare that are relevant to the content of this article.

References

1. IDTA - Der Standard für den digitalen Zwilling. https://industrialdigitaltwin.org/
2. Textlint, a natural language linter for text and markdown. https://github.com/textlint/textlint, MIT license [Software]
3. Vale. https://vale.sh/. [Software]
4. SKOS: Simple Knowledge Organization System (2024). https://www.w3.org/2004/02/skos/
5. Adersberger, J., Philippsen, M.: ReflexML: UML-based architecture-to-code traceability and consistency checking. In: Crnkovic, I., Gruhn, V., Book, M. (eds.) Software Architecture, pp. 344–359. Springer, Berlin Heidelberg, Berlin, Heidelberg (2011)
6. Aghajani, E., et al.: Software documentation: the practitioners' perspective. In: Proceedings of the ACM/IEEE 42nd International Conference on Software Engineering, pp. 590–601. ICSE 2020. Association for Computing Machinery, New York (2020)
7. Agrawal, G., Kumarage, T., Alghamdi, Z., Liu, H.: Can knowledge graphs reduce hallucinations in LLMs? : a survey. In: Proceedings of the 2024 Conference of the North American Chapter of the Association for Computational Linguistics: Human Language Technologies, vol. 1, pp. 3947–3960. Association for Computational Linguistics (2024)
8. Ali, N., Baker, S., O'Crowley, R., Herold, S., Buckley, J.: Architecture consistency: state of the practice, challenges and requirements. Empirical Softw. Eng. **23**, 1–35 (2018)
9. Atil, B., Chittams, A., Fu, L., Ture, F., Xu, L., Baldwin, B.: LLM stability: a detailed analysis with some surprises (2024)
10. Beck, K., et al.: Manifesto for agile software development (2001). http://www.agilemanifesto.org/
11. Bernd Bertsche, M.D.: Zuverlässigkeit im Fahrzeug- und Maschinenbau, 4th edn. Springer, Heidelberg (2022)

12. Buse, R., Weimer, W.R.: Learning a metric for code readability. IEEE Trans. Software Eng. **36**(4), 546–558 (2010)
13. Cullyer, W., Storey, N.: Tools and techniques for the testing of safety-critical software. Comput. Control Eng. J. **5**(5), 239–244 (1994)
14. Cámara, J., Troya, J., Burgueño, L., Vallecillo, A.: On the assessment of generative AI in modeling tasks: an experience report with ChatGPT and UML. Softw. Syst. Model. **22**, 1–13 (2023)
15. Eisenreich, T., Speth, S., Wagner, S.: From requirements to architecture: an AI-based journey to semi-automatically generate software architectures. In: Proceedings of the 1st International Workshop on Designing Software, pp. 52–55 (2024)
16. Farley, D., Humble, J., Farley, D.: Continuous Delivery: Reliable Software Releases Through Build, Test, and Deployment Automation, 1st edn. Addison-Wesley Professional (2010)
17. Firesmith, D.G.: Common System and Software Testing Pitfalls: How to Prevent and Mitigate Them Descriptions, Symptoms, Consequences, Causes, and Recommendations, 1st edn. Addison-Wesley Professional (2013)
18. Foote, B., Yoder, J.: Big ball of mud. In: Harrison, N., Foote, B., Rohnert, H. (eds.) Pattern Languages of Program Design, vol. 4, pp. 654–692. Addison-Wesley (2000)
19. Fung, J., Li, Z., Stephens, D., Mao, A., Goel, P., Walpole, E., Dima, A.A., Boyd-Graber, J.: Human-in-the-loop Technical Document Annotation: Developing and Validating a System to Provide Machine-Assistance for Domain-Specific Text Analysis (2024)
20. Gentle, A.: Docs Like Code: Collaborate and Automate to Improve Technical. Lulu Press, Incorporated (2022)
21. Gesellschaft für Technische Kommunikation - tekom Deutschland e. V.: iiRDS - the intelligent information Request and Delivery Standard. https://www.iirds.org/
22. Grieves, M.: Digital twin: manufacturing excellence through virtual factory replication. White paper **1**(2014), 1–7 (2014)
23. Gruber, T.: Toward principles for the design of ontologies used for knowledge sharing. Int. J. Hum. Comput. Stud. **43**, 907–928 (1994)
24. Häring, I.: Requirements for Safety-Critical Systems, pp. 209–226. Springer, Singapore (2021)
25. Hunt, A., Thomas, D.: Software archaeology. IEEE Softw. **19**(2), 20–22 (2002)
26. ISO/IEC/IEEE 15289:2019: Systems and software engineering - content of life-cycle information items (documentation) (2019)
27. ISO/IEC/IEEE 26514:2022: Systems and software engineering - design and development of information for users (2022)
28. Kim, G., Humble, J., Debois, P., Willis, J.: The DevOps Handbook: How to Create World-Class Agility, Reliability, and Security in Technology Organizations. IT Revolution Press, ITpro collection (2016)
29. Krunic, M.: Documentation as code in automotive system/software engineering. Elektronika ir Elektrotechnika **29**, 61–75 (2023)
30. Laadhar, A., Acharya, N., Wagner, J., Ley, M.: Product information management systems powered by knowledge graphs. In: European Semantic Web Conference ESWC (2024)
31. Lakatos, D.: Crafting Docs for Success - An End-to-End Approach to Developer Documentation, 1st edn. Apress (2023)
32. Laplante, P.: Technical Writing: A Practical Guide for Engineers, Scientists, and Nontechnical Professionals. What Every Engineer Should Know, 2nd edn. CRC Press (2018)

33. Martin, R.C., Coplien, J.O.: Clean Code: A Handbook of Agile Software Craftsmanship. Prentice Hall, Upper Saddle River, NJ [etc.] (2009)

34. Martraire, C.: Living Documentation: Continuous Knowledge Sharing by Design, 1st edn. Addison-Wesley Professional (2019)

35. Mastropaolo, A., et al.: On the robustness of code generation techniques: an empirical study on github copilot. In: Proceedings of the 45th International Conference on Software Engineering, pp. 2149-2160. ICSE 2023. IEEE Press (2023)

36. Meinke, K., Bennaceur, A.: Machine learning for software engineering: models, methods, and applications. In: Proceedings of the 40th International Conference on Software Engineering: Companion Proceeedings, pp. 548–549. ICSE 2018. Association for Computing Machinery, New York (2018)

37. Munro, R.: Human-in-the-Loop Machine Learning: Active learning and annotation for human-centered AI. Manning (2021)

38. Nunes, D., Zhang, P., Sá Silva, J.: A survey on human-in-the-loop applications towards an internet of all. IEEE Commun. Surv. Tutor. **17**, 1 (2015)

39. OMG: Business Process Model and Notation (BPMN), Version 2.0.2 (2014). https://www.omg.org/spec/BPMN/2.0.2/

40. OMG: Unified Modeling Language (UML), Version 2.5.1 (2017). https://www.omg.org/spec/UML/2.5.1/

41. Ortmeier, F., Reif, W., Schellhorn, G.: Deductive cause-consequence analysis (DCCA). In: Proceedings of IFAC World Congress. Elsevier (2006)

42. Peng, C., Xia, F., Naseriparsa, M., Osborne, F.: Knowledge graphs: opportunities and challenges. Artif. Intell. Rev. **56**(11), 13071–13102 (2023)

43. Priestley, M.: DITA XML: a reuse by reference architecture for technical documentation, pp. 152–156 (2001)

44. Rivera-Trigueros, I.: Machine translation systems and quality assessment: a systematic review. Lang. Resour. Eval. **56**(2), 593–619 (2022)

45. Sequeda, J., Allemang, D., Jacob, B.: A benchmark to understand the role of knowledge graphs on large language model's accuracy for question answering on enterprise SQL databases. In: Proceedings of the 7th Joint Workshop on Graph Data Management Experiences & Systems (GRADES) and Network Data Analytics (NDA). GRADES-NDA 2024. Association for Computing Machinery, New York (2024)

46. Theunissen, T., van Heesch, U., Avgeriou, P.: A mapping study on documentation in continuous software development. Inf. Softw. Technol. **142**, 106733 (2022)

47. Yetistiren, B., Ozsoy, I., Tuzun, E.: Assessing the quality of github copilot's code generation. In: Proceedings of the 18th International Conference on Predictive Models and Data Analytics in Software Engineering, pp. 62–71. PROMISE 2022, Association for Computing Machinery, New York (2022)

48. Ziegler, A., et al.: Productivity assessment of neural code completion. In: Proceedings of the 6th ACM SIGPLAN International Symposium on Machine Programming, pp. 21–29. MAPS 2022. Association for Computing Machinery, New York (2022)

Towards Automatic Structured Inference of Module Abstractions

Gidon Ernst[✉] and Marian Lingsch-Rosenfeld

LMU Munich, Munich, Germany
`gidon.ernst@lmu.de, marian.lingsch-rosenfeld@sosy.ifi.lmu.de`

Abstract. We describe and exemplify an approach to infer specifications that soundly and completely represent the behavior of a given software module, without referring to the representation of the internal state.

Prelude

A group of computer scientists heard that a strange software system had been installed, but none of them were aware of its shape and form. Out of curiosity, they said: "We must inspect and know it by observing, of which we are capable". So, they sought it out, and when they found it they called the operation defined in its system interface. The first person, who invoked the constructor of the system, said, "I'm not sure whether anything had happened at all". For another one entered some data, and it seemed the system was crunching along happily. As for another person, who also entered their data, remarked, how sometimes they would be able to access their data later but sometimes not—the pattern seemed unclear. Yet another one who tried to upload pictures of cats was disappointed, as all they got were cryptic error messages and the number 42.

After a while, a wise man joined and said: Look within. You will find the answers to your questions in the implementation of the system, but also numerous bugs that may need to be rectified. Write down what you find out, but so that the next person can understand it easily. I give you the tools: Expressive structured algebraic specifications and a powerful verification system called KIV, by which you may gain ultimate trust in your judgement. But once your curiosity is satisfied, close the lid, not to be opened again, and leave this hardship behind, as then you will know the truth.

The computer scientists set out to do as they were advised. And certainly they found the answers to their questions. The bugs were even more numerous than anticipated, but finally, they had proudly produced a concise specification and had verified that it adequately describes the behavior of the system. It had indeed been a great deal of hardship and as they celebrated their achievements in the Room of Scrum, they pondered if they could build a machine that could do all this work on its own, without much guidance of a human. Imagine the benefits such a machine would bring to software engineers! This paper is the beginning of the story of some of them set out to construct a machine as described.

© The Author(s), under exclusive license to Springer Nature Switzerland AG 2025
G. Ernst et al. (Eds.): Wolfgang Reif Festschrift, LNCS 15765, pp. 303–324, 2025.
https://doi.org/10.1007/978-3-031-92196-4_15

1   Introduction

Deductive verification [4,33] of software systems typically makes use of *abstractions* [37,45], over which requirements of behaviors can be formulated. The virtue of such abstractions is that on one hand they are *precise*, i.e., they capture desired functionality in full detail, but that on the other hand that they are *easy to understand*, which enables validation and facilitates reasoning. The traditional view and presentation is that such abstractions are *provided*, typically by a human engineer, as an alternative representation of a given system that explains its behavior in a self-contained way, i.e., decoupled from implementation-level algorithms and data structures.

Well-established theory, e.g., that of Data Refinement [23], explains how the correspondence between an encapsulated stateful implementation of a software module and its abstraction is established during initialization and maintained via proofs throughout possible dynamic evolution of the system as a representation invariant that is shown to be a forward simulation [46,61]. This view is centered around finding a safety certificate that expresses that any possible momentary interaction with the module complies with the requirements outlined by the specification and moreover that this guarantee is inductively and therefore indefinitely preserved throughout the system's evolution.

Crafting good formal specifications is challenging [7,60]: It is not always easy to find a suitable degree of abstraction which allows one to express the actual behaviors of the system while at the same time avoiding to leak irrelevant implementation details. Moreover, the specification languages that achieve such abstractions are often not familiar to software engineers in practice as they emphasize mathematical rigor, use obscure notation, and require knowledge of concepts beyond those found in programming. Conducting the verification itself is even more challenging [28,31,72]: Expert knowledge and experience is required to make effective use of proof tools and to find arguments underpinning the correctness in the form of inductive invariants and auxiliary lemmas. Nevertheless, a wide range of mature tools are available, including powerful interactive verification systems like KIV [24,57,58,63], KeY [1], Isabelle/HOL [52], and Coq [8], and auto-active verification tools like Dafny [43], VerCors [11], Why 3 [27], and Frama-C [21]. Such tools have been applied to large case studies like seL4 [40], CompCert [44], and Flashix [12,25,62,64].

It is appealing to automate these two aspects. Assuming that a specification is given, there is a wide range of methods to automate the first-order aspects (discharging proof obligations, e.g., with SMT solvers [6,50] or ATPs [41]) and the second-order aspects (inference of invariants and coupling relations, e.g., with abstract domains [20] or interpolation [13,17]) as well as methods to discover auxiliary insights (lemma synthesis through goal-directed methods like rippling [15,38,68], recently in combination with E-graphs [42,51], as well as bottom-up theory exploration [18,26,65,66]).

Likewise, there are methods to discover the specifications themselves [2,54], which is more ambitious as one has to construct the proof of correctness alongside. Specification inference methods in practice typically aim to summarize the effect of code such as procedures at the level of program variables. While this may offer some potential to simplify away computational details, it does not address the main challenge of fundamentally raising the degree of abstraction that underpins good specifications. As an example, we may summarize operations such as lookup or insertion of a hash table data structure using quantified formulas over the array that stores the hash table, but coming up with the insight that a hash table is best described by a mathematical map requires human insight and understanding of the problem domain.

Contribution: In this paper, we aim to tackle that fundamental challenge: Given a system or module implementation, how can we describe its behavior *independently of the representation of its internal state*? We describe an approach that may discover such a description automatically, without human insight and creativity. Theoretically, this approach is rooted in well-known principles of analyzing axiomatic systems with respect to their model classes, and the conditions under which it is possible to describe these models uniquely. We showcase several examples that emphasize the importance of automatic lemma discovery as a key building block of reasoning across abstraction boundaries like those found in Data Refinement proofs and we discuss some promising ideas for effective algorithms to do so.

2   Motivation and Overview

Goal of this paper is to infer high-level, abstract descriptions of behaviors of stateful encapsulated software modules. To represent both modules as well as their requirements, we rely on *algebraic specifications Sp* [5,59,71], comprised of definitions of data types as well as functional realizations of interface operations. Software modules in this view are semantically described by algebras $\mathcal{A}$, in which the operations obey the constraints laid out by the specification as $\mathcal{A} \models Sp$. Algebras $\mathcal{A}$ that are *models* of a specification in this way can be seen as a thinking aid and as surrogates for the physical reality (e.g., program running on actual computers), whereas specifications Sp are syntactic objects that may be processed by a compiler and reasoned about mechanically by a theorem prover [56]. Taking such a model-centric perspective allows us to differentiate between descriptions of behavior and the behaviors themselves.

Here, implementations are characterized by *behavioral specifications* [19,32, 34] $I = Sp(S, \Delta)$, in which the internal state space S, the system operations, as well as all definitions Δ that characterize S and govern I's behavior are singled out.

Goal. Given an implementation of a software module as an algebraic specification $I = Sp(S, \Delta)$ we seek to infer a specification $Sp^*(S, \Phi)$, in which S remains an abstract sort, and all definitions Δ are replaced by axioms Φ that make no reference to any implementation detail. We aim to guarantee two properties:

- **Soundness:** The axiomatization Φ is faithful to the implementation and therefore encodes facts only that hold for the actual system, i.e., Δ implies Φ.
- **Completeness:** The discovered axiomatization captures all behavioral aspects of the implementation, i.e., describes the same class of models $\mathcal{A}$.

Table 1. Overview of modeling and specification approaches.

	implementation	behavioral model	axiomatic spec.
representation of state	concrete types: structs, classes, arrays, memory	mathematical types: records, sets, sequences, maps, bags, ...	*implicitly*
realization of operations	functional and imperative programs	transition relations, abstract programs	*implicitly*
verification approach		simulation proofs	observer completeness
main proof artifact		coupling invariants	lemmas connecting observations and updates

To explain our approach, we draw a distinction between what we call here *behavioral models* and *axiomatic specifications* as two possible approaches for formalizing requirements of stateful systems, as summarized in Table 1. In both cases, trust in the system's correct behavior is gained by establishing a formal correspondence between the two complementary explanations of the system's functionality, once in terms of the potentially complex but efficiently executable implementation, and once in terms of a much more abstract and therefore easy to understand mathematical description.

The key characteristic of implementations as well as behavioral models is the presence of an *explicit representation of state* in terms of which the computations of the operations are explained. The foundations in terms of data refinement [23] are well-established. In its emphasis on state representations, this approach offers a focal point by which the degree of abstraction is raised, and thus by which trust is generated. However, the same aspect that makes this idea worthwhile, namely ease of understanding, also suggests that it is ambitious to have such "true" abstractions be introduced automatically and algorithmically. From personal experience, retaining just as much detail as necessary to capture behavioral nuances while avoiding irrelevant complexity is usually a fine design point that relies on human judgement and creativity. Moreover, taking specific design choices can have somewhat unforeseen consequences on the ease and success of proofs. Overall, we are not aware of methods to automatically make judgements about designs of behavioral specifications in such regards.

In this paper, we focus instead on *axiomatic specifications*. The key difference is that they can be expressed without explicit reference to states (see Table 1 right-most column): Instead of explaining the behavior of individual operations in terms of state, the idea is to instead relate system operations among each other. This is reminiscent of how mathematical data types are presented as axioms like $x \in (s_1 \cup s_2) \iff x \in s_1 \lor x \in s_2$ the semantics of "update" $\cup$ is precisely determined with the help of "observer" operation $\in$.

Approach. In this paper, we construct axiomatic specifications from behavioral counterparts (implementations or models). The main idea is to express possible behaviors of the system by the relationship of observations that can be made about the system at different points during its computation. Concretely-represented states are replaced by the equivalence classes induced by the observations that we can make. State changes are then represented by the differences in observations that can be detected. It is well-known, however, that for some systems, state changes may become observable only later and only under certain future evolutions of the system. To express references to those future system states abstractly, we introduce *synthetic observers* that can generically describe the relationship of *sequences of operations*.

In this paper, we focus on deterministic systems, which is a proper restriction of our approach that can potentially be lifted by careful introduction of choice functions. We present the operations of a module implementation as total functional programs, which is a choice made for ease of presentation. The examples make use of standard functional data types, notably optionals and lists, where Constructors of the data types are written in green and are typically capitalized.

$$\textbf{data } \mathtt{Option}\langle\alpha\rangle = \mathtt{None} \mid \mathtt{Some}(\alpha)$$
$$\textbf{data } \mathtt{List}\langle\alpha\rangle = \mathtt{[]} \mid \alpha :: \mathtt{List}\langle\alpha\rangle$$

Example 1 (Implementation of Stacks with Lists). A realization of stacks using functional lists is comprised of a type definition for the representation for the internal state space S and the operations empty, push, pop, and top,

$$\textbf{type } S\langle\alpha\rangle := \mathtt{List}\langle\alpha\rangle \tag{1}$$

$$\mathtt{empty}() := \mathtt{[]} \qquad\qquad \mathtt{push}(x, xs) := x :: xs$$

$$\mathtt{pop}(\mathtt{[]}) := \mathtt{[]} \qquad\qquad \mathtt{top}(\mathtt{[]}) := \mathtt{None}$$
$$\mathtt{pop}(x :: xs) := xs \qquad\qquad \mathtt{top}(x :: xs) := \mathtt{Some}(x)$$

where we distinguish between **update** operations that change the internal state (written in blue) and **observer** operations with outputs (written in purple). ♡

Example 2 (Partial Axiomatization of Stacks). The intention behind making the type definition explicit here is that this is precisely the part that we want to avoid. An axiomatic specification is constructed from lemmas that determine the result of the single observer top after any update. We can easily find the first two equations, but the third one remains elusive.

$$\textbf{type } S\langle\alpha\rangle \quad \text{(left abstract)}$$

$$\mathtt{top}(\mathtt{empty}()) = \mathtt{None}$$
$$\mathtt{top}(\mathtt{push}(x, st)) = \mathtt{Some}(x)$$
$$\mathtt{top}(\mathtt{pop}(st)) = ? \tag{2}$$

In order to determine the result of $\mathtt{top}(\mathtt{pop}(st))$ it may be necessary to traverse the stack st even further and the number of pop operations we might encounter is not statically bounded.   ♡

Even though such partial system may be useful in practice, finding an answer to (2) is essential to achieve completeness (Sect. 3).

Example 3 (Ad-Hoc Axiomatization of Stacks). A well-known idea is to manually introduce a more powerful observer $\mathtt{nth}$ that allows to access all information stored in the data structure:

$$\mathtt{nth}(n, [\,]) := \mathsf{None}$$
$$\mathtt{nth}(0, x :: xs) := \mathsf{Some}(x)$$
$$\mathtt{nth}(n + 1, x :: xs) := \mathtt{nth}(n, xs) \tag{3}$$

It is easy to determine the laws connecting this function $\mathtt{nth}$ to all update operations. The previously-problematic pop can be handled simply by reaching one element further into the list in the pre-state as shown in (4). We also get the right-hand side for (2) in terms of the more general observer as shown in (5).

$$\mathtt{nth}(n, \mathtt{pop}(st)) = \mathtt{nth}(n + 1, st) \tag{4}$$
$$\mathtt{top}(\mathtt{pop}(st)) = \mathtt{nth}(1, st) \tag{5}$$

However, note that the definition of $\mathtt{nth}$ as shown above is yet again defined with reference to the concrete representation of state, and we think it is not really clear how to do this systematically in the general case.   ♡

Why $\mathtt{nth}$ does not get stuck like $\mathtt{top}$ when encountering a pop? It happens that $\mathtt{nth}$ already generalizes over an *arbitrary* number of pops, and so handling another one comes essentially for free. This correspondence is neatly described by recurrence Eq. (4) once we regard parameter n as a sequence of n deferred pop operations. It suggests a general principle to switch away from the data-centric view promoted by Example 3 towards an explicit representation of *sequences of operations*. In our approach, for some observer operation we introduce variants that are allowed to perform an some hypothetical updates before making the actual observation. This generalization of observer operations is driven by the failure to discover suitable lemmas. Here, $\mathtt{top}(st)$ in (2) is extended with an additional parameter c as $\mathtt{top}^*(c, st)$, where c ranges over a data type with one constructor corresponding to each problematic update operation.

Example 4 (Generalized Observers). For the stack, we add the following synthetic definitions that are a canonical part of inferred specifications $Sp^*(S, \Phi)$.

$$\mathbf{data}\ \Gamma := \square \mid \mathrm{Pop}(\Gamma)$$

$$\mathtt{top}^*(\square, st) := \mathtt{top}(st)$$
$$\mathtt{top}^*(\mathrm{Pop}(c), st) := \mathtt{top}^*(c, \mathtt{pop}(st)) \tag{6}$$

Note how the data type Γ is isomorphic to natural numbers, i.e., parameter n of nth roughly corresponds to $\mathrm{Pop}^n(\square)$, but in contrast to nth, the generalized observer top* is not realized by direct reference to the internal state, but solely with the system's interface. ♡

This construction allows us to canonically deal with all previously problematic cases for the original observers (7) by deferring them to their generalized variants. Also by construction, reading (6) *backwards* yields the required laws of the generalized observer for all operations captured by Γ. Of course, in order to arrive at a complete system description, we still need to discover all other laws.

Example 5 (Axiomatization of Stacks with Generalized Observers). Together with the first two equations from Example 2 the set of lemmas Φ of $Sp^*(S,\Phi)$ becomes:

$$\mathrm{top}(\mathrm{pop}(st) = \mathrm{top}^*(\mathrm{Pop}(\square), st)$$

$$\mathrm{top}^*(c, \mathrm{empty}()) = \mathrm{None}$$
$$\mathrm{top}^*(\square, \mathrm{push}(x, st)) = \mathrm{Some}(x)$$
$$\mathrm{top}^*(\mathrm{Pop}(c), \mathrm{push}(x, st)) = \mathrm{top}^*(c, st)$$
$$\mathrm{top}^*(c, \mathrm{pop}(st)) = \mathrm{top}^*(\mathrm{Pop}(c), st) \tag{7}$$

Covering all combinations between observers and update operations is the main ingredient in achieving goal "completeness". The other ingredient is to avoid cyclic arguments, which is ensured by the existence of a suitable well-founded order, here the number of operation calls in the second argument. ♡

The rest of the paper is structured as follows. In Sect. 3, we briefly summarize established notions, specifically "observable equivalence" as a criterion for completeness. In Sect. 4, we describe how the synthetic observers are constructed in a generic way. We assemble these ideas into an inference algorithm in Sect. 5, which in turn relies on a lemma discovery oracle, some initial work towards the latter is described in Sect. 6.

3   Implementations, Specifications, and Observations

We briefly summarize notions from the literature that describe behavioral specifications in terms of the "Hidden Algebra" paradigm [32], observational logic [36], and criteria for sound and complete axiomatizations [9,35,53], based on the notion of "behavioral equivalence" [22,55,61].

Definition 1 (Algebraic Specifications). *A specification $Sp = (\Sigma, Ax)$ consists of a signature $\Sigma = (T, F)$ of sorts T and function symbols F, and a set of axioms Ax which characterize properties of the symbols declared by the signature.*

In this paper, we allow axioms that define abbreviations for types, such as such as **type** $S\langle\alpha\rangle := \mathrm{List}\langle\alpha\rangle$ in Example 2. The theoretical presentation will be in terms of many-sorted signatures without parametric types for simplicity.

Definition 2 (Models of Specifications). *An algebra* $\mathcal{A} = (T^{\mathcal{A}}, F^{\mathcal{A}})$ *over a signature* $\Sigma = (T, F)$ *assigns carrier sorts* $t^{\mathcal{A}}$ *to all sorts* $t \in T$ *and functions* $f^{\mathcal{A}}$ *to all function symbols* $f \in F$. *An algebra* $\mathcal{A}$ *is called a* model *of a specification* $Sp = (\Sigma, Ax)$ *if it satisfies all axioms, i.e.,* $\mathcal{A} \models Ax$, *where the satisfaction relation* $\models$ *is defined in the standard way. We write this as* $\mathcal{A} \models Sp$.

As motivated in Sect. 2, behavioral specifications are distinguished from axiomatic ones by the explicit designation of certain sorts as representing internal states.

Definition 3 (Behavioral Signature). *A behavioral signature* $\Sigma(S) = (T \uplus \{S\}, F \uplus Ops)$ *where* $Ops = \{\mathtt{init}\} \uplus \{\mathtt{upd}_n\}_{n \in N} \uplus \{\mathtt{obs}_m\}_{m \in M}$ *over hidden sorts* S *distinguishes a constant* $\mathtt{init} \colon S$, *observer operations* $\mathtt{obs}_m \colon I_m \times S \to O_m$, *and update operations* $\mathtt{upd}_n \colon J_n \times S \to S$, *where* $I_m, O_m, J_n \subseteq T$.

Note that the hidden sort S occurs at most once as an argument and is passed on "linearly" to describe executions of the system. We refrain from generalizing this notion to multiple hidden sorts, as we are for now interested in describing a single module only. If necessary, S can be a tuple type with several individual components, to represent that the system is implemented with more than one state variable. Sorts I_m, O_m resp. J_n are the respective inputs and outputs of observer $\mathtt{obs}_m$ resp. update $\mathtt{upd}_n$.

Definition 4 (Behavioral Specification). *An algebraic behavioral specification* $Sp(S, \Delta) = (\Sigma(S), \Delta \uplus Ax)$ *is a specification that is defined over a behavioral signature* $\Sigma(S)$, *in which* Δ *are exactly those axioms that define the hidden sort* S *and the operations* $\mathtt{init}$, $\mathtt{obs}_m$, *and* $\mathtt{upd}_n$.

From this notation, which we leave somewhat informal, we expect that if we remove the details related to the hidden sort and the implementation of the operations, written $Sp(S, \varnothing)$, these will be entirely unconstrained:

Proposition 1. *By convention on the notation* $Sp(S, \varnothing)$, *given* $\mathcal{A} \models Sp(S, \Delta)$ *we also have* $\mathcal{A}' \models Sp(S, \varnothing)$ *for all* $\mathcal{A}'$ *that coincide with* $\mathcal{A}$ *but assign an arbitrary carrier set* $S^{\mathcal{A}'}$ *to the hidden sort* S *and any interpretation* $\mathtt{init}^{\mathcal{A}'}$, $\mathtt{obs}_m^{\mathcal{A}'}$, *and* $\mathtt{upd}_n^{\mathcal{A}'}$ *to the behavioral part of the signature.*

Subsequently we equip behavioral specifications with the means to characterize the observations we can make about terms of the hidden sort S, i.e., observations about the internal states of the system. We will do so syntactically by introducing new predicate and function symbols with suitable definitions.

Definition 5 (Observable Equivalence). *Two expressions denoting states* $s_1, s_2 \colon S$ *are observably equivalent, if they cannot be differentiated by immediate observations. We capture this by relation* $s_1 \approx s_2$, *syntactically defined as*

$$s_1 \approx s_2 \quad := \quad \bigwedge_{m \in M} \forall\, i \colon I_m.\ \mathtt{obs}_m(i, s_1) = \mathtt{obs}_m(i, s_2)$$

The notion of "observer-complete definitions" [9] provides a sufficient criterion for equations Φ which if they exist, fully characterize the behavior of $Sp(S, \Delta)$ in terms of $Sp(S, \Phi)$. Here we give a somewhat weaker version than possible, as we will generalize it to more complex scenarios in Sect. 4 anyway. Note that the equations shown in Example 2 are instances of this definition.

Definition 6 (Observer-Complete Descriptions). *A set of equations Φ for an observable specification $Sp(S, \Delta)$ is called* observer-complete, *if Φ follows from $Sp(S, \Delta)$ and contains an equation for each pair of observer $m \in M$ and the constructor* init *as well as updates $n \in N$ that explains their commutation:*

$$\mathsf{obs}_m(i, \mathtt{init}) = \iota(i) \quad \in \quad \Phi \tag{8}$$

$$\mathsf{obs}_m(i, \mathsf{upd}_n(j, s)) = \chi_m(i, j, \{\mathsf{obs}_{m'}(i', s)\}) \quad \in \quad \Phi \tag{9}$$

where $\iota(i)$ is a state-independent expression that explains observations over the initial state; and χ_m is an expression that explains observations over any reachable state. It may refer to the inputs i and j and arbitrary observations about the pre-state, but χ_m may syntactically refer to arbitrary observations $\mathsf{obs}_{m'}(i', s)$ about the predecessor state s but not to any $\mathsf{upd}_{n'}$.

Definition 7 (Soundness). *An (observer-complete set) of equations Φ is* sound, *if $\mathcal{A} \models Sp(S, \Delta) \implies \mathcal{A} \models Sp(S, \Phi)$, for all $\mathcal{A}$.*

Note, Def. 6 already requires that Φ follows from $Sp(s, \Delta)$.

Lemma 1. *Given an observer-complete set of equations Φ for $Sp(S, \Delta)$, then Def. 5 yields a congruence relation $\approx^{\mathcal{A}}$ on $S^{\mathcal{A}}$ when $\mathcal{A} \models Sp(S, \Phi)$ is given.* □

Corollary 1. *Given an observer-complete set of equations Φ for an observable specification $Sp(S, \Delta)$ and a model $\mathcal{A} \models Sp(S, \Delta)$, the quotient algebra $\mathcal{A}/_{\approx^{\mathcal{A}}}$, which identifies congruent elements of $S^{\mathcal{A}}$, is well-defined.* □

Theorem 1 (Completeness). *For two models $\mathcal{A} \models Sp(S, \Phi)$ and $\mathcal{A}' \models Sp(S, \Phi)$ their quotients $\mathcal{A}/_{\approx^{\mathcal{A}}}$ and $\mathcal{A}'/_{\approx^{\mathcal{A}'}}$ wrt. observably-equivalent states are isomorphic.*

This property follows from a helper lemma, cf. [9, Thm 4.6].

Lemma 2. *For all expressions $s : S$, observers $m \in M$, and input terms $i : I_m$ there is an expression $r_{m,i,s} : O_m$ so that $\mathcal{A} \models \mathsf{obs}_m(i, s) = r_{m,i,s}$.*

Proof. By structural induction on the subterm order of s, we have $r_{m,i,\mathtt{init}} = \iota(i)$ for $s = \mathtt{init}$ and $r_{m,i,\mathsf{upd}_n(j,s')} = \chi_m(i, j, \{r_{m',i',s'}\})$ for $s = \mathsf{upd}_n(j, s')$. As s' is a subterm of s, $\mathcal{A} \models \mathsf{obs}_{m'}(i', s') = r_{m',i',s'}$. □

4   Behavioral Completion

Observer-complete sets of equations may not exists over the original signature of $Sp(S, \Delta)$, as shown in Example 1. Depending on whether all information stored in the system can be accessed at any time, as it is the case, e.g., with map-like modules or a file system, or whether some information may be temporarily inaccessible, as it is the case, e.g., with stacks or queues, where we can see only the top resp. front/rear elements. A standard idea is to add further (synthetic) observer operations:

Definition 8 (Behavioral Completion). *For a given behavioral specification $Sp(S, \Delta) = (\Sigma(S), \Delta \uplus Ax)$ with $\Sigma(S) = (T \uplus \{S\}, F \uplus Ops)$, we define its behavioral completion $Sp^*(S, \Delta)$ by*

$$\Sigma^*(S) = \big(T \uplus \{S, C\}, \; F \uplus Ops \uplus \{obs_m^*\}_{m \in M} \uplus \{Upd_n\}_{n \in N}\big)$$
$$Sp^*(S, \Delta) = \big(\Sigma^*(S), \; \Delta \uplus Ax \uplus \{ \; Eqs. \; (10) \; and \; (11), (12) \; \}\big)$$

where

$$\textbf{data } C := \Box \; \big| \; \{ \; \mathrm{Upd}_n(J_n, C) \; \}_{n \in N} \tag{10}$$

$$\mathbf{obs}_m^*(i, \Box, s) := \mathbf{obs}_m(i, s) \tag{11}$$
$$\mathbf{obs}_m^*(i, \mathrm{Upd}_n(j, c), s) := \mathbf{obs}_m^*(i, c, \mathbf{upd}_n(j, s)) \tag{12}$$

The observable completion introduces generalized observer operations $\mathbf{obs}_m^*(c, s)$, which take an additional argument $c \colon C$ over their original counterparts $\mathbf{obs}_m(s)$. Data type C reifies sequences of operations as shown in Example 4. We consider the axioms that characterize C as a freely generated data type to be part of Sp^*. The definition of $\mathbf{obs}_m^*$ simply unwinds its extra argument and applies the sequence of operations that it specifies.

Note that the outermost constructor corresponds to the first operation applied and finally the original observer $\mathbf{obs}_m$ is invoked, and recall that our goal is to ultimately explain Sp without relying on a model for $\mathbf{upd}_n$ in the right-hand side of Eq. (12).

Lemma 3. *In the behaviorally completed specification $Sp^*(S, Ax)$, the generalized observers alone are sufficient to characterize uniqueness of models because*

$$s_1 \approx s_2 \iff \bigwedge_{m \in M} \forall \, i \colon I_m, c \colon C. \; \mathbf{obs}_m^*(i, c, s_1) = \mathbf{obs}_m^*(i, c, s_2)$$

Proof. Key to understanding the $\Longrightarrow$-direction is that Def. 5 on the left hand side unfolds into conjuncts for all observers, which contains not only the original ones but also each $\mathbf{obs}_m^*$. The $\Longleftarrow$-direction follows from Eq. (11) by instantiating the universal quantifier with $c = \Box$. $\Box$

Therefore, to construct an axiomatic specification, we may simply explain the original observers in terms of the generic ones through (11) and perform the specification inference on the latter only. Moreover, as seen in Example 4, interpreting Eq. (12) from right-to-left yields a canonical way to dispatch observations after updates $\mathsf{upd}_n(j, s)$ that can not easily be reduced to the pre-state s.

However, we have to pay attention not shift back and forth the deferred operations *indefinitely*—at some point we have to make reliable progress. This is achieved as usual by requiring the existence of a well-founded order that ensures that the model of obs^*_m in Sp^* is described uniquely even in the absence of its normative definition by Eqs. (11) and (12).

Definition 9 (Behavior-Complete Descriptions). *A set of equations Φ for a specification $Sp^*(S, \Delta)$ is* behavior-complete, *if Φ follows from $Sp^*(S, \Delta)$ and contains an equation for each $m \in M$ that explains an observation with $\mathsf{obs}_m(i, \_)$ after evaluating an additional construction context c:*

$$\mathsf{obs}^*_m(i, c, s) = \chi_m(i, c, \{\mathsf{obs}^*_{m'}(i', c', s')\}) \quad \in \quad \Phi \tag{13}$$

as long as the right-hand side refers to combinations $(m', i', c', s') \prec (m, i, c, s)$ only that decrease according to some well-founded order $\prec$ over the operation name and terms for the input, construction context, and denotation of state

We emphasize that the validity checks and therefore the construction of equations Eq. (13) can be be performed inside the logic, thanks to the reification supported by Def. 8. In contrast, the well-founded order $\prec$ is at the meta-level as it analyzes the syntactic structure of how the equation is presented. Equations (8) and (9) are an instances of Eq. (13), in which $\prec$ compares the terms s, s' by structural recursion. Theorem 1 can be proved with an analogue lemma by induction over the well-founded order $\prec$.

Example 6 (Priority Queues). The simplest behavioral explanation of priority queues is with sets. We present the corresponding specification where elements are natural numbers and their priority is simply given by their value—with smaller numbers being preferred. We assume a type $\mathsf{Set}\langle\alpha\rangle$ that is provided in the standard library with operators $\varnothing, \in, \{\,\}, \cup, \cap$ and function min that satisfies $\mathsf{min}(s) \in s$ and $\forall\, x \in s.\ \mathsf{min}(s) \leq x$ whenever $s \neq \varnothing$.

$$\begin{aligned}
&\mathbf{type}\ S\langle\alpha\rangle := \mathsf{Set}\langle\alpha\rangle \\
&\quad \mathsf{init}() := \varnothing \\
&\quad \mathsf{enq}(x, q) := q \cup \{x\} \\
&\qquad \mathsf{deq}(q) := \mathbf{if}\ q = \varnothing\ \mathbf{then}\ \varnothing\ \mathbf{else}\ q - \{\mathsf{min}(q)\} \\
&\qquad \mathsf{top}(q) := \mathbf{if}\ q = \varnothing\ \mathbf{then}\ \mathsf{None}\ \mathbf{else}\ \mathsf{Some}(\mathsf{min}(q))
\end{aligned}$$

As with the stack in Example 1, there are lemmas connecting **top** with **init** and **enq**:

$$\text{top}(\text{init}) = \text{None}$$

$$\begin{aligned}
\text{top}(\text{enq}(x, q)) = \ &\textbf{match } \text{top}(q) \\
&\textbf{case } \text{None } \textbf{then } \text{Some}(x) \\
&\textbf{case } \text{Some}(y) \ \textbf{if } x \leq y \ \textbf{then } \text{Some}(x) \\
&\textbf{case } \text{Some}(y) \ \textbf{if } x > y \ \textbf{then } \text{Some}(y)
\end{aligned} \tag{14}$$

Analogously to Example 2 we get stuck with $\text{top}(\text{deq}(q))$. We let Γ to have a corresponding constructor Deq and we introduce generalized top^* observer for which we discover a complete set of lemmas as required by Def. 9.

$$\textbf{data } \Gamma := \square \mid \text{Deq}(\Gamma) \tag{15}$$

$$\text{top}(q) = \text{top}^*(\square, q)$$

$$\text{top}^*(c, \text{init}()) = \text{None}$$
$$\text{top}^*(c, \text{deq}(q)) = \text{top}^*(\text{Deq}(c), q)$$
$$\text{top}^*(\square, \text{enq}(x, q)) = \text{top}(\text{enq}(x, q)) \qquad \text{(can unfold as in (14))}$$

$$\begin{aligned}
\text{top}^*(\text{Deq}(c), \text{enq}(x, q)) = \ &\textbf{match } \text{top}(q) \\
&\textbf{case } \text{None } \textbf{then } \text{None} \\
&\textbf{case } \text{Some}(y) \ \textbf{if } x < y \ \textbf{then } \text{top}^*(c, q) \\
&\textbf{case } \text{Some}(y) \ \textbf{if } x = y \ \textbf{then } \text{top}^*(c, \text{deq}(q)) \\
&\textbf{case } \text{Some}(y) \ \textbf{if } x > y \ \textbf{then } \text{top}^*(c, \text{enq}(x, \text{deq}(q)))
\end{aligned} \tag{16}$$

The last Eq. (16) processes one **deq** operation. There are four cases. If the queue contained one element before, i.e. $q = \varnothing$, then the queue becomes empty and remains so while processing the remaining operations in c. In the second and third case, the most-recently added element x is indeed the topmost one. If the element was not part of the queue q before, then the Deq and the **enq** cancel out. Otherwise, f the element *was* already present before, $\text{enq}(x, q)$ does not cause a state change and we propoagate the **deq** further to the point in the history when x was added first. Lastly, if x is not the topmost element, the result is described by first removing the topmost element of q recursively but then adding x again before further processing any pending Deq in c. This is because we do not know which priority x takes in the entire queue, and consequentially, it remains unclear whether that x will be removed at all, at least for now. In other words, Deq and **enq** commute, but we make progress because the queue becomes smaller.

We believe that this operation is terminating on all inputs, based on a termination measure that is in $O(|c| \times |q|)$ where $|c| = n$ if $c = \text{Deq}^n(\square)$ measures the size of c and $|q|$ measures the number of elements in the queue. The argument is that deleting $|c|$ elements by linear search in a list should have that asymptotic complexity. However, we have not yet worked out the full details of the termination order. Note that both arguments are increased sometimes, and $\mathsf{enq}(x, \mathsf{deq}(q))$ is smaller in cardinality than $\mathsf{enq}(x, q)$ while other operations decrease that argument in terms of operation calls, and it appears that the details of how we count exactly do matter. ♡

5 Inference Algorithm

In Sect. 3 we have seen the criteria on how to construct axiomatic abstractions of a given module implementation $I = Sp(S, \Delta)$, where hidden sort S represents the internal state space and Δ are behavioral definitions of the interface operations. The form of the abstraction is a specification $Sp^*(S, \Phi)$ in which observer-complete equations replace those definitions, according to Defs. 6 and 9. We describe an *algorithm* to construct such axiomatic specifications below, focusing on the general observers obs_m^* and defining the original ones by Eq. (11) from right to left.

Input. Given a module implementation, represented as a behavioral specification $I = Sp(S, \Delta)$ with hidden sorts S and internal definitions Δ (cf. Def. 4).

State. The algorithm maintains two sets of equations as instances of Eq. (13) The first set of equations Λ contains all *lemmas* that have been found so far, with the required property that Λ follows from Δ and the intended property that the left-hand sides of $\Lambda \cup \Theta$ always cover all cases, where equations in Θ are the *todo* list of left-hand sides that remain to be considered. Moreover, we track a set $\Gamma_m \subseteq C$ that constrains which contexts need to be considered in left-hand sides for obs_m^* in Θ.

$$\mathsf{obs}_m^*(i, c, s) = \chi_m(i, c, \{\mathsf{obs}_{m'}^*(i', c', s')\}) \qquad \in \Lambda$$
$$\mathsf{obs}_m^*(i, c, s) = ? \quad (\text{where } c \in \Gamma_m) \qquad \in \Theta$$

Initially, we start with $\Theta = \{\mathsf{obs}_m^*(i, c, s) = ?\}_{m \in M}$ where i, c, s are just variables. Below, Γ_m is represented by its grammar, starting with **data** $\Gamma_m := \square$. This reflects the initial assumption that observational equivalence is sufficient (Def. 6), which is gradually refined by the algorithm. More complex types of constraints to represent Γ_m could be interesting in the future.

Result. To construct the axiomatization $Sp^*(S, \Phi)$, the algorithm must select a set of equations $\Phi \subseteq \Lambda$ that satisfies the requirements of Def. 9, i.e., the left-hand sides $\mathsf{obs}_m^*(i, c, s) = \cdots \in \Phi$ are a partitioning of possible operations m and arguments i, c, s and there exists a suitable well-founded ordering $\prec$.

Calculation Steps. The main algorithm is a refinement loop:

(a) It incrementally discerns cases of the most recent operation applied to the state s in an equation of Θ and asks a lemma synthesis oracle to find the corresponding right-hand side. Initially, this will solve Eqs. (8) and (9). The lemma synthesis oracle operates over the full specification $Sp^*(S, \Delta \cup \Lambda)$ and may take into account all lemmas that have been found in prior steps.

(b) For the cases $\mathsf{obs}^*_m(i, c, \mathsf{upd}_n(j, s)) = ?$ for which we cannot find a lemma, we enlarge $\Gamma_m := \Gamma_m \mid \mathsf{Upd}_n(j, \Gamma_m)$ to contain the corresponding constructor. We define a default case that shifts the problematic update into the context

$$\mathsf{obs}^*_m(i, c, \mathsf{upd}_n(j, s)) = \mathsf{obs}^*_m(i, \mathsf{Upd}_n(j, c), s) \qquad \in \Lambda$$

and add the requirement that the now-enlarged context is considered with respect to initialization and all other updates n',

$$\mathsf{obs}^*_m(i, c, \mathsf{init}) = ? \qquad\qquad \in \Theta$$
$$\mathsf{obs}^*_m(i, c, \mathsf{upd}_{n'}(j, s)) = ? \qquad \text{for all } n' \neq n \qquad \in \Theta$$

where c is either a plain variable, or we can already construct a case distinction over the outermost constructors, for example including a pattern $c = \mathsf{Upd}_n(j, c')$ for each n recently added to Γ_m.

(c) If we find lemmas for all cases in step (a) we attempt to construct a result as described above.

Note that we do not make assumptions on the shape of lemmas constructed by the oracle, instead, only at the end we check for the existence of a well-founded order. Moreover, it surely is possible that lemma synthesis proposes multiple candidates in step (a), which is fine even if some of these syntactically still refer to internal implementation details, as that requirement is also checked only at the end. Different approaches are of course possible and may be more efficient.

Example 7. We illustrate a potential run of the algorithm for the stack (Example 5).

1. Step (a) of the first iteration for the unique observer top^* with **data** $\Gamma^1 := \square$ produces new lemmas Λ^1

$$\mathsf{top}^*(\square, \mathsf{empty}()) = \mathsf{None} \qquad\qquad (\dagger)$$
$$\mathsf{top}^*(\square, \mathsf{push}(x, s)) = \mathsf{Some}(x)$$

Step (b) finds that we have no lemma for **pop**, sets **data** $\Gamma^2 := \square \mid \mathsf{Pop}(\Gamma^2)$, and adds two further queries to Θ, where $c \in \Gamma^2$ ranges over the now-generalized context of deferred operations:

$$\mathsf{top}^*(\mathsf{Pop}(c), \mathsf{empty}()) = ? \qquad\qquad (\ddagger)$$
$$\mathsf{top}^*(\mathsf{Pop}(c), \mathsf{push}(x, s)) = ?$$

where we have done the case distinction over the innermost constructor of the context.

2. In the second iteration, lemma synthesis first unfolds these new equations using (12) and then simplifies them using discovered properties from the implementation, i.e., $\Delta \models$

$$\cdots = \text{top}^*(c, \text{pop}(\text{empty}())) = \text{top}^*(c, \text{empty}())$$
$$\cdots = \text{top}^*(c, \text{pop}(\text{push}(x, s))) = \text{top}^*(c, s)$$

$\heartsuit$

6   Calculational Lemma Synthesis

We briefly showcase a few approaches to lemma synthesis that can provide solutions to the queries asked by the algorithm from Sect. 5.

Definition 10 (Equational Lemma Synthesis). *Given a specification Sp, a lemma synthesis query is of the form lhs = ? A solution is an expression rhs with free(rhs) $\subseteq$ free(lhs) and all models $\mathcal{A} \models Sp$ also validate $\mathcal{A} \models lhs = rhs$.*

Enumeration. The classic solution is based on enumeration [14,18,65], sometimes called syntax-guided synthesis [3], where candidates for *rhs* are taken from the search space of all well-formed expressions of the correct type, over the variables *free(lhs)* in scope. This search space is rather large, in particular, if many function symbols are present in *Sp*, which can be a significant problem for the analysis of complex systems. Moreover, the validity check whether candidates are correct typically involves an automated and inductive proof oracle, which is often costly or may fail due to the undecidability of the general problem. Nevertheless, this approach has proved to be effective for exploring smaller theories [39], in particular when combined with data-driven reasoning (i.e., bounded counterexample search to quickly reject false candidates or accept candidates based on testing only [48,66]).

Calculation. However, we can do better, and proceed by forward symbolic reasoning, starting from the left-hand side and transforming it gradually towards solutions that are correct by construction. For instance, the lemma synthesis in the second iteration of Example 7 may simply rewrite subterms pop(empty) and pop(push(x, s)) by unfolding the definitions.

Calculational techniques based on unfold-fold transformations [10,16] approach equational rewriting modulo induction, i.e., we intertwine the discovery of right-hand side candidates with an inductive proof over some structure of the left-hand side. These techniques have been used for program transformations in optimizing compilers for manual program development [49], and also for proofs [67]. The idea to leverage transformations for lemma discover has been pursued by the first author [26], by specifically applying two transformations, fusion [47,69,70] and parameter removal [29,30], which produce lemmas that fit well with the shape of distributive laws like those of Def. 6.

$$f(x, g(y)) = fg^?(x, y) \qquad \text{(fusion)}$$
$$h(x, a) = e^?(a, x, h^?(x)) \qquad \text{(parameter removal)}$$

It combines an exploration procedure that, in contrast to enumeration-based lemma synthesis, does not work on the term level but rather on the level of function definitions. Here, functions $fg^?$ and $h^?$ are calculated from the definitions of f, g, and h, respectively, whereas expression $e^?$ may be found by heuristics or enumeration. In particular, if a is an accumulator, or even passed down to bases cases without modification, solutions for $e^?$ often make use of laws for neutral elements and associativity of some functions involved.

Lemmas are generated whenever one of these synthetic functions is discovered to be structurally identical to another one that is already known, and similarly if one function is recognized to be constant, i.e.,

$$f(x) = g(\pi(x)) \qquad\qquad f(x) = k$$

where π is a permutation of the arguments and k is "constant" expression over variables x that omits the recursion of f.

The advantage of this approach is that the structure of the functions involved already guides a large part of the proof, notably, it governs the underlying induction argument that semantically validates these equations. Therefore, it does not rely as much on powerful automated proof oracles. Similarly, enumeration is tamed, because much of the construction of $fg^?$ and $h^?$ is canonical from the already existing structure.

Not all lemmas resp. examples are in scope of the current prototype implementation for lemma discovery (called LEMMACALC [26]). We are currently working on improving the algorithms and integrating enumeration into the search procedure in a principled way.

Of the lemmas shown in Example 5, Eq. (7) is the only one that requires an inductive proof. We illustrate how a calculational approach can find it with a combination of fusion for $f = \texttt{top}^*$ and $g = \texttt{empty}()$ and recognizing that the resulting functions is constant. However, we emphasize that this lemma merely simplifies the axiomatization, it is just as good to keep the case distinction over the context as with (†) and (‡) in Example 7.

Example 8. (Fusing $fg(c) = \texttt{top}^*(c, \texttt{empty}())$for the Stack). Fusion is based on the unfold-fold paradigm, pioneered by [16]. We proceed by case analysis on c

to unfold the definition of $\texttt{top}^*$ from Example 4.

$$\begin{aligned}
fg(\Box, \texttt{empty}()) &:= \texttt{top}^*(\Box, \texttt{empty}()) &&\text{(unfold } fg) \\
&= \texttt{top}(\texttt{empty}()) &&\text{(Def. } \texttt{top}^*) \\
&= \texttt{None} &&\text{(Def. } \texttt{top}) \\[6pt]
fg(\texttt{Pop}(c), \texttt{empty}()) &:= \texttt{top}^*(\texttt{Pop}(c), \texttt{empty}()) &&\text{(unfold } fg) \\
&= \texttt{top}^*(c, \texttt{pop}(\texttt{empty}())) &&\text{(Def. } \texttt{top}^*) \\
&= \texttt{top}^*(c, \texttt{empty}()) &&\text{(Def. } \texttt{pop}) \\
&= fg(c) &&\text{(fold. } fg)
\end{aligned}$$

Reading the equations in the above calculations end-to-end gives a self-contained recursive definition of a new synthetic function fg. ♡

The advantage of fg over the explicit composition of $\texttt{top}^*$ and $\texttt{empty}()$ is that some computations can be inlined symbolically. The crucial step in the calculation is from the third to fourth line in the recursive case, where $\texttt{pop}(\texttt{empty}())$ gets simplified to $\texttt{empty}()$.

Example 9. (Recognizing $fg(c)$ as constant). The base case result $\texttt{None}$ of the calculated definition of fg is independent of the recursive traversal. Moreover, it does not depend on any computation performed by fg (which is trivial anyway). Therefore, we can structurally recognize fg to return a constant result, analogously to hoisting constant expressions over loops. Remembering how fg was fused in the first place, we have:

$$\texttt{top}^*(c, \texttt{empty}()) \overset{\text{fusion}}{=} fg(c) \overset{\text{constant}}{=} \texttt{None}$$

♡

7 Conclusion

Programming is fun, and for some people[1] so is writing the formal specifications that accompany the programs. But for most software engineers, writing formal specifications is a difficult and tedious task. *This need not be the case.* We strongly believe that it is possible to use automatic techniques and formal methods to infer an appropriate specification from the code and proof that it models the program. In particular this should be possible knowing only the interface of the module and the code that implements it.

In this paper, we presented some first thoughts on how to infer specifications from code. Generalizing and automating our approach would be a first step in the right direction. This paper furthermore emphasizes the importance of lemma synthesis as a key ingredient for reasoning about data abstractions. We have

---
[1] This includes the authors of this paper.

presented preliminary ideas to achieve efficiency beyond exclusively enumerative approaches to lemma discovery, and we envision other promising applications for such methods. Future steps could extend the approach to non-deterministic systems, or to ones with bugs in the implementation whose existence are made apparent during the inference of the specification or when showing that the program conforms to it.

Acknowledgement. We thank Rolf Hennicker for an exciting and insightful discussion on the issue of observer-(in-)completeness of the stack example, which has led to the development of the presented approach.

References

1. Ahrendt, W., Beckert, B., Bubel, R., Hähnle, R., Schmitt, P.H., Ulbrich, M. (eds.): Deductive Software Verification - The KeY Book - From Theory to Practice, LNCS, vol. 10001. Springer (2016). https://doi.org/10.1007/978-3-319-49812-6
2. Albarghouthi, A., Dillig, I., Gurfinkel, A.: Maximal specification synthesis. In: POPL, pp. 789–801. ACM (2016). https://doi.org/10.1145/2837614.2837628
3. Alur, R., et al.: Syntax-guided synthesis. IEEE (2013)
4. Appel, A.W., et al.: Position paper: the science of deep specification. Philos. Trans. R. Soc. A Math. Phys. Eng. Sci. **375**(2104), 20160331 (2017). https://doi.org/10.1098/rsta.2016.0331
5. Astesiano, E., et al.: CASL: the common algebraic specification language. Theoret. Comput. Sci. **286**(2), 153–196 (2002). https://doi.org/10.1016/S0304-3975(01)00368-1
6. Barbosa, H., et al.: cvc5: a versatile and industrial-strength SMT solver. In: TACAS, pp. 415–442. Springer, Cham (2022). https://doi.org/10.1007/978-3-030-99524-9_24
7. Baumann, C., Beckert, B., Blasum, H., Bormer, T.: Lessons learned from microkernel verification–specification is the new bottleneck. In: SSV. EPTCS, vol. 102, pp. 18–32. Elsevier (2012). https://doi.org/10.4204/EPTCS.102.4
8. Bertot, Y., Castéran, P.: Coq'Art: Interactive theorem proving and program development. In: Texts in Theoretical Computer Science. An EATCS Series, Springer, Heidelberg (2004). https://doi.org/10.1007/978-3-662-07964-5
9. Bidoit, M., Hennicker, R.: Observer complete definitions are behaviourally coherent. OBJ/CafeOBJ/Maude Formal Methods **99**, 83–94 (1999)
10. Bird, R.S.: Algebraic identities for program calculation. Comput. J. **32**(2), 122–126 (1989). https://doi.org/10.1093/COMJNL/32.2.122
11. Blom, S., Darabi, S., Huisman, M., Oortwijn, W.: The VerCors tool set: verification of parallel and concurrent software. In: IFM, pp. 102–110. Springer, Cham (2017). https://doi.org/10.1007/978-3-319-66845-1_7
12. Bodenmüller, S., Schellhorn, G., Bitterlich, M., Reif, W.: Flashix: modular verification of a concurrent and crash-safe flash file system. Logic, Computation and Rigorous Methods: Essays Dedicated to Egon Börger on the Occasion of His 75th Birthday, pp. 239–265 (2021)
13. Bradley, A.R.: SAT-based model checking without unrolling. In: VMCAI, pp. 70–87. Springer, Heidelberg (2011). https://doi.org/10.1007/978-3-642-18275-4_7

14. Buchberger, B., et al.: Theorema: towards computer-aided mathematical theory exploration. J. Appl. Log. **4**(4), 470–504 (2006). https://doi.org/10.1016/J.JAL.2005.10.006

15. Bundy, A., Stevens, A., Van Harmelen, F., Ireland, A., Smaill, A.: Rippling: a heuristic for guiding inductive proofs. Artif. Intell. **62**(2), 185–253 (1993). https://doi.org/10.1016/0004-3702(93)90079-Q

16. Burstall, R.M., Darlington, J.: A transformation system for developing recursive programs. J. ACM **24**(1), 44–67 (1977). https://doi.org/10.1145/321992.321996

17. Cimatti, A., Griggio, A.: Software model checking via IC3. In: CAV, pp. 277–293. Springer, Heidelberg (2012). https://doi.org/10.1007/978-3-642-31424-7_23

18. Claessen, K., Johansson, M., Rosén, D., Smallbone, N.: Automating inductive proofs using theory exploration. In: International Conference on Automated Deduction, pp. 392–406. Springer, Heidelberg (2013). https://doi.org/10.1007/978-3-642-38574-2_27

19. Clavel, M., et al.: Maude: specification and programming in rewriting logic. Theor. Comput. Sci. **285**(2), 187–243 (2002). https://doi.org/10.1016/S0304-3975(01)00359-0

20. Cousot, P., Cousot, R.: Abstract interpretation: a unified lattice model for static analysis of programs by construction or approximation of fixpoints. In: POPL, pp. 238–252 (1977). https://doi.org/10.1145/512950.512973

21. Cuoq, P., Kirchner, F., Kosmatov, N., Prevosto, V., Signoles, J., Yakobowski, B.: Frama-C. In: Eleftherakis, G., Hinchey, M., Holcombe, M. (eds.) SEFM 2012. LNCS, vol. 7504, pp. 233–247. Springer, Heidelberg (2012). https://doi.org/10.1007/978-3-642-33826-7_16

22. De Nicola, R., Hennessy, M.C.: Testing equivalences for processes. Theoret. Comput. Sci. **34**(1–2), 83–133 (1984). https://doi.org/10.1016/0304-3975(84)90113-0

23. De Roever, W.P., Engelhardt, K., Buth, K.H.: Data Refinement: Model-Oriented Proof Methods and Their Comparison, vol. 47. Cambridge University Press (1998)

24. Ernst, G., Pfähler, J., Schellhorn, G., Haneberg, D., Reif, W.: KIV–overview and verify this competition. Softw. Tools Technol. Transfer (STTT) **17**(6), 677–694 (2015)

25. Ernst, G., Pfähler, J., Schellhorn, G., Reif, W.: Modular, crash-safe refinement for ASMs with submachines. Sci. Comput. Program. (SCP) **131**, 3–21 (2016)

26. Ernst, G., Fedyukovich, G., Sögtrop, R.: Quick theory exploration for algebraic data types via program transformations (2023). https://www.sosy-lab.org/research/pub/2023-Draft.LemmaCalc.pdf

27. Filliâtre, J.C., Paskevich, A.: Why3—where programs meet provers. In: ESOP, pp. 125–128. Springer, Heidelberg (2013). https://doi.org/10.1007/978-3-642-37036-6_8

28. Garavel, H., Beek, M.H.t., Pol, J.v.d.: The 2020 expert survey on formal methods. In: FMICS, pp. 3–69. Springer, Cham (2020). https://doi.org/10.1007/978-3-030-58298-2_1

29. Giesl, J.: Context-moving transformations for function verification. In: LOPSTR, pp. 293–312. Springer, Heidelberg (1999). https://doi.org/10.1007/10720327_17

30. Giesl, J., Kühnemann, A., Voigtländer, J.: Deaccumulation techniques for improving provability. J. Logic Algebraic Program. **71**(2), 79–113 (2007). https://doi.org/10.1016/J.JLAP.2006.11.001

31. Gleirscher, M., Marmsoler, D.: Formal methods in dependable systems engineering: a survey of professionals from Europe and North America. Empir. Softw. Eng. **25**(6), 4473–4546 (2020). https://doi.org/10.1007/S10664-020-09836-5

32. Goguen, J.: Hidden algebra for software engineering. In: Proceedings Conference on Discrete Mathematics and Theoretical Computer Science, Auckland, New Zealand, pp. 35–59. Citeseer (1999)

33. Hähnle, R., Huisman, M.: Deductive software verification: from pen-and-paper proofs to industrial tools. Computing and Software Science: State of the Art and Perspectives pp. 345–373 (2019). https://doi.org/10.1007/978-3-319-91908-9_18

34. Hatcliff, J., Leavens, G.T., Leino, K.R.M., Müller, P., Parkinson, M.: Behavioral interface specification languages. ACM Comput. Surv. **44**(3), 16:1–16:58 (2012). https://doi.org/10.1145/2187671.2187678

35. Hennicker, R.: Observational implementation of algebraic specifications. Acta Informatica **28**(3), 187–230 (1991). https://doi.org/10.1007/BF01178505

36. Hennicker, R., Bidoit, M.: Observational logic. In: AMAST, pp. 263–277. Springer, Heidelberg (1999). https://doi.org/10.1007/3-540-49253-4_20

37. Hoare, P.: Proof of correctness of data representations. Acta Informatica **1**, 271–281 (1972). https://doi.org/10.1007/BF00289507

38. Hutter, D., Sengler, C.: INKA: the next generation. In: CADE. LNCS, vol. 1104, pp. 288–292. Springer, Heidelberg (1996). https://doi.org/10.1007/3-540-61511-3_92

39. Johansson, M., Dixon, L., Bundy, A.: Case-analysis for rippling and inductive proof. In: Interactive Theorem Proving, pp. 291–306. Springer, HeidelbergD (2010). https://doi.org/10.1007/978-3-642-14052-5_21

40. Klein, G.: Comprehensive formal verification of an OS microkernel. TOCS **32**(1), 1–70 (2014). https://doi.org/10.1145/2560537

41. Kovács, L., Voronkov, A.: First-order theorem proving and Vampire. In: CAV, pp. 1–35. Springer, Heidelberg (2013). https://doi.org/10.1007/978-3-642-39799-8_1

42. Kurashige, C., et al.: CCLemma: e-graph guided lemma discovery for inductive equational proofs. Proc. ACM Program. Lang. **8**(ICFP), 818–844 (2024). https://doi.org/10.1145/3674653

43. Leino, K.R.M.: Dafny: An automatic program verifier for functional correctness. In: LPAR, pp. 348–370. Springer, Heidelberg (2010). https://doi.org/10.1007/978-3-642-17511-4_20

44. Leroy, X.: Formal verification of a realistic compiler. Commun. ACM **52**(7), 107–115 (2009). https://doi.org/10.1145/1538788.1538814

45. Liskov, B., Guttag, J., et al.: Abstraction and Specification in Program Development, vol. 20. MIT Press Cambridge (1986)

46. Lynch, N., Vaandrager, F.: Forward and backward simulations. Inf. Comput. **121**(2), 214–233 (1995). https://doi.org/10.1006/INCO.1995.1134

47. Meijer, E., Fokkinga, M.M., Paterson, R.: Functional programming with bananas, lenses, envelopes and barbed wire. In: FPCA, vol. 91, pp. 124–144 (1991). https://doi.org/10.1007/3540543961_7

48. Miltner, A., Padhi, S., Millstein, T., Walker, D.: Data-driven inference of representation invariants. In: PLDI, pp. 1–15 (2020). https://doi.org/10.1145/3385412.3385967

49. de Moor, O., Sittampalam, G.: Generic program transformation. In: Advanced Functional Programming. LNCS, vol. 1608, pp. 116–149. Springer, Heidelberg (1998). https://doi.org/10.1007/10704973_3

50. Moura, L.D., Bjørner, N.: Z3: An efficient SMT solver. In: TACAS. LNCS, vol. 4963, pp. 337–340. Springer, Heidelberg (2008). https://doi.org/10.1007/978-3-540-78800-3_24

51. Nandi, C., et al.: Rewrite rule inference using equality saturation. Proc. ACM Program. Lang. **5**(OOPSLA), 1–28 (2021). https://doi.org/10.1145/3485496

52. Nipkow, T., Wenzel, M., Paulson, L.C. (eds.): Isabelle/HOL. LNCS, vol. 2283. Springer, Heidelberg (2002). https://doi.org/10.1007/3-540-45949-9
53. Orejas, F., Nivela, P.: Constraints for behavioural specifications. In: ADT. LNCS, vol. 534, pp. 220–245. Springer, Heidelberg (1990). https://doi.org/10.1007/3-540-54496-8_12
54. Park, K., Peng, X., D'Antoni, L.: LOUD: synthesizing strongest and weakest specifications. CoRR **abs/2408.12539** (2024). https://doi.org/10.48550/ARXIV.2408.12539
55. Reichel, H.: Behavioural equivalence-a unifying concept for initial and final specifications. In: Proceedings of Third Hungarian Computer Science Conference. Akademiai Kiado (1981)
56. Reif, W.: Correctness of generic modules. In: Logical Foundations of Computer Science-Tver 1992: Second International Symposium Tver, Russia, July 20–24, 1992 Proceedings 2, pp. 406–417. Springer (1992). https://doi.org/10.1007/BFB0023893
57. Reif, W.: The KIV system: Systematic construction of verified software. In: CADE. Lecture Notes in Computer Science, vol. 607, pp. 753–757. Springer, Heidelberg (1992). https://doi.org/10.1007/3-540-55602-8_218
58. Reif, W.: The KIV-approach to software verification. KORSO: Methods, Languages, and Tools for the Construction of Correct Software: Final Report, pp. 339–368 (2005). https://doi.org/10.1007/BFB0015471
59. Reif, W., Schellhorn, G., Stenzel, K., Balser, M.: Structured specifications and interactive proofs with KIV. Automated Deduction-A Basis for Applications: Volume II: Systems and Implementation Techniques, pp. 13–39 (1998). https://doi.org/10.1007/978-94-017-0435-9_1
60. Rozier, K.Y.: Specification: the biggest bottleneck in formal methods and autonomy. In: VSTTE, pp. 8–26. Springer, Cham (2016). https://doi.org/10.1007/978-3-319-48869-1_2
61. Sannella, D., Tarlecki, A.: On observational equivalence and algebraic specification. J. Comput. Syst. Sci. **34**(2/3), 150–178 (1987). https://doi.org/10.1016/0022-0000(87)90023-7
62. Schellhorn, G., Ernst, G., Pfähler, J., Haneberg, D., Reif, W.: Development of a verified Flash file system. In: Proceedings of Alloy, ASM, B, TLA, VDM, and Z (ABZ). LNCS, vol. 8477, pp. 9–24. Springer (2014), invited Paper
63. Schellhorn, G., Bodenmüller, S., Bitterlich, M., Reif, W.: Software & system verification with KIV. In: The Logic of Software. A Tasting Menu of Formal Methods: Essays Dedicated to Reiner Hähnle on the Occasion of His 60th Birthday, pp. 408–436. Springer (2022). https://doi.org/10.1007/978-3-031-08166-8_20
64. Schierl, A., Schellhorn, G., Haneberg, D., Reif, W.: Abstract specification of the ubifs file system for flash memory. In: FM 2009: Formal Methods: Second World Congress, Eindhoven, The Netherlands, November 2–6, 2009. Proceedings 2, pp. 190–206. Springer, Heidelberg (2009). https://doi.org/10.1007/978-3-642-05089-3_13
65. Singher, E., Itzhaky, S.: Theory exploration powered by deductive synthesis. In: CAV, pp. 125–148. Springer, Cham (2021). https://doi.org/10.1007/978-3-030-81688-9_6
66. Smallbone, N., Johansson, M., Claessen, K., Algehed, M.: Quick specifications for the busy programmer. J. Funct. Program. **27**, e18 (2017). https://doi.org/10.1017/S0956796817000090
67. Sonnex, W.: Fixed point promotion: taking the induction out of automated induction, Tech. rep. University of Cambridge, Computer Laboratory (2017)

68. Sonnex, W., Drossopoulou, S., Eisenbach, S.: Zeno: an automated prover for properties of recursive data structures. In: Flanagan, C., König, B. (eds.) TACAS 2012. LNCS, vol. 7214, pp. 407–421. Springer, Heidelberg (2012). https://doi.org/10.1007/978-3-642-28756-5_28
69. Turchin, V.F.: The concept of a supercompiler. TOPLAS **8**(3), 292–325 (1986). https://doi.org/10.1145/5956.5957
70. Wadler, P.: Deforestation: transforming programs to eliminate trees. In: ESOP. pp. 344–358. Springer, Heidelberg (1988). https://doi.org/10.1007/3-540-19027-9_23
71. Wirsing, M.: Algebraic specification. Formal Models and Semantics, pp. 675–788 (1990). https://doi.org/10.1016/B978-0-444-88074-1.50018-4
72. Woodcock, J., Larsen, P.G., Bicarregui, J., Fitzgerald, J.: Formal methods: practice and experience. ACM Comput. Surv. **41**(4), 1–36 (2009). https://doi.org/10.1145/1592434.1592436

Mirror is Not Strong: Discovery of a Persistent Memory Bug using Refinement in KIV

Gerhard Schellhorn[1]($\boxtimes$)(iD), Stefan Bodenmüller[1](iD), Brijesh Dongol[2](iD), and Heike Wehrheim[3](iD)

[1] University of Augsburg, Augsburg, Germany
schellhorn@informatik.uni-augsburg.de
[2] University of Surrey, Guildford, UK
[3] Carl von Ossietzky Universität Oldenburg, Oldenburg, Germany

Abstract. Designing concurrent objects for non-volatile memory (NVM) is challenging since designers must take both inter-thread synchronization and durability into account. To simplify development, a number of libraries have been developed to support *transformations to durability*. In this paper, we study one such library, Mirror, which offers a technique for transforming *linearizable* (lock-free) concurrent data structures into a *durably linearizable* version. We aimed to prove soundness of Mirror using the interactive proof assistant KIV, but in doing so discovered a subtle bug: Mirror can only be shown to correctly implement a durable `compare_exchange_weak` instruction rather than a durable `compare_exchange_strong`, as claimed in the original paper. Unlike the `compare_exchange_strong` instruction, `compare_exchange_weak` may fail spuriously. It turns out that the bug in Mirror's implementation is difficult to fix, thus, we only provide a formal proof that Mirror implements `compare_exchange_weak`. Our proof is via refinement against an abstract persistency library, PLib, that we developed in prior work. We split the proof into two steps via a new intermediate specification IMirror: we prove forward simulation between Mirror and IMirror and backward simulation between IMirror and PLib. Since any implementation of PLib is guaranteed to ensure correct transformations to durability, we conclude that Mirror continues to support this property.

Keywords: Durable Linearizability · Persistency Libraries · Mirror · Interactive Verification · KIV

1 Introduction

Formal methods are techniques for the formal specification and analysis of hardware and software systems. The development of formal methods has been the

Schellhorn and Bodenmüller are supported by DFG project RE 828/26-1. Dongol is supported by VeTSS and EPSRC grants EP/X037142/1, EP/X015149/1, EP/V038915/1, and EP/R025134/2. Wehrheim is supported by DFG project WE 2290/14-1.

G. Ernst et al. (Eds.): Wolfgang Reif Festschrift, LNCS 15765, pp. 325–355, 2025.
https://doi.org/10.1007/978-3-031-92196-4_16

longest running topic pursued at the chair of Professor Wolfgang Reif. Work already started in 1986 with research on Dynamic Logic (DL) [17] at the chair of Professor Menzel in Karlsruhe. One of the key contributions at the chair of Prof. Reif which also started in Karlsruhe is the interactive prover KIV (Karlsruhe Interactive Verifier).

KIV has since then been developed to a theorem prover focusing on the verification of sequential and concurrent software systems. Modularization using refinement as we employ here has always been a key concept. In 2007, research on concurrent systems led to the start of the cooperation among (part of) the authors of this paper and their investigation of concurrent systems. Within this cooperation, research has mainly been based on the IO Automata formalism and its refinement definition of Lynch and Vaandrager [21]. To reason about such refinements, support in KIV has been developed. It uses step-local and thread-local proof obligations to prove invariants [7] as well as forward simulations [4] for refinement. Such refinement proofs have in particular been employed to show *linearizability* [18] of concurrent objects, e.g. in [8–11,30,31,34].

In this paper, we employ KIV and its built-in IOA refinement support to prove soundness of the library Mirror [16] used in the context of hardware architectures with non-volatile memory (NVM). NVM is a new storage technology which offers higher performance than flash memory while providing resilience to system-wide crashes. Unlike DRAM, the content of NVM survives system failures (e.g., a power outage). While this brings advantages, it also poses new challenges. In particular, programmers need to guarantee that their programs are still in a consistent state after a failure.

To help programmers with this challenge, there have been several proposals in recent years [5,16,28,35] providing *transformations to durability*. In such transformations, program operations such as reads and writes to memory are replaced by calls to library operations that manage durability. In particular, given a (concurrent) library that is correct in a setting *without* crashes, the transformation mechanism can be used to ensure correctness *with* crashes.

The approach we study here is that of the library Mirror [16]. Mirror transforms linearizable [18] concurrent objects (viz., data structures like stacks or queues) into *durable linearizable* [19] objects provided that the object in question is non-blocking. We prove soundness of this transformation to durability. Our proof employs the abstract persistency library PLib [3] which we have previously developed for proving soundness of yet another transformation to durable linearizability, namely the library FliT [35]. The soundness proof of FliT in [3] proceeds by (1) showing PLib to guarantee correct transformation to durable linearizability and (2) proving FliT to refine PLib. Here, we build on the existing result (1), and for soundness of Mirror only need to prove that Mirror refines PLib. Still, the proof presents us with specific challenges: while FliT works by reading ahead data which is not yet persisted, Mirror always first persists data and then allows reading from a volatile copy which never gets persisted. The design of PLib, constructed for the purpose of showing soundness of FliT, reflects the persistency strategy of FliT. Hence – at first sight – there seems to be a mis-

```
1  put(x, arg){                    5  get(x){
2    write(x.val, arg);            6    r := read(x.flag)
3    write(x.flag, true);          7    if r {
4  }                               8      s := read(x.val);
                                   9      return s; }
                                  10    return ⊥; }
```

Fig. 1. Object O with operations put and get

match between Mirror and PLib implying no refinement relationship. However, refinement does hold and we manage to show it by employing an intermediate specification IMirror, proving IMirror to refine PLib (upper refinement) and Mirror to refine IMirror (lower refinement). The seeming mismatch can be resolved by using a *backward simulation* for the upper refinement.

Initial proof attempts for the lower refinement furthermore revealed a subtle bug in Mirror. In [16], Mirror is stated to overload existing operations of the C++ standard std::atomic, among others the method

bool compare_exchange_strong(T& expected, T newVal),

(T being a type). This method is supposed to compare the value of the current object with expected. If successful, the object's current value is replaced by newVal. If not successful, the current value is loaded into expected. The latter is not achieved in all executions, thus "Mirror is not strong." Instead we now prove Mirror to implement compare_exchange_weak. This does, however, not change the overall soundness of Mirror: It still guarantees correct transformation to durable linearizability.

Overview. This paper is structured as follows. Section 2 motivates our work with an example and Sect. 3 provides the necessary background definitions including the specification of PLib. Section 4 describes the persistency library Mirror and Sect. 5 gives the proof of its soundness via refinement. Finally, Sects. 6 and 7 discuss related work and give a conclusion.

Supplementary Material. The KIV mechanisation corresponding to the proofs in Sect. 5 may be found at [29].

2   Motivation

We start by explaining the memory model that our work assumes (Sect. 2.1), in particular, we give details on non-volatile memory (NVM). We then describe a motivating example which illustrates the challenges on NVM (Sect. 2.2).

2.1   Memory Model

Like Friedman et al. [16], we assume that operations of concurrent objects are executed by a finite (but unbounded) number of n processes $p_1, p_2, \ldots, p_n$. For simplicity, we assume the PSC memory model, which comprises a sequentially consistent memory model with persistency [20]. Since the bug in Mirror manifests even under this stronger memory model, we can conclude that it is not caused by weak memory effects. Of course, since weaker memory models such as PTSO [20, 27] permit all PSC behaviors, the bug we describe will also be present under weak memory.

Under PSC, memory comprises two parts: (1) a *volatile memory* (e.g., DRAM), and (2) a *persistent memory* (e.g., NVM). Both memories can be read-from and written-to in a byte-addressable manner. However, the contents of DRAM do not survive a crash (e.g., a power failure), whereas those of NVM are preserved. Writes first take place in volatile memory and are later transferred to persistent memory either *implicitly* or *explicitly* using programmer-controlled instructions (e.g., FLUSH). We then talk about *persisted writes*. Reads always read from the volatile copy and in case a crash occurs, all contents from the persistent memory are first transferred to volatile memory before execution resumes (typically through a recovery operation).

2.2   Durably Linearizable Concurrent Objects

We illustrate the challenges involved in concurrent programming on NVM on an example. Consider a concurrent object O implementing put and get operations as given in Fig. 1. Concurrency here means that several threads can execute these operations concurrently. The object uses a variable x.flag to determine whether a value for a given location x.val has been set. Initially, x.flag is *false* and x.val is 0. Operation get(x) only reads from x.val after x.flag has been set, otherwise it returns the value $\bot$. Note that the put operation would in principle need to reset flag to *false* before writing a new value. For simplicity we elide this here.

We are interested in finding out whether this object is (durable) *linearizable*. In general, linearizability [18] requires operations to behave as if they were executed instantaneously, some time in between its invocation and response. In refinement proofs of linearizability [7,8,11,30,31], we thus compare concurrent objects against abstract specifications which perform each operation atomically. A formal definition of linearizability (as given in [18]) compares concurrent *histories* (sequences of invocations and responses of operations) to sequential histories. In this comparison, we allow reorderings as long as they preserve the real-time order of operations (i.e., responses and later invocations cannot be put in a different order). Without NVM, the operations of object O are linearizable (with the above mentioned modification of put) since the fine-grained implementation only sets x.flag to true after writing to x.val. Thus any (concurrent) history of put and get operations can be reordered into a valid sequential history.

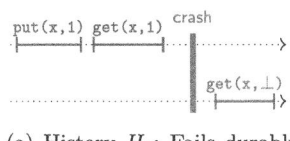 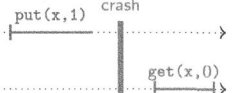 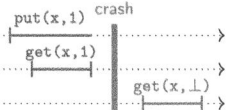

(a) History H_1: Fails durably linearizability because the second **get** returns $\bot$ instead of **1**

(b) History H_2: Fails durably linearizability because the **get** returns 0 instead of **1** or $\bot$

(c) History H_3: Fails durably linearizability because the second **get** returns $\bot$ instead of **1**

Fig. 2. Three histories (dotted lines denote operations of different threads, x-axis is time, a bar depicts the beginning and end of an operation)

Let us now consider the behavior of object O under NVM. After a crash, only the contents of persistent memory is preserved; thus a crash action resets the volatile memory to persistent memory. *Durable linearizability* [19] requires all histories – now potentially including crashes – to be linearizable when all crash actions are removed from the histories. It is straightforward to see that executing **put** and **get** in such a memory model results in histories that are not durably linearizable. In particular, it is possible to generate the history in Fig. 2a, where the **get** operation before the crash can read from unpersisted writes of the **put** operation, and the crash can occur before these writes are persisted (so that the get after the crash returns $\bot$). This history is not linearizable after removing the crash, thus is not durably linearizable. Similarly, since writes do not persist in the order of their occurrence, histories such as H_2 (Fig. 2b) would also be possible. Here, the **get** operation performs a read and sees the flag to be *true* (i.e., x.flag is persisted), but still returns the old (initial) value of x.val. A naive approach to addressing this issue is to modify the program so that each write is flushed to persistent memory immediately after it occurs. However, this approach is not only inefficient, it is also incorrect. For instance, under concurrency, it would be possible to generate the history in Fig. 2c since the second write to x.flag in put may have occurred, but not yet persisted, yet be read by the first **get**.

The key to achieving durable linearizability is to ensure that a client (i.e., a thread executing an object's operation (e.g., **get**) that reads from another client operation (e.g., **put**) should *not* complete unless it can be assured that all writes it has read from have been persisted. With this approach, histories such as H_3 (Fig. 2c) *cannot* occur even when the put(x,1) executes concurrently with get(x). In particular, the **get** itself would then ensure that any unpersisted writes that it has read from have been persisted *before* returning. Thus it would be impossible for the second get(x) to return $\bot$. Further, to prevent writes from being persisted out-of-order, we must ensure prior writes executed by a thread have been persisted before starting a new write by that thread. This prevents histories such as H_2 (Fig. 2b).

The purpose of libraries such as Mirror [16] and FliT [35] is now to provide read/write methods to clients that – when used instead of standard reads and writes – avoid such problems leading to non-linearizability. Mirror basically provides these durability guarantees by keeping two replicas of each shared location, rep_p and rep_v. Reading is only performed on the volatile replica rep_v. Writing first takes place on (volatile) rep_p, then persists rep_p and only then puts the value into rep_v. The main contribution of this paper is to prove that this construction of reads and writes always ensures durable linearizability. We do so by showing that Mirror refines the library PLib (presented as an input/output automaton in Fig. 5) that we have already proven to guarantee durable linearizability in [3].

3   Background

This section provides some standard definitions and known results, providing background for the rest of the paper. In particular, we introduce input/output automata (Sect. 3.1), which we use (in Sect. 3.2) to specify the abstract persistency library PLib and state its soundness result [3].

In the following, we let Loc be the set of shared locations, Tid the set of thread identifiers and Val the values of locations.

3.1   Input/Output Automata (IOA)

We use input/output automata (IOA) [23] to model the libraries at three levels: the abstract persistency library PLib (see Sect. 3.2), an intermediate specification IMirror that is used in the correctness proof (see Sect. 5.2) and the formal model of Mirror (see Sect. 4.3). Each of these models has been encoded in KIV to support mechanized correctness proofs.

Definition 1 (Input/Output Automaton (IOA)). *An* Input/Output Automaton (IOA) *is a labeled transition system A with a set of* states $states(A)$, *a set of* actions $acts(A)$, *a set of* start states $start(A) \subseteq states(A)$, *and a transition relation* $trans(A) \subseteq states(A) \times acts(A) \times states(A)$ *(so that the actions label the transitions).*

The set $acts(A)$ is partitioned into internal actions $internal(A)$ that represent events of the system that are hidden from the environment, and external actions $external(A)$ (typically atomic steps or invocations/responses to calling an operation) representing the IOA's interactions with its environment. We sometimes also call the internal actions τ actions. For a sequence of actions γ, we let $\widehat{\gamma}$ denote the sequence of external actions in γ. We write $s \xrightarrow{a} s'$ for $(s, a, s') \in trans(A)$ and $s \xRightarrow{a} s'$ if there is some $\gamma = a_1 \ldots a_n$ and $s = s_0 \xrightarrow{a_1} s_1 \xrightarrow{a_2} \ldots \xrightarrow{a_n} s_n = s'$ and $\widehat{\gamma} = a$.

We specify IOAs (like the one in Fig. 5) by giving the states in terms of state variables and their initial values. For every action $a \in acts(A)$ we give a precondition Pre on the state s that enables a step $(s, a, s') \in trans(A)$, and specify the result state s' by assignments given under Eff.

Our proof strategy is showing the existence of a refinement relationship between IOAs which we define next. An *execution* of an IOA A is a sequence

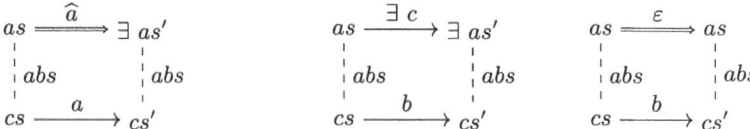

Fig. 3. Left: General condition for a forward simulation *abs* between abstract IOA *A* and concrete IOA *C*, where $a \in external(C) \cup internal(C)$, *as, as'* abstract and *cs, cs'* concrete states. The existentially quantified state must be found such that the diagram commutes. Right: Two special cases with $b \in internal(C)$ and $c \in internal(A)$.

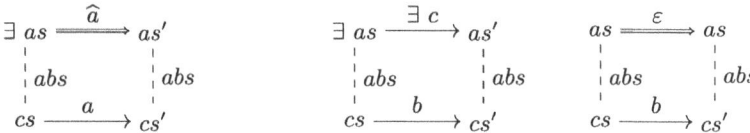

Fig. 4. Left: General condition for a backward simulation *abs* between abstract IOA *A* and concrete IOA *C*, where $a \in external(C) \cup internal(C)$, *as, as'* abstract and *cs, cs'* concrete states. The existentially quantified state must be found such that the diagram commutes. Right: Two special cases with $b \in internal(C)$ and $c \in internal(A)$.

$\sigma = s_0 a_0 s_1 a_1 \ldots s_n a_n s_{n+1}$ of alternating states and actions such that $s_0 \in start(A)$ and for all states s_i, $(s_i, a_i, s_{i+1}) \in trans(A)$. A *trace* of *A* (an element of *traces(A)*) is any sequence of (external) actions obtained by projecting the external actions of any execution of *A*. For IOAs *C* and *A* with $external(C) = external(A)$, we say that *C* is a *refinement* of *A*, denoted $C \leq A$, iff $traces(C) \subseteq traces(A)$. Refinement can be proven by establishing forward or backward simulations between IOAs (see e.g., [21]) using *abstraction relations* relating the states of the two IOAs. Given that the methodology is well-known, we elide the formal definitions here, and explain the proof obligations for forward and backward simulation diagrammatically in Figs. 3 and 4 (letting *abs* be the abstraction relation).

3.2 Abstract Persistency Library PLib

Next, we present the abstract persistency library PLib introduced in [3][1], which guarantees correct transformation to durability for concurrent linearizable objects. A concurrent object therein is a concurrent data structure (like a stack) providing operations (like push and pop) to clients. The idea behind the library PLib is the following: Instead of directly accessing shared locations and executing read, write or CAS[2] instructions on them, an object O has to call libWrite, libRead and libCAS actions of PLib for each memory access and additionally call action complete before returning from its operations.

---

[1] We have extended it with a CAS instruction here.

[2] CAS is a compare-and-swap instruction, atomically making a comparison and – if successful – setting a new value for a location.

State variables:

$vmem, pmem : Loc \rightarrow Val$ (initially: $\forall\, x \in Loc.\ vmem(x) = pmem(x) = 0$)

$log : ((\mathsf{Rd} \cup \mathsf{Wr}) \times \mathbb{B})^*$ (initially: $log = \varepsilon$)

$pc : Tid \rightarrow \{idle, ready, crashed, inR(x), outR(v),$
$\qquad\qquad\quad inC(x, v_1, v_2), outC(bv, v)\ |$
$\qquad\qquad\quad v, v_1, v_2 \in Val, bv \in \{true, false\}\}$ (initially: $\forall\, t \in Tid.\ pc(t) = idle$)

Actions:

$\mathsf{inv}_t(\mathsf{libRead}(x))$
Pre: $pc(t) \in \{idle, ready\}$
Eff: $pc(t) := inR(x)$

$\mathsf{do}_t(\mathsf{libRead})$
Pre: $pc(t) = inR(x)$
Eff: $log := log \cdot \langle \mathsf{R}_t(x, v), lwp(x, log)\rangle$
$\qquad\quad pc(t) := outR(vmem(x))$

$\mathsf{inv}_t(\mathsf{libCAS}(x, exp, newVal))$
Pre: $pc(t) \in \{idle, ready\}$
Eff: $pc(t) := inC(x, exp, newVal)$

$\mathsf{do}_t(\mathsf{libCAS}^T)$
Pre: $pc(t) = inC(x, exp, newVal)$
$\qquad\quad log_{|t,false} = \varepsilon$
$\qquad\quad lwp(x, log) = true$
$\qquad\quad exp = vmem(x)$
Eff: $vmem(x) := newVal$
$\qquad\quad log := log \cdot \langle \mathsf{R}_t(x, vmem(x)), true\rangle$
$\qquad\qquad\qquad\quad \cdot \langle \mathsf{W}_t(x, newVal), false\rangle$
$\qquad\quad pc(t) := outC(true, vmem(x))$

$\mathsf{do}_t(\mathsf{libCAS}^F)$
Pre: $pc(t) = inC(x, exp, newVal)$
$\qquad\quad vmem(x) \neq exp$
Eff: $log := log \cdot$
$\qquad\qquad\quad \langle \mathsf{R}_t(x, vmem(x)), lwp(x, log)\rangle$
$\qquad\quad pc(t) := outC(false, vmem(x))$

$\mathsf{res}_t(\mathsf{libRead}(v))$
Pre: $pc(t) = outR(v)$
Eff: $pc(t) := ready$

$\mathsf{res}_t(\mathsf{libCAS}(bv, v))$
Pre: $pc(t) = outC(bv, v)$
Eff: $pc(t) := ready$

$\mathsf{complete}_t$
Pre: $pc(t) \in \{idle, ready\}$
$\qquad\quad log_{|t,false} = \varepsilon$
Eff: $pc(t) := ready$

$\mathsf{persist}(x)$
Pre: $\exists\, log_1, log_2.$
$\qquad\qquad log = log_1 \cdot \langle \mathsf{W}_t(x, v), false\rangle \cdot log_2$
$\qquad\qquad \wedge\ log_{1|\mathsf{Wr},x,false} = \varepsilon$
Eff: $pmem(x) := v$
$\qquad\quad log := log_1 \cdot$
$\qquad\qquad\quad \langle \mathsf{W}_t(x, v), true\rangle \cdot pdr(x, log_2)$

crash
Eff: $pc := \lambda\, t \in Tid.\ \textbf{if}\ pc(t) \neq idle\ \textbf{then}\ crashed\ \textbf{else}\ pc(t)$
$\qquad\quad vmem := pmem$
$\qquad\quad log := \varepsilon$

Fig. 5. PLib^{IR} specification (where $log_{|\mathsf{Wr}}$ restricts log to all Wr entries, $log_{|t,false}$ to all entries of t with flag $false$ and so on; crash and persist are global transitions). In red, we highlight the part which needs to be dropped for the weak version. (Color figure online)

The key requirement on PLib to ensure durability is that it tracks certain dependencies (see [35]) and persists writes in this order. Namely,

1. writes of the same thread must be persisted in the order in which they occur, and

2. when a thread, say t, executes complete, all of t's writes as well as the writes t has read from must have been persisted.

Figure 5 gives the IOA of PLib. We elide write instructions here as they can be implemented via libCAS. We moreover directly give the IOA in a form required for verification, namely as a canonical automaton. A *canonical automaton* [22] splits every operation into separate external *invocation* and *response* actions and an internal *do*-action that occurs between the invocation and response. We thus refer to the IOA of Fig. 5 as $\mathsf{PLib^{IR}}$ (Invocations, Reponses). The specification of $\mathsf{PLib^{IR}}$ employs the auxiliary functions *pdr* (persist dependent reads) and *lwp* (last write persisted) given in Fig. 6.

$$pdr(x, \varepsilon) = \varepsilon$$
$$pdr(x, \langle \mathsf{W}_t(x,v), \mathit{flag} \rangle \cdot log) = \langle \mathsf{W}_t(x,v), \mathit{flag} \rangle \cdot log$$
$$pdr(x, \langle \mathsf{R}_t(x,v), \mathit{flag} \rangle \cdot log) = \langle \mathsf{R}_t(x,v), \mathit{true} \rangle \cdot pdr(x, log)$$
$$pdr(x, \langle \mathsf{W}_t(x',v), \mathit{flag} \rangle \cdot log) = \langle \mathsf{W}_t(x',v), \mathit{flag} \rangle \cdot pdr(x, log) \quad \text{for} \quad x \neq x'$$
$$pdr(x, \langle \mathsf{R}_t(x',v), \mathit{flag} \rangle \cdot log) = \langle \mathsf{R}_t(x',v), \mathit{flag} \rangle \cdot pdr(x, log) \quad \text{for} \quad x \neq x'$$

$$lwp(x, log) = \mathit{false} \leftrightarrow \exists\, log_1, log_2.\ log = log_1 \cdot \langle \mathsf{W}_t(x,v), \mathit{false} \rangle \cdot log_2 \wedge log_{2|\mathsf{Wr},x} = \varepsilon$$

Fig. 6. Auxiliary functions *pdr* and *lwp*.

The state of $\mathsf{PLib^{IR}}$ assumes a volatile memory *vmem*, a persistent memory *pmem* and a *log* of all *read/write events* together with their persistency status and a program counter. The program counter tracks the control flow of the library. When a crash occurs, all currently running threads are stopped and cannot continue. To comply with durable linearizability [19], a thread that crashes is never restarted. Note that $\mathsf{PLib^{IR}}$ specifies a concurrent system since each thread can independently execute its actions.

The key mechanism of $\mathsf{PLib^{IR}}$ for ensuring durable linearizability of objects can be found in the usage of state variable *log*. It records the read and writes events which have taken place so far (CAS instructions are reads directly followed by writes when successful). We let Rd and Wr be the set of all read and write events, respectively. Each read event has the form $\mathsf{R}_t(x,v)$, where t is a thread identifier, $x \in Loc$ is a shared location and $v \in Val$ is a value. Similarly, a write event has the form $\mathsf{W}_t(x,v)$. The *log* pairs write events with a boolean flag that determines whether the operation (viz. its value) has been persisted. We call a write *persisted* if its effect is written to persistent memory and it has flag *true*, and a read *persisted* if the write that it reads from is persisted.

A libCAS and complete executed by thread t may only proceed if it has neither unpersisted writes nor unpersisted reads in *log*, i.e., when $log_{|t,\mathit{false}} = \varepsilon$. We believe the condition is as weak as possible to allow a proof of durable linearizability for arbitrary data structures: A write is disallowed only if there is a previous read or write by the same thread that has yet to be persisted. Returning from an operation must thus have persisted its values. The fact that we here prove Mirror to refine $\mathsf{PLib^{IR}}$ despite mismatches in the persistency strategies supports this conjecture.

Execution of libRead reads the value from volatile memory. It adds the read to the log together with a flag that is computed by checking whether the last write to this location (if there is any; otherwise the flag is set to true) has already been persisted using predicate lwp. Persisting a write also persists dependent reads (i.e., those which have read this value) using function pdr. The recursive definition of pdr ensures that the reads affected are those in the log before the next write to the same location. Execution of libCAS in the successful case adds a read *and* a write event to log and – in case the comparison fails – just a read event. Operation libCAS has three values in its parameters, the first is the expected value (to be compared with the current value), the second is the new value for the location if the comparison is successful, and the third is the old value of the location to be returned by the CAS operation.

Now, as Mirror is not implementing `compare_exchange_strong`, we have to adapt PLib here as well. Mirror only implements

```
bool compare_exchange_weak(T& expected, T newVal).
```

This instruction is allowed to spuriously fail, i.e., return *false* although the current value of the object is equal to the expected value. To account for this, we provide a weak version of PLib here (Fig. 5) which drops the red line within libCAS^F.

Soundness of PLib has been proven in [3]. Despite moving to a weak version here, PLib still guarantees a correct transformation to durable linearizability for linearizable objects.

Theorem 1 (Bodenmüller et al. [3]). *Let* O *be (an IOA of) a concurrent object. If* O *is linearizable, then* O *using* PLib *is durable linearizable.*

By standard results on refinement, we then get:

Corollary 1. *If a library* L *refines* PLib *and the object* O *is linearizable, then* O *using* L *is durable linearizable.*

In this paper, we intend to apply this corollary to the library Mirror. To do so, we need to prove Mirror to refine (the weak version of) PLib^{IR}.

4   Mirror

In this section, we present Mirror's CAS operation [16] (which is the most complex of its operations) and explain its overall design (Sect. 4.1). We describe the bug in Sect. 4.2, then in Sect. 4.3, we present the KIV model corresponding to Mirror.

4.1   C Implementation of Mirror

For each location x, Mirror maintains a volatile and a persistent copy, which we will refer to as $x.rep_v$ and $x.rep_p$, respectively. The persistent copy $x.rep_p$ exists in both volatile and non-volatile memory, whereas the volatile copy is never explicitly persisted, so we only refer to the volatile copy. Read operations only read from the volatile copy, and write operations write to both the persistent and volatile copies. Each variable x is associated with a sequence number, x.*seq* that is incremented with each update to x. We omit the write operation in this paper for brevity since is implemented by repeatedly retrying the CAS operation until it succeeds.

The C-like pseudocode for Mirror's CAS operation is presented in Listing 1.1. Later in Sect. 4.3, we also provide the KIV model of the code, where each line of the program must be distinguished by a label denoting the possible values for a thread's program counter. To simplify the mapping from the C code to the KIV model, we include line numbers in Listing 1.1 in orange.

Moreover, in describing the algorithm, we assume that the location being updated (i.e., this) is x with an expected value expected and a new value newVal. Some line numbers (e.g., CE9f) have a suffix "f", which is used to stress the fact that the KIV model does not have a continue statement. Control flow in these cases is modeled by refactoring the program by negating and only executing beyond the continue if the negation of the guard holds.

The algorithm is based on a primitive DWCAS ("double-word compare and swap") instruction, which is supported natively by many modern processors[3]. DWCAS(addr, old, new) compares the value (whose size is double the word-size of the processor) stored at addr to old. If they are equal, the value at addr is atomically replaced with new and the call returns old. Otherwise the current value stored at addr is not changed, but returned. The result is used only at line CE22, all other calls ignore it. The algorithm basically should behave the same way, using this, expected and newVal. However, in addition, it should look to a caller of the algorithm, as if a modified value is persisted atomically too.

At line CE1, the address of x.*rep$_p$* is calculated. At lines CE3–CE4, the persistent sequence number and value of x are first read, and then the persistent sequence number is read again at line CE5. If the values read at CE3 and CE5 are different, then there must be another update of x in progress, so the algorithm later immediately retries via the test at line CE9f. Similarly, the volatile sequence number and value is read at lines CE6 and CE7, with a second read of the volatile sequence number at line CE8.

If the test at line CE9f fails (i.e., the sequence numbers are consistent), the algorithm next checks whether there is another concurrent update to x at line CE10f, namely whether the persistent sequence number is one greater than the volatile sequence number. If so, there must be another write in progress that has managed to update the new (to be persisted) value, but not yet persisted it, e.g., another CAS operation that has successfully executed the DWCAS at CE22. Since this other update would first persist the address and then update the volatile copy, helping proceeds by first persisting the location (CE11), constructing the current and new value/sequence-number pairs (CE12 and CE13), then attempting to update x.*rep$_v$* using a DWCAS (CE14). Regardless of whether this helping via DWCAS is successful, the CAS operation restarts.

---

[3] We note that DWCAS is different from DCAS ("double compare and swap"), which performs a compare-and-swap across two distinct locations.

```
template <typename T>
bool patomic <T>::
  compare_exchange_strong(T& expected, T newVal) {
CE1     patomic <T>* rep_p_addr = REP_V_2_REP_P(this);
CE2     while (true) {
CE3         rep_p_seq = rep_p_addr->seq; // Read rep_p
CE4         rep_p_val = rep_p_addr->val;
CE5         rep_p_seq_again = rep_p_addr->seq;
CE6         rep_v_seq = this->seq; // Read rep_v
CE7         rep_v_val = this->val;
CE8         rep_v_seq_again = this->seq;
            // Restart if seq and val inconsistent
CE9f        if (rep_p_seq_again != rep_p_seq ||
                rep_v_seq_again != rep_v_seq)
              continue;
            // Help to complete another ongoing write
CE10f       if (rep_p_seq == rep_v_seq+1) {
CE11            FLUSH(rep_p_addr);
CE11            FENCE();
CE12            before = ⟨rep_v_val, rep_v_seq⟩;
CE13            after = ⟨rep_p_val, rep_p_seq⟩;
CE14            DWCAS(this, before, after);
              continue;
            }
            // Make sure we have the same versions
CE15f       if (rep_p_seq != rep_v_seq) continue;
            // If value on rep_p is not expected, fail
CE16        if (rep_p_val != expected) {
CE17            expected = rep_p_val;
CE19            return false;
            }
            // Update rep_p
CE20        before = ⟨rep_p_val, rep_p_seq⟩;
CE21        after = ⟨newVal, rep_p_seq+1⟩;
CE22        bool res = DWCAS(rep_p_addr, before, after);
CE23        FLUSH(rep_p_addr);
CE23        FENCE();
CE24        if res {
CE25            DWCAS(this, before, after);
CE27            return res;
            } else {
CE28f           if (before.val == expected) continue;
CE29            DWCAS(this, ⟨rep_v_val, rep_v_seq⟩, before);
CE31            return res;
            }
        }
    }
```

Listing 1.1. compare_and_exchange_strong implementation in Mirror

If no helping is needed, the CAS double-checks that the persisted and volatile values are the same (CE15f), and restarts if they are not. Next, at CE16, the operation checks whether the value of the persistent copy is the expected value, and if it is not it updates the expected value to the value read (CE17), then returns false (CE19) to indicate that the CAS has failed. As we shall discuss later, the KIV model requires an additional line (CE18) to set the value of the result followed by a "return" that effectively jumps to the program counter signifying termination of the CAS operation.

From line CE20, the CAS is finally ready to perform its own updates. This occurs in three steps:

(W_1) the volatile version of x.rep_p is updated to the new value and sequence number using a DWCAS at CE22,
(W_2) x.rep_p is flushed to persistent memory at CE23, and
(W_3) if the DWCAS update was successful, then x.rep_v is updated using another DWCAS at CE25.

This implies that memory at address x is always in one of three *stages*: Either, in stage 1, all three memories are equal (before step W_1 and after W_3). In stage 2 the volatile copy x.rep_p has been updated (after W_1), and in stage 3, both copies of x.rep_p (stage 3), but x.rep_v has not. We will formally define this as an invariant in Sect. 5.1.

Note that the update to x.rep_v must also occur through a DWCAS because another thread may have helped update the volatile copy after the persistent copy (CE22) has been successfully updated. If the DWCAS at CE22 fails (i.e., the persistent copy could not be updated), the thread checks whether it is possible to return false (to indicate failure) at line CE28f. If returning false is not possible, it retries the entire operation. Otherwise, we know that before holds the latest version of x.rep_p, (provided that the value of x.rep_v is unchanged since it was read at lines CE6–CE7) since before is updated as a side-effect of the DWCAS at CE22. Thus, the CAS attempts to help the other in-flight CAS operation via the DWCAS at line CE29 by updating the volatile copy, before returning the result (i.e., false) at line CE31.

4.2 Bug Against Expected Abstract Specification

Mirror's CAS implements a durable version of the compare_exchange_strong from C++ standard (see [6]), as announced in [16]. As discussed earlier, this property does not hold.

The problem can be illustrated with the following scenario, where we assume all operations execute a CAS on the same location, x, and that all values and sequence numbers are 0.

1. Assume that thread 1 starts the CAS algorithm with expected = 0 and new value 1 in the initial state and executes up to line CE22.
2. Thread 1 is interrupted by thread 2, which starts a CAS with expected = 0 and new value 2, executes the algorithm fully, and returns successfully.
3. Thread 1 resumes, but its DWCAS at line CE22 now fails (since the sequence number for x is now 1). Helping at line CE28f fails since after CE22, we have before.val = 2, and thread 1 returns with expected still being 0.

This execution is incompatible with two atomic `compare_exchange_strong` operations in any order. If the CAS of thread 1 linearizes first, it should succeed and not fail. If the CAS of thread 2 executes first, then its result is correct, but the value of `expected` should have been set to 2 instead of 0.

At first, this seemed to be a simple oversight in the algorithm that could be repaired by reloading `before.val` into `expected` before returning at CE31 (since the side-effect of line CE22 ensures that `before` has the latest version of `rep_p_addr`). For the scenario above, thread 1 correctly returns with `expected` set to 2 instead of 0. However, this "fix" also turned out to be incorrect, as demonstrated by the following execution.

1. The program is initialized as above, and thread 1 executes its CAS up to CE22, but is interrupted by thread 2.
2. Thread 2 executes a complete CAS operation with value 2.
3. Thread 2 starts a second CAS operation with value 3, but stops immediately after its successful DWCAS at line CE22. At this point, the volatile copy of $x.rep_p$ is set to $\langle 3, 2 \rangle$, while the persistent copy of $x.rep_p$ (as well as $x.rep_v$) is still set to $\langle 2, 1 \rangle$.
4. Thread 1 resumes. This time the failed DWCAS for thread 1 will read $\langle 3, 2 \rangle$ into `before`, and the FLUSH/FENCE at lines CE23/CE24 will set the persistent copy of $x.rep_p$ to $\langle 3, 2 \rangle$. However, helping at CE29 now fails since the value of $x.rep_v$ is set to $\langle 2, 1 \rangle$, but thread 1's copy of $x.rep_v$ is $\langle 0, 0 \rangle$. In the modified algorithm, thread 2 would now return with `expected` = 3.
5. Thread 1 executes a read operation, returning the value 2 since it reads from $x.rep_v$.

Once again, there is no linearization order for the execution above. Thread 1's CAS must be after thread 2's second CAS since it returns the value 3 for `expected`, but the final read is now incorrect since it returns 2 instead of 3.

We therefore conclude that the algorithm cannot be easily repaired to implement the strong version of CAS, but only implements the weak version, which allows the CAS algorithm to fail spuriously even if it sees the correct expected value. In the next section, we provide a KIV model for the algorithm that allows us to verify this claim.

Note that a weak CAS is sufficient to implement a write (aka store) operation with a loop repeating the CAS algorithm of Listing 1.1 until it succeeds as described in the original paper [16]. Thus such a write operation would never witness the bug that we have described.

```
automaton specification MIRROR
using MIRROR-preds
variables
  buf: L → Patomic;
  ram: L → Patomic;
  pm: L → Patomic;
  locStates: L → LocState;
  t: Tid;
  addr, vaddr, paddr: L;
  pseq, vseq, pseq_again, vseq_again: Nat;
  pval, vval, expected, val, newVal, newExpected: T;
  res: Bool;

initial state
  (pm = λ addr. ⟨0, 0⟩) ∧ (ram = λ addr. ⟨0, 0⟩) ∧
  (buf = λ addr. ⟨0, 0⟩) ∧ (pcf = λ tid. idle);
```

Listing 1.2. State model for Mirror in KIV

4.3 KIV Model

The KIV model of Mirror's state is given in Listing 1.2. The translation from program to automaton is as detailed in our prior work [7]. Here, we only outline some key elements. The full state of the automaton consists of the *global state*, together with *local states* and *pc* values for all threads. As discussed in Sect. 2.1, we assume the PSC memory model which comprises a (volatile) persistence buffer and a (non-volatile) memory [20]. Since each location contains two replicas rep_v and rep_p with rep_p further replicated in both the volatile and non-volatile memories, our KIV model uses three mappings:

1. *ram*, which tracks the rep_v copy,
2. *pm*, which tracks the persistent rep_p copy, and
3. *buf*, which tracks the volatile rep_p copy.

The local state is generated from the algorithm as the tuple of all local variables for each thread. The possible program counter values of algorithms are the labels before each program step. Each labeled instruction of an algorithm defines one possible step of the automaton, that may be executed by any thread. Programs start with label *idle* and the return instruction shows the program counter after the return. For each program step the generated automaton has steps for every thread that modify the global state together with the local state and pc of the executing thread.

The action of invoking/returning steps (e.g. from `idle` to `CE1` and from `CE19` back to `ready`) is generated automatically as an invoke/response action that has the thread as well as the input/output parameters of the algorithm as parameters. All other program steps have the empty action τ by default. The default can be overridden by providing a `with` clause after the line number that specifies a user-defined action for the step. We use this feature to override the empty action at linearization points, where we want a step of the algorithm to implement a step of the abstract specification. The `with` clause can be of the conditional form ($\phi \supset a_1; a_2$), which results in two steps with actions a_1 and a_2, depending on whether condition ϕ is true or not. Details how this feature is used for the verification are given in Sect. 5.3.

The main operations of the algorithms are defined using code-like syntax (see Listing 1.3). Note that KIV does not have a `continue` statement, and hence statements such as CE9f must be refactored so that the test is negated. The `if*` statement is used to model atomic compare-and-swap primitives, where the test and the first statement after the test are executed as a single atomic step. Finally, KIV does not have assignment via reference, and hence, old and new values of `expected` must be expressed as two separate parameters. For example, the COMPARE_EXCHANGE_STRONG operation has input parameter `expected` and output parameters `newExpected` to model the old and new values of `expected` in Listing 1.1. This also means that we have additional assignments, e.g., CE33 that are not in the C version of the algorithm (Listing 1.1).

Apart from the operations corresponding to the program, it is also possible to specify other atomic operations to describe system-level steps. These can be thread-local or global steps (specified with corresponding keywords). Mirror assumes two global operations CRASH and FLUSH.

5   Soundness of Mirror

The use of the Mirror library allows to transform a linearizable algorithm to a durably linearizable one. To be applicable, the original, linearizable algorithm must use read, write and CAS instructions over volatile memory. The transformed algorithm calls the libRead and the libCAS methods of the library instead, resulting in a durably linearizable algorithm over persistent memory.

To verify that this is the case, we verify that the Mirror library refines the abstract PLib^{IR} specification. To do so, we must find steps of the algorithms that implement the do-steps of the PLib^{IR} specification. These steps are in essence the *linearization points* (LPs) of the algorithms. Invokes and returns as well as crashes are external events, so they must refine their abstract counterparts in every case. For the read-algorithm there is little choice: reading from *ram* must implement do(libRead). This directly implies for the abstraction relation *abs* that the persistent memory of PLib^{IR} (i.e., the abstract *pmem*) must always be equal to the *ram* of Mirror, *not* to the persistent memory *pm* of the algorithm.

This adds several difficulties to the refinement proof: first, the CAS instruction at line CE22 that successfully writes a new value of Mirror to the buffer cannot be the linearization point that implements do(libCAS^T). Otherwise, if the thread is stopped directly after executing CE22, and another thread runs a complete Read this thread should already return the updated value. It however still receives the old value from *ram*. The same argument rules out the flush in the next line CE23. Instead the LP must be the subsequent helping instruction at CE25 that moves the value to *ram* or – alternatively – an earlier helping CAS by another thread (at CE29).

```
LOAD:
idle :- LOAD(vaddr; ;val) {
L1    with do_t(Read) :- val := ram(vaddr).val;
L2    :- return idle :- };

COMPARE_EXCHANGE_STRONG:
idle :- COMPARE_EXCHANGE_STRONG(vaddr, newVal, expected; ;
                                newExpected, res) {
CE1   :- let paddr = vaddr in
CE2   :-   while true do {
CE3   :-      let pseq = buf(paddr).seq in
CE4   with do_t(CAS^F)
      :-       let pval = buf(paddr).val in
CE5   :-       let pseq_again = buf(paddr).seq in
CE6   :-       let vseq = ram(vaddr).seq in
CE7   :-       let vval = ram(vaddr).val in
CE8   :-       let vseq_again = ram(vaddr).seq in
CE9   :-       if pseq_again = pseq ∧ vseq_again = vseq then {
CE10  :-          if pseq = vseq + 1 then {
CE11  with Flush(paddr) :-
                   pm(paddr) := buf(paddr);
CE12  :-             let before = ⟨vval, vseq⟩ in
CE13  :-             let after = ⟨pval, pseq⟩ in
CE14  with (ram(vaddr) = before ⊃ Help(vaddr); τ) :-
                   if* (ram(vaddr) = before)
                   then ram(vaddr) := after
                   else skip; }
                else
CE15  :-          if vseq = pseq then {
CE16  :-             if pval ≠ expected then {
CE17  :-                newExpected := pval;
CE18  :-                res := false;
CE19  :-                return idle :- };
CE20  :-             let before = ⟨pval, pseq⟩ in
CE21  :-             let after = ⟨newVal, pseq + 1⟩ in {
CE22  with (buf(paddr) = before ⊃ do_t(CAS^T); τ) :-
                   if* buf(paddr) = before
                   then res := true, buf(paddr) := after
                   else res := false, before := buf(paddr);
CE23  with Flush(paddr) :-
                   pm(paddr) := buf(paddr);
CE24  :-             if res then {
CE25  with (ram(vaddr) = before ⊃ Help(vaddr); τ) :-
                   if* ram(vaddr) = before
                   then ram(vaddr) := after;
CE26  :-                newExpected := expected;
CE27  :-                return idle :-; }
                   else
CE28  :-                if before.val ≠ expected then {
CE29  with (ram(vaddr) = ⟨vval, vseq⟩ ⊃ Help(vaddr); τ) :-
                   if* (ram(vaddr) = ⟨vval, vseq⟩)
                   then ram(vaddr) := before;
CE30  :-                   newExpected := expected;
CE31  :-                   return idle :-; } } } } };
CE32  :-   res := false;
CE33  :-   newExpected := expected;
CE34  :-   return idle :-; };

CRASH: CRASH() global
{ pcf := λ t. (pcf(t) = idle ⊃ idle; crashed),
  ram := pm,
  buf := pm };

FLUSH: FLUSH(addr) global
{ pm(addr) := buf(addr); };
```

Listing 1.3. Model of Mirror's LOAD and CAS operations

Second, the crash step of Mirror that updates the content of *ram* with persistent memory (as done in recovery) implements the do(libCAS^T) step of *all* those threads t that have executed a successful CAS, where the buffered value has been flushed (either by the algorithm or the operating system) to persistent memory already.

Finally, the algorithm has to decide which step of a failed CAS run should implement do(libCAS^F). This poses the problem of a *potential linearization point* [11]: loading a value at CE4 that is different from the expected one must implement do(libCAS^F) when the algorithm later on reaches line CE19 where it returns false. No later step is possible, since the value in persistent memory as well as the one in *ram* may be changed directly after loading it by other threads. However, whether the algorithm will continue to CE19 is not decided at this point: if persistent memory and *ram* are changed by another thread immediately after loading at CE4, then the algorithm will restart the loop, since the check at CE9 fails. Then the algorithm may still execute both a successful as well as a failed CAS. Therefore, loading at CE4 is what we call a potential linearization point: whether it becomes a real one (i.e., refines do(libCAS^F)) depends on how the execution proceeds afterwards.

The problem is similar to proving linearizability of the dequeue operation of the Michael-Scott queue [25]. There, a queue is represented as a singly-linked list, with a dummy cell that contains no data at the start (to avoid a special representation for the empty queue). A head and tail pointer point to the start and end of the queue, respectively. The dequeue algorithm loads the head cell, and this again is a potential LP for the result that the queue is empty: if a subsequent check that head and tail point to the same cell succeeds, it is. But this check first requires to load the tail cell, which may have already been changed by other threads enqueuing elements. In this case loading the head is not the LP (and the algorithm then successfully dequeues an element).

Dealing with potential LPs where the decision depends on future execution requires the use of backward simulation (or equivalently: of prophecy variables) for proving refinement. We have shown in earlier work that linearizability proofs then can be done with backward simulation alone [31].

However, since the proofs are *global* proofs (over all threads) then and get very complex and non-modular, we decided to roughly follow the strategy of [12], which splits the proof into two refinements:

- A lower one, that refines an intermediate model IMirror to Mirror. IMirror is designed to abstract from all the intricate details of the CAS algorithm. This refinement can be verified by proving the step-local, thread-local forward simulation conditions we derived in [4].
- The IMirror model consists of 10 simple atomic steps, which makes a backward simulation proof that it refines PLib feasible.

The intermediate model is designed to make the three stages of a successful CAS explicit as steps of an abstract automaton. Before showing that this is a faithful abstraction, it is necessary to prove that the algorithm indeed keeps each memory location in one of the three stages described at the end of Sect. 4.1.

This is expressed in the next subsection as a suitable invariant which we can show via step-local, thread-local proof obligations. Then Sect. 5.2 will define the intermediate model IMirror. Section 5.3 then shows the results of trying to prove the lower refinement. This section discusses the problem we found in the algorithm, since it was discovered when trying to prove the refinement correct. Finally, Sect. 5.4 is about the proof of the upper refinement by backward simulation.

5.1   Invariant Proof

As a first step to proving Mirror correct, it has to be shown that a successful run of the CAS algorithm indeed updates memory in three stages. The following invariant expresses the fact that the three values stored at an address x are indeed in one of three stages:

$$\begin{aligned}
\mathsf{inv}(\mathit{buf}, \mathit{pm}, \mathit{ram}) \equiv \\
\forall\ x.\ \mathit{buf}(x) = \mathit{pm}(x) = \mathit{ram}(x)\ \vee & \qquad /\!\!* \text{ stage 1 } *\!/ \\
\mathit{buf}(x).\mathsf{seq} = \mathit{pm}(x).\mathsf{seq} + 1 \wedge \mathit{pm}(x) = \mathit{ram}(x)\ \vee & \qquad /\!\!* \text{ stage 2 } *\!/ \\
\mathit{buf}(x) = \mathit{pm}(x) \wedge \mathit{pm}(x).\mathsf{seq} = \mathit{ram}(x).\mathsf{seq} + 1 & \qquad /\!\!* \text{ stage 3 } *\!/
\end{aligned}$$

However, it is crucial too that the algorithm only cycles forward through the three stages, since it only increments sequence numbers. This is expressed as a rely condition, that holds for all steps of Mirror from a state to another:

$$\begin{aligned}
\forall\ x.\ \mathit{buf}(x).\mathsf{seq} \leq \mathit{buf}'(x).\mathsf{seq}\ \wedge \\
\mathit{pm}(x).\mathsf{seq} \leq \mathit{pm}'(x).\mathsf{seq}\ \wedge \mathit{ram}(x).\mathsf{seq} \leq \mathit{ram}'(x).\mathsf{seq}
\end{aligned}$$

In the formula (like in the specification language Z) primed variables indicate the state after a step.

To verify the global invariant, it is necessary to give a lot of assertions that characterize individual states that occur during the executions of the CAS algorithm (a full listing is given in the KIV specification at [29]). Some are simple to figure out. Others are quite tricky, e.g., at CE25 we prove that

$$\begin{aligned}
\phi_{\mathrm{CE25}} \equiv \mathit{pseq} + 1 \leq \mathit{pm}(\mathit{paddr}).\mathsf{seq} \wedge \mathit{pseq} + 1 \leq \mathit{buf}(\mathit{paddr}).\mathsf{seq}\ \wedge \\
(\mathit{pseq} + 1 = \mathit{buf}(\mathit{paddr}).\mathsf{seq} \rightarrow \mathit{buf}(\mathit{paddr}).\mathtt{val} = \mathit{newVal})
\end{aligned}$$

If this formula would be proved in a global approach, all assertions ϕ_{pcval} for specific program counter values pcval, like ϕ_{CE25} would be part of a large invariant that collects all the assertions into one formula. This formula would state that for all threads t where $\mathit{pcf}(t) = \mathrm{CE25}$, ϕ_{CE25} holds (where paddr, pseq, newVal denote the respective fields $\mathit{lsf}(t).\mathit{paddr}$ etc.). KIV however generates thread-local step-local proof obligations that allow to avoid reasoning about the large invariant and all threads. These fall into three categories (see [7] for details):

- Each step from $pcval$ to $pcval'$ of a thread establishes the global invariant inv and the assertion $\phi_{pcval'}$, given that inv and ϕ_{pcval} held before the step.
- All steps (of other threads) satisfy the rely condition.
- All assertions ϕ_{pcval} are stable, i.e., preserved by steps that satisfy the rely.

For our algorithm, the assertions given for the first steps of the algorithm must ensure that when the algorithm does a successful CAS at CE22, then the memory at the address $addr$ is in stage 1. This requires that at CE10, when the test at CE9 has succeeded we can prove

$$paddr = vaddr \wedge pseq \leq buf(paddr).\mathtt{seq} \wedge vseq \leq ram(vaddr).\mathtt{seq}$$
$$\wedge \; (pseq = buf(paddr).\mathtt{seq} \rightarrow pval = buf(paddr).\mathtt{val})$$
$$\wedge \; (vseq = ram(paddr).\mathtt{seq} \rightarrow vval = ram(vaddr).\mathtt{val})$$

This assertion is then propagated to the CAS instruction at CE22, which implies that the test $vseq = pseq$ at CE15 must have succeeded. When the CAS is successful, i.e. when $vseq = buf(vaddr).\mathtt{seq}$, then $buf(vaddr).\mathtt{seq} = vseq = pseq \leq ram(vaddr).\mathtt{seq}$ can be deduced. Together with the global invariant it is implied that $ram(vaddr)$ must be equal to $buf(vaddr)$ in this case, i.e. that memory at $vaddr$ is indeed in stage 1, when the successful CAS happens.

Figuring out the assertions is supported by the thread-local proof obligations. If an assertion for an early program counter is still too weak, the proof of one of the steps before reaching CE22 will fail. When all the assertions are correct (we get ca. 50 lines of assertions), the proof obligations are all rather simple to verify (19 interactions are needed overall for the 83 obligations generated).

5.2   Intermediate Layer

The intermediate model IMirror is given in Fig. 7. Every thread t now has a status $status(t)$ which is a program counter with information about locations and values[4]. The status is $idle$ before the first invocation of an operation, and $ready$ in between operations[5].

The model essentially abstracts the CAS algorithm of Mirror to the three stages a modification of a location x passes through.

Invocation by thread t sets the status of this thread to $inC(x, e, v)$ which stores the location updated, the expected value and the new value. Then the main step of the algorithm is a successful CAS at CE22. All the steps of the algorithm before this step just ensure success only happening if all three values $buf(x)$, $pm(x)$ and $ram(x)$ are equal, i.e., in stage 1. This is now enforced simply by the precondition $buf(x) = ram(x)$ for the $do_t(\mathsf{CAS}^T)$ step.

After a successful CAS memory is in stage 2, and the status is $outC(x, v, sqno)$ where $sqno = buf(v).\mathtt{seq} + 1$ and v are the sequence number and value just

---

[4] The formal KIV specification separates $status$ into a pure program counter and data stored in the local state of the thread.

[5] The KIV model has an extra step that starts a thread by changing $status$ from $idle$ to $ready$. We simplify this detail here.

State variables:

$pm, ram, buf : Loc \rightarrow (Val \times \mathbb{N})$ (initially: $\forall x \in Loc.pm(x) = ram(x) = buf(x) = (0,0)$)

$status : Tid \rightarrow PCVal$ (initially: $\forall t \in Tid.\ status(t) = idle$)

Actions:

$inv_t(Read(x))$
Pre: $status(t) = idle$
Eff: $status(t) := inR(x)$

$do_t(Read)$
Pre: $status(t) = inR(x)$
      $ram(x).val = v$
Eff: $status(t) := outR(v)$

$res_t(Read(v))$
Pre: $status(t) = outR(v)$
Eff: $status(t) := idle$

$flush(x)$
Pre: $true$
Eff: $pm(x) := buf(x)$

$help(x)$
Pre: $pm(x).\mathbf{seq} = ram(x).\mathbf{seq} + 1$
Eff: $ram(x) := pm(x)$

$crash$
Pre: $true$
Eff: $status := \lambda t : Tid.$
      $\quad$ if $status(t) \neq idle$
      $\quad$ then $crashed$ else $idle$
      $buf := pm$
      $ram := pm$

$inv_t(CAS(x,e,v))$
Pre: $status(t) = idle$
Eff: $status(t) := inC(x,e,v)$

$do_t(CAS^T)$
Pre: $status(t) = inC(x,e,v)$
      $\lor\ status(t) = inoutC(x,e,v,\cdot,\cdot)$
      $ram(x) = buf(x)$
      $buf(x).\mathbf{val} = e$
Eff: $status(t) := outC(x,v,buf(x).\mathbf{seq} + 1)$
      $buf(x) := \langle v, buf(x).\mathbf{seq} + 1 \rangle$

$do_t(CAS^F)$
Pre: $status(t) = inC(x,e,v)$
      $\lor\ status(t) = inoutC(x,e,v,\cdot,\cdot)$
      $buf(x).\mathbf{val} \neq e$
Eff: $status(t) := inoutC(x,e,v,$
      $\qquad\qquad\qquad buf(x).\mathbf{val}, buf(x).\mathbf{seq})$

$res_t(CAS(bv, newE))$
Pre: $(bv \land status(t) = outC(v, newE, sqno))$
      $\lor\ (\neg bv \land status(t) = inoutC(\cdot,\cdot,\cdot, newE, sqno))$
      $ram(x).\mathbf{seq} \geq sqno$
Eff: $status(t) := idle$

Fig. 7. Specification of intermediate layer IMirror (with global transitions crash, help, flush). Again, the red color indicates the condition which was dropped for the weak version. (Color figure online)

written. The code of Mirror after a successful CAS ensures that this value reaches *ram* before it is returned as the new expected value. Since the necessary flushing and helping can happen in other threads (or in case of the flush even by the operating system) the steps are abstracted to global steps flush and help that update *pm* and then *ram* to stage 3 and finally back to stage 1.

That the thread cannot return before these steps (and possibly more) have happened is now enforced with a precondition for the response action $res_t(CAS)$: it is checked that the sequent number of $ram(x).\mathbf{seq}$ must be at least the *sqno* that CAS updated to. This is the reason, why the $do_t(CAS^T)$ step saves this sequence number as the last field of the status *outC*.

It was explained earlier that the CAS algorithm of Mirror has a potential linearization point (LP) for CAS when loading a value at CE4, where the decision to place one depends on future execution. In IMirror, this is realized by allowing that the $do_t(CAS^F)$ step may be repeated several times nondeterministically (not executing an LP), followed by a final $\mathsf{do}_t(\mathsf{CAS}^T)$ or $\mathsf{do}_t(\mathsf{CAS}^F)$. Only this final do-step determines the LP of the CAS. Technically, the status $inoutC(x, e, v, newE, sqno)$ resulting from $\mathsf{do}_t(\mathsf{CAS}^F)$ allows two behaviors: first, another do-step may be executed as if the status still were $inC(x, e, v)$. Then the LP has not happened, corresponding to the algorithm restarting the loop. It is however possible that an LP has been executed by allowing a response similar to the one for $outC(x, newE, sqno)$, except that the returned boolean bv that indicates a failed CAS is then false.

Before a failed CAS can return it is crucial again, that the algorithm ensures by flushing and helping in lines CE23 and CE29 that the modified expected value stored in the $newE$ field of the status has been copied to ram. Again this is abstracted to saving the $buf(x).sqno$ into the last field of the status, and by checking that the response step $\mathsf{res}_t(\mathsf{CAS})$ is possible only when the sequence number in ram is at least this saved number.

Finally, a crash of Mirror aborts all running threads with a status that is not $idle$ by setting their status to $crashed$. On a crash, Mirror executes a recovery routine that will copy the current content of pm to ram, which is part of the abstract crash transition.

Altogether, this gives us the following status values of the thread's program counters:

$$PCVal = \{idle, crashed, inR(x), outR(v),$$
$$inC(x, v, e), outC(x, newE, sqno), inoutC(x, v, e, newE, sqno) \mid$$
$$x \in Loc, v, e, newE \in Val, sqno \in \mathbb{N}\}$$

5.3   Lower Refinement

The proof of the lower refinement assumes that the invariant from Sect. 5.1 is already proved. It also generates thread-local, step-local forward simulation conditions. The automatic generation of such proof obligations is supported by KIV under two conditions:

- the refinement can be proved with a forward simulation, where all concrete steps (of the algorithm) either refine one abstract step of IMirror or none (the general forward simulation condition of Fig. 3 allows a sequence of steps).
- The abstract step which a step of thread t can implement must either be an abstract step of the same thread or a global step.

The case where the abstract step is a global one is used to allow that successful helping steps of the CAS algorithm can implement help($paddr$) of IMirror.

To specify the steps of the Mirror algorithm that should refine steps of IMirror we specify a mapping between the actions of both automata. For invoking and returning steps the mapping must be 1:1, since these steps are external. To specify which steps of Mirror implement do steps of IMirror, we use the with clause in the algorithm (see again Listing 1.3), and override the default action

τ with the action of IMirror[6] it should implement. All the other steps that keep their τ action 'implement skip', i.e., an empty sequence of abstract steps.

Before we found the bug, CE4 could only be a potential linearization point (LP), when the loaded value is different from *expected*, so the with clause at CE4 was[7]

with $(expected \neq buf(x).val \supset \text{do}_t(\text{CAS}^F); \tau)$

A failed CAS at CE22 then also had to implement $\text{do}_t(\text{CAS}^F)$, except when *expected* is the right value, so the test at CE28 succeeds and restarts the loop. Therefore the clause was

with $(buf(paddr) = before \supset \text{do}_t(\text{CAS}^T);$
$(buf(paddr).\text{val} \neq pval \supset \text{do}_t(\text{CAS}^F); \tau))$

Verification then failed while trying to prove that the failed CAS indeed implements $\text{do}_t(\text{CAS}^F)$. The reason for this is that the wrong expected value is returned (it should be the one which is read at this point). The counter example given in Sect. 4.2 demonstrates this problem. The attempted correction that sets *newExpected* to *before*.val at CE30 instead of the old *expected* failed, since the assumption that this value has reached *ram* failed, when helping is not successful. This led to discovering the second run given in Sect. 4.2.

Just assuming that the Mirror algorithm implements a weak CAS makes verification much simpler, since then CE4 is now the only instruction that implements $\text{do}_t(\text{CAS}^F)$, even in the case where the expected value is equal to the loaded one. The failed CAS at CE22 no longer refines $\text{do}_t(\text{CAS}^F)$. The with clauses are simplified accordingly in the algorithm given in Listing 1.3. All states between CE4 and the main CAS at CE22 now refine status *inoutC* of IMirror, while before in many cases *inC* was possible as the status of IMirror, too.

For proving a refinement, we need to specify an abstraction relation. Like the invariant for Mirror, this relation is split into a global abstraction relation that relates the global states of Mirror and IMirror and *local abstract relations* that relate the local states of individual threads t (these relations have access to the two global states too).

The global relation here is simply saying that the three memories of both models always agree. The difficult assertions are again the thread local ones, which are again specified to hold for specific *pc* ranges of Mirror.

As two examples, when the algorithm is at CE10 or at one of CE23, CE24 we have

$$CE10 : status = inoutC(vaddr, expected, vval, newVal, expected)$$
$$CE23 \rightarrow CE24 : status = (res \supset outC(vaddr, newVal, vseq + 1);$$
$$inoutC(vaddr, expected, newVal, vval, vseq, expected))$$

---

[6] Technically, the custom actions defined are not identical to those of IMirror (since they are not available in Mirror), but are mapped 1:1 when defining the refinement.

[7] Recall that $(\phi \supset e_1; e_2)$ returns e_1, when ϕ holds, e_2 otherwise.

Note that *status* in the formula is the status of the considered thread t of IMirror, and that *vaddr*, *res* etc. are the local variables of the same thread in Mirror.

Given the split into global and local abstraction relation, and the mapping that gives one or no abstract action for every concrete action, KIV again generates proof obligations for every step of the Mirror algorithm. These proof obligations define commuting diagrams for each concrete step of thread t, when the corresponding abstract step (or no step if the action is τ) is executed. They show that the concrete step of thread t preserves a) the global abstraction relation, b) the local abstraction relation of t, c) the local abstraction relation of an arbitrary different thread t'. Together, the proof obligations allow to reconstruct a proof for the full abstraction relations. Various simplifications are possible that save some of these proof obligations, full details are given in [4].

The KIV proofs available at [29] still show the failed proof attempt for the modified version that returns *before*.val. Unfortunately the problem was detected only with the last proof obligation for step from CE29 to CE30, where it must be ensured that *before*.val has reached *ram*. The final version that shows that weak CAS is implemented generates 73 refinement proof obligations. Again, the main effort is in getting the ca. 30 lines of assertions for the local abstraction relation right. Once these are correct, proofs were almost automatic (27 interactions).

5.4 Upper Refinement

Mechanized verification of the upper refinement from IMirror to PLib^{IR} is work in progress. However, the main definitions needed for the proof have been worked out already, so we give them here. First, one needs to define a global abstraction relation that maps the memories *buf*, *pm* and *ram* to the two memories *vmem* and *pmem* of PLib^{IR}. A first observation is that simply identifying the two persistent memories *pm* and *pmem* will not work, since reading an address x in PLib^{IR} will return $pmem(x)$ while IMirror will return $ram(x)$, which are different, when the location is in stage 1 or 2 in IMirror. Therefore we are forced to use *ram* to implement *pmem*. For the implementation of *vmem* there is more choice. One could consider *buf* and *pm* as two caches for the *ram* and choose e.g., *buf* to implement *vmem*. However it is simpler to ignore them altogether and to choose *ram* as the implementation of *vmem*, too.

The reasoning for this choice is as follows: PLib^{IR} allows writing a value to *vmem* and to finish the libWrite operation, before it is flushed to *pmem*. Under specific conditions such a non-persisted value may be read by another libRead already. The *log* of PLib^{IR} marks the writes and reads of such non-persisted values with a flag *false* and gives general conditions that ensure that this does not lead to non-linearizable behavior of the operations calling a library implementing PLib^{IR}. The FliT library [35] which we verified in [3] to be a refinement of PLib^{IR} indeed exploits that reading a nonpersisted value is sometimes possible. In Mirror, we have no such unpersisted writes at the end of a CAS operation (and, since writing is implemented using CAS at the end of a write operation too), since a CAS of IMirror cannot finish, before the value has reached *ram*. If we choose the abstraction relation to equate *vmem* to be equal to *buf*, then there is still a state, where *log* contains a last write entry with flag 'false'. If we choose *vmem* to be *ram* too, then all *log* entries have flag *true*, leading to the simple invariant $log_{|t,false} = \varepsilon$. This has two consequences: first of all, $\mathsf{complete}_t$ of PLib^{IR} can be implemented with an empty program in IMirror (and consequently in Mirror), since its precondition is trivially true. This explains, why mirror in consequence does not need a call to $\mathsf{complete}_t$ at the end of an operation at all when the Mirror library is used. Second, moving a value from *pm* to *ram* in the help transition then is the only transition in a successful CAS run that has to implement abstract steps: it implements a sequence of $\mathsf{do}_t(\mathsf{libCAS}^T)$ and $\mathsf{persist}(x)$ for the relevant address x. The action $\mathsf{do}_t(\mathsf{CAS}^T)$ itself simply implements skip, since it has no effect on *ram*, and therefore none in PLib^{IR}.

Together the global abstraction relation is

$$buf.\mathtt{vals} = vmem \wedge ram.\mathtt{vals} = pm \wedge log_{|t,false} = \varepsilon$$

where $\mathtt{vals}$ discards the sequence numbers of IMirror which are not present in PLib^{IR}.

Careful consideration is now necessary for crashes. In Mirror (and IMirror) a crash copies *pm* to *ram* when executing the recovery program. For the given abstraction relation this means, that the "cache" *pm* is not discarded on a crash, but instead all the values that have not yet reached *ram* (i.e., all addresses in stage 2) are "persisted" instead. Therefore a crash of IMirror refines a sequence persistActions followed by the crash step of PLib^{IR} as shown by the commuting diagram given in Fig. 8, to the upper left. persistActions is a sequence of persist actions for all addresses x_i that are currently in stage 2 followed by the crash of PLib^{IR}:

$$\mathsf{persistActions} = \mathsf{persist}(x_1) \ldots \mathsf{persist}(x_n)$$

Note that we exploit here that in IO automata refinement an external action (crash here) can refine a sequence of internal steps (the persist steps) followed by the same external action.

Finally, the most complex issue of the refinement is that we need to deal with the potential linearization points caused by the possibility of executing $\mathsf{do}_t(\mathsf{CAS}^F)$ several times, before the final one really decides whether a successful or failed CAS is executed on the abstract level. For the backward simulation this means that the local abstraction relation for thread t relates a state with $status(t) = inoutC$ to two abstract states of PLib^{IR}: one, where $pc(t) = outC$ and one where $pc(t) = inC$. Simulating a $\mathsf{do}_t(\mathsf{libCAS}^F)$ step for address x of thread t backwards then decides on whether an abstract step must be executed with a case split that looks at the abstract state as' that is related to the state cs' that is reached after the $\mathsf{do}_t(\mathsf{libCAS}^F)$ step. The two cases result in the two commuting diagrams that are shown in the bottom row of Fig. 8. If the state related to cs' has $pc(t) = inC$, the $\mathsf{do}_t(\mathsf{libCAS}^F)$ step is not the last one (the operation can not return before executing $\mathsf{do}_t(\mathsf{libCAS}^T)$ or $\mathsf{do}_t(\mathsf{libCAS}^F)$), so

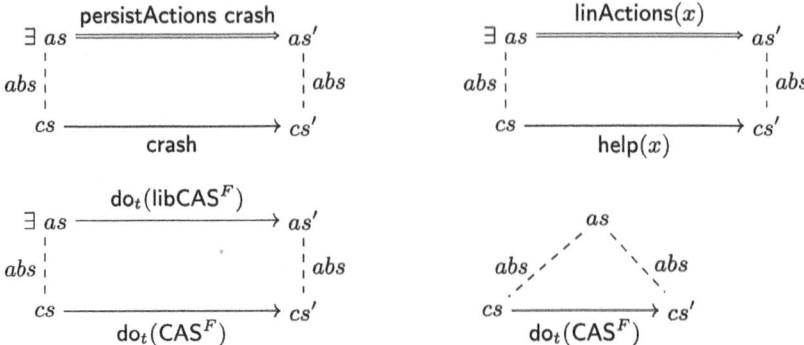

Fig. 8. Important commuting diagrams of the backward simulation. The last row shows the two cases for executing a failed CAS in IMirror. The left diagram is proved, when the state of t is $outC$ in state as' and $buf(x) = ram(x)$ for the location x that thread t accesses. The sequences persistActions and linActions(x) are explained in the text.

the step refines skip. This gives the diagram, where no abstract step is executed, and $as = as'$ on the right. The other case, where $pc(t) = outC$ suggests that the abstract step should be the corresponding $do_t(\mathsf{libCAS}^F)$ of $\mathsf{PLib}^{\mathsf{IR}}$ (as shown on the left). However, this is correct only when the failed CAS reads an expected value, that has already been moved to ram, i.e., when address x is in stage 1 and $ram(x) = buf(x)$. Otherwise, the abstract dolibCAS^F step must again be delayed to the help(x) step. Otherwise the expected values computed by $do_t(\mathsf{CAS}^F)$ and $do_t(\mathsf{libCAS}^F)$ would differ.

As a result the help(x) transition of IMirror refines a sequence linActions(x) of abstract actions shown as the upper right commuting diagram in Fig. 8. First, it refines the successful CAS for address x together with a persist(x). Second, it refines all those failed CAS instructions that have loaded $buf(x)$.val as their expected value that is now persisted to ram. These are precisely the ones that will try to help at line CE29 with moving the value to ram. Letting $t_1, \ldots, t_n$ be those threads we have

$$\mathsf{linActions}(x) = do_t(\mathsf{libCAS}^T)\ \mathsf{persist}(x)\ do_{t_1}(\mathsf{libCAS}^F)\ \ldots\ do_{t_n}(\mathsf{libCAS}^F)$$

Mechanized verification of the refinement needs to precisely define the sequences $t, t_1, \ldots, t_n$ of threads used in linActions and the sequence $x_1, \ldots, x_n$ of locations used in persistActions. Since no order is easily available when inspecting a state of IMirror, we have already formally specified two lists of threads as auxiliary variables in IMirror that fix an order.

The first auxiliary variable waitForHelp(x, $sqno$) directly stores the list $t, t_1, \ldots, t_n$. The list is nonempty, iff some thread t has just executed a successful CAS, that has written $sqno$ at the given address x, has not yet returned (so its status is $outC(x, v, sqno)$) and the value has not reached ram, so $ram(x) \neq buf(x)$ (the location x is currently in stage 2 or 3). The list then starts with t. The remaining elements t_i of the list are all those threads that have executed a failed CAS for the same x and $sqno$. These also have not returned, and store the same $sqno$ in their $inoutC$-status. The variable is updated in the $\mathsf{do}_t(\mathsf{CAS}^T)$ step, where it sets the variable to $List(t)$. A thread t' attaches itself to the end of the list $\mathsf{waitForHelp}(x, buf(x).\mathsf{seq})$ when it executes $\mathsf{do}_{t'}(\mathsf{CAS}^F)$ for a location x, but only if $ram(x) \neq buf(x)$. The list $\mathsf{waitForHelp}(x, buf(x).\mathsf{seq})$ is emptied, when $\mathsf{help}(x)$ is executed. A crash empties all lists.

The second auxiliary variable $\mathsf{linOnCrash}$ is a list of threads, too. It stores those t_i that have executed a successful CAS, where the value has already been flushed, but has not reached ram, i.e., memory at the updated location x is in stage 2. It should be clear that for each location x_i in stage 2 a single thread t_i exists, that is responsible for this stage and that the thread stores the location x_i needed for $\mathsf{persistActions}$ as $status(tid_i) = outC(loc_i, \ldots)$. Variable $\mathsf{linOnCrash}$ is updated on the $\mathsf{flush}(x)$ step, by adding the head of $\mathsf{waitForHelp}(x, buf(x).\mathsf{seq})$ (which is the thread that has executed the successful CAS). The $\mathsf{help}(x)$ step removes the head of $\mathsf{waitForHelp}(x, buf(x).\mathsf{seq})$ from the list again. A crash empties the variable.

As a first step towards verification of the backward simulation, we have mechanized the proof that the two auxiliary variables always contain the right threads affected by a help resp. crash step as an invariant. Although the proofs are not difficult, they are somewhat more interactive than the previous ones, since they involve global arguments over all threads and all locations that require lots of quantifier instances (177 interactions). Mechanizing the upper refinement with KIV is current work in progress.

6   Related Work

Our proof technique is based on our prior work [11], which provides a verification method for concurrent data structures with *potential* linearization points. Here, status fields are used to determine whether a linearization point has been executed. In particular, status values inC and $outC$ hold before and after the linearization point, while a status value $inoutC$ describes a state in which it is unclear whether the linearization point has been executed. The abstract library PLib is from our earlier work [3], where we show that (1) PLib (and by extension its refinements) ensure transformation of a linearizable object to a durably linearizable object (Sect. 3.2), and (2) another library called FliT [35] is a refinement of PLib. By applying PLib to Mirror, we show that it is a sufficient persistency library abstraction for more than one implementation.

In concurrent work, Stefanesco et al. have also verified refinement of both FliT and Mirror [33] under weak memory models using a declarative weak-memory framework, rather than our operational approach. Interestingly, they have not reported on the bug in Mirror and claim that it is correct. On closer inspection, it turns out that they use an abstract specification that does *not* track the value of $\mathsf{expected}$, causing them to miss this bug in their proofs.

Besides Mirror and the already mentioned FliT, many other works have been developed to support transformations to durability. NVTraverse [15] is a technique for making linearizable concurrent objects durably linearizable, but is limited to node-based tree-like data structures. Pronto [24] supports transformations for sequential and lock-based data structures, while approaches such as Persimmon [37] and Mangosteen [14] build sophisticated techniques for *transparent* approaches to durability, with little to no programmer overhead. Fully verifying these systems is a subject for future work.

For weaker consistency conditions, Izraelevitz and Scott [19] propose a technique for converting linearizable objects into (buffered) durable linearizable objects by inserting pfence and pwb instructions. They prove correctness of their transformations. Montage [36] is a library for use by programmers when intending to achieve buffered durable linearizability [19]. To this end, programmers need to identify so-called payloads and follow several additional rules. Neither of these approaches is accompanied by a fully formal proof of its soundness.

Apart from that, there are other approaches studying transformation of durability and their soundness. D'Osualdo, Raad and Vafeiadis [13] pursue the idea of reusing proofs of linearizability for showing concrete data structures to be durable linearizable. They formulate a so-called *Pathway Theorem* which presents a method for proving durable linearizability. This method is modular in the sense of separating the proof of linearizability from issues concerning volatile and persistent memory. A similar approach (but for durable opacity) has been investigated by Vafeiadi Bila et al. [1] who aim at reusing proofs of opacity [2]. Durable opacity is the NVM analogue of the correctness criterion *opacity* employed for Transactional Memory [32] (not concurrent objects). Finally, Raad et al. show that for software transactional memory (STM), it is possible to seamlessly combine STMs with sequential failure atomic transactions (e.g., PMDK) [26], resulting in a system that is durably opaque.

7 Conclusion

This paper provides a case study for the use of the interactive prover KIV within refinement proofs. The case study has once again shown that in particular within the field of concurrency the rigor and precision of mechanized proving is indispensable for finding subtle bugs in implementations. More specifically, with the help of KIV we found a subtle bug in Mirror, which does not concern its overall soundness, but shows that Mirror disagrees with the specification it is supposed to implement. In future work, we will work on finding a repair for this bug. Up to now, the refinement could only be completed by changing the abstract specification.

This work has also provided evidence for two presumptions we made in earlier work. In [3], we conjectured that PLib is also an adequate abstract specification for proving refinement to Mirror which we now know to actually be true. In [4], we claimed that thread-local proof obligations can significantly simplify proofs which has turned out to be case here once more.

As future work, we see finishing the mechanized proof of the upper refinement. Furthermore, we are interested in fixing the bug and then providing a refinement proof for the strong version. We currently think, that returning *before.lVAL* as in the failed proof attempt could be a possible solution, when making sure that this value has been moved to *ram* already.

Acknowledgment. The authors of this paper are very grateful for Prof. Reif's continuous support and help with the topic over many years.

References

1. Bila, E., Derrick, J., Doherty, S., Dongol, B., Schellhorn, G., Wehrheim, H.: Modularising verification of durable opacity. Log. Methods Comput. Sci. **18**(3) (2022). https://doi.org/10.46298/lmcs-18(3:7)2022
2. Bila, E., Doherty, S., Dongol, B., Derrick, J., Schellhorn, G., Wehrheim, H.: Defining and verifying durable opacity: correctness for persistent software transactional memory. In: Gotsman, A., Sokolova, A. (eds.) FORTE, LNCS, vol. 12136, pp. 39–58. Springer (2020). https://doi.org/10.1007/978-3-030-50086-3_3
3. Bodenmüller, S., Derrick, J., Dongol, B., Schellhorn, G., Wehrheim, H.: A fully verified persistency library. In: VMCAI (2), vol. 14500 of Lecture Notes in Computer Science, pp. 26–47. Springer (2024). https://doi.org/10.1007/978-3-031-50521-8_2
4. Bodenmüller, S., Schellhorn, G., Reif, W.: Verification of forward simulations with thread-local, step-local proof obligations. Sci. Comput. Program. **241**, 103–227 (2024). https://doi.org/10.1016/j.scico.2024.103227
5. Chakrabarti, D.R., Boehm, H.-J., Bhandari, K.: Atlas: leveraging locks for non-volatile memory consistency. ACM SIGPLAN Not. **49**(10), 433–452 (2014)
6. compare_exchange as defined in the C++ standard library (2024). https://en.cppreference.com/w/cpp/atomic/atomic/compare_exchange
7. Derrick, J., Doherty, S., Dongol, B., Schellhorn, G., Wehrheim, H.: Verifying correctness of persistent concurrent data structures: a sound and complete method. Formal Aspects Comput. 547–573 (2021). https://doi.org/10.1007/s00165-021-00541-8
8. Derrick, J., Schellhorn, G., Wehrheim, H.: Proving linearizability via non-atomic refinement. In: Davies, J., Gibbons, J. (eds.) IFM 2007. LNCS, vol. 4591, pp. 195–214. Springer, Heidelberg (2007). https://doi.org/10.1007/978-3-540-73210-5_11
9. Derrick, J., Schellhorn, G., Wehrheim, H.: Mechanizing a correctness proof for a lock-free concurrent stack. In: Barthe, G., de Boer, F.S. (eds.) FMOODS 2008. LNCS, vol. 5051, pp. 78–95. Springer, Heidelberg (2008). https://doi.org/10.1007/978-3-540-68863-1_6
10. Derrick, J., Schellhorn, G., Wehrheim, H.: Mechanically verified proof obligations for linearizability. ACM Trans. Program. Lang. Syst. **33**(1), 1–43 (2011)
11. Derrick, J., Schellhorn, G., Wehrheim, H.: Verifying linearisability with potential linearisation points. In: Butler, M.J., Schulte, W. (eds.) FM, volume 6664 of LNCS, pp. 323–337. Springer (2011). https://doi.org/10.1007/978-3-642-21437-0_25
12. Doherty, S., Groves, L., Luchangco, V., Moir, M.: Formal verification of a practical lock-free queue algorithm. In: FORTE, volume 3235 of LNCS, pp. 97–114. Springer (2004)
13. D'Osualdo, E., Raad, A., Vafeiadis, V.: The path to durable linearizability. Proc. ACM Program. Lang. **7**(POPL), 748–774 (2023). https://doi.org/10.1145/3571219

14. Egorov, S., Chockler, G.V., Dongol, B., O'Keeffe, D., Keshavarzi, S.: Mangosteen: fast transparent durability for linearizable applications using NVM. In: Bagchi, S., Zhang, Y. (eds.) Proceedings of the 2024 USENIX Annual Technical Conference, USENIX ATC 2024, Santa Clara, CA, USA, July 10–12, 2024, pp. 799–815. USENIX Association (2024). https://www.usenix.org/conference/atc24/presentation/egorov

15. Friedman, M., Ben-David, N., Wei, Y., Blelloch, G.E., Petrank, E.: NVTraverse: in NVRAM data structures, the destination is more important than the journey. In: Donaldson, A.F., Torlak, E. (eds.) PLDI, pp. 377–392. ACM (2020). https://doi.org/10.1145/3385412.3386031

16. Friedman, M., Petrank, E., Ramalhete, P.: Mirror: making lock-free data structures persistent. In: Freund, S.N., Yahav, E. (eds.) PLDI, pp. 1218–1232. ACM (2021). https://doi.org/10.1145/3453483.3454105

17. Hähnle, R., Heisel, M., Reif, W., Stephan, W.: An interactive verification system based on dynamic logic. In: Proceedings of the 8th International Conference on Automated Deduction, pp. 306-315. Springer (1986)

18. Herlihy, M., Wing, J.M.: Linearizability: a correctness condition for concurrent objects. ACM TOPLAS **12**(3), 463–492 (1990)

19. Izraelevitz, J., Mendes, H., Scott, M.L.: Linearizability of persistent memory objects under a full-system-crash failure model. In: Gavoille, C., Ilcinkas, D. (eds.) DISC, volume 9888 of LNCS, pp. 313–327. Springer (2016). https://doi.org/10.1007/978-3-662-53426-7_23

20. Khyzha, A., Lahav, O.: Taming ×86-TSO persistency. Proc. ACM Program. Lang. **5**(POPL), 1–29 (2021). https://doi.org/10.1145/3434328

21. Lynch, N., Vaandrager, F.: Forward and backward simulations. Inf. Comput. **121**(2), 214–233 (1995)

22. WDAG 1996. LNCS, vol. 1151. Springer, Heidelberg (1996). https://doi.org/10.1007/3-540-61769-8_9

23. Lynch, N.A., Tuttle, M.R.: Hierarchical correctness proofs for distributed algorithms. In: PODC, pp. 137–151. ACM, New York, NY, USA (1987)

24. Memaripour, A.S., Izraelevitz, J., Swanson, S.: Pronto: easy and fast persistence for volatile data structures. In: Larus, J.R., Ceze, L., Strauss, K. (eds.) ASPLOS '20: Architectural Support for Programming Languages and Operating Systems, Lausanne, Switzerland, March 16–20, 2020, pp. 789–806. ACM (2020). https://doi.org/10.1145/3373376.3378456

25. Michael, M.M., Scott, M.L.: Simple, fast, and practical non-blocking and blocking concurrent queue algorithms. In: PODC, pp. 267–275. ACM (1996)

26. Raad, A., Lahav, O., Wickerson, J., Balcer, P., Dongol, B.: Intel PMDK transactions: specification, validation and concurrency. In: Weirich, S. (ed.) ESOP, volume 14577 of LNCS, pp. 150–179. Springer (2024). https://doi.org/10.1007/978-3-031-57267-8_6

27. Raad, A., Wickerson, J., Neiger, G., Vafeiadis, V.: Persistency semantics of the intel-x86 architecture. Proc. ACM Program. Lang. **4**(POPL), 11:1–11:31 (2020). https://doi.org/10.1145/3371079

28. Scargall, S.: Programming Persistent Memory: A Comprehensive Guide for Developers. Springer Nature (2020)

29. Schellhorn, G., Bodenmüller, S., Dongol, B., Wehrheim, H.: Verification of correctness of the MIRROR persistence library (2024). http://www.informatik.uni-augsburg.de/swt/projects/MIRROR.html

30. Schellhorn, G., Derrick, J., Wehrheim, H.: A sound and complete proof technique for linearizability of concurrent data structures. ACM Trans. Comput. Log. **15**(4), 31:1–31:37 (2014). https://doi.org/10.1145/2629496

31. Schellhorn, G., Wehrheim, H., Derrick, J.: How to prove algorithms linearisable. In: Madhusudan, P., Seshia, S.A. (eds.) CAV 2012. LNCS, vol. 7358, pp. 243–259. Springer, Heidelberg (2012). https://doi.org/10.1007/978-3-642-31424-7_21

32. Shavit, N., Touitou, D.: Software transactional memory. Distrib. Comput. **10**(2), 99–116 (1997)

33. Stefanesco, L., Raad, A., Vafeiadis, V.: Specifying and verifying persistent libraries. In: ESOP (2), volume 14577 of LNCS, pp. 185–211. Springer (2024)

34. Tofan, B., Travkin, O., Schellhorn, G., Wehrheim, H.: Two approaches for proving linearizability of multiset. Sci. Comput. Program. **96**, 297–314 (2014)

35. Wei, Y., Ben-David, N., Friedman, M., Blelloch, G.E., Petrank, E.: FliT: a library for simple and efficient persistent algorithms. In: Lee, J., Agrawal, K., Spear, M.F. (eds.) PPoPP, pp. 309–321. ACM (2022). https://doi.org/10.1145/3503221.3508436

36. Wen, H., Cai, W., Du, M., Jenkins, L., Valpey, B., Scott, M.L.: A fast, general system for buffered persistent data structures. In: Sun, X.-H., Shende, S., Kalé, L.V., Chen, Y. (eds.) ICPP, pp. 73:1–73:11. ACM (2021). https://doi.org/10.1145/3472456.3472458

37. Zhang, W., Shenker, S., Zhang, I.: Persistent state machines for recoverable in-memory storage systems with NVRAM. In: 14th USENIX Symposium on Operating Systems Design and Implementation, OSDI 2020, Virtual Event, November 4–6, 2020, pp. 1029–1046. USENIX Association (2020). https://www.usenix.org/conference/osdi20/presentation/zhang-wen

On the Quest for Criticality Searching for Interactions that Matter When Hunting for Bugs

Franz Wotawa[✉][iD]

Graz University of Technology, Institute of Software Engineering and Artificial Intelligence, Inffeldgasse 16b/2, 8010 Graz, Austria
wotawa@tugraz.at

Abstract. Quality assurance of mobile and autonomous systems is essential for ensuring trust and avoiding harm, especially when such systems are safety-critical, heavily interacting with humans in our physical environment. Considering only requirements or models of systems for generating test cases or formal verification may not be sufficient due to the huge number of possible interactions that the requirements or system model might not cover. Therefore, this paper discusses a testing approach that utilizes environmental models for test case generation to mitigate this challenge. For representing the environmental models, we suggest using formal ontologies from which we can automatically generate test input models. The latter can be used directly for test case generation. Although the proposed approach was initially closely paired with combinatorial testing, it is agnostic regarding the underlying testing methodology. It can also be combined with search-based or random testing, as discussed in this paper. Besides introducing the foundations and presenting an algorithm that allows us to compile ontologies into test input models, we show an application scenario from mobile robots and summarize already reported results of an experimental evaluation in the context of automated driving assistance systems.

Keywords: Autonomous system testing · Test automation · Ontology-based testing · Finding critical scenarios

1 Introduction

Testing automated and autonomous systems has always been challenging, leading to developing various testing methodologies, like model-based testing [21,22]. With the rise of integrating artificial intelligence technology and methods like machine learning and self-adaptivity into safety-critical systems like autonomous cars or automated driving assistant functions, testing bears additional challenges ranging from adversarial attacks [15] to the more general question of how to test smart autonomous systems to assure meeting safety requirements? Earlier work [7] in the area of self-adaptive systems considered utilizing fault-trees for

G. Ernst et al. (Eds.): Wolfgang Reif Festschrift, LNCS 15765, pp. 356–373, 2025.
https://doi.org/10.1007/978-3-031-92196-4_17

identifying critical scenarios and conditions. For autonomous driving, several studies, including [8,11] have been conducted to deal with this testing challenge. However, obviously, only considering the number of autonomously driven miles, as indicated by [8], is not a good termination criterion for testing in this domain unless assuming that sufficient critical scenarios have been experienced in this fixed number of miles. Criticality in this context means finding scenarios that may bring an autonomous system into an unexpected state that might lead to a crash or other severe safety violations. Similarly, there has been a lot of other research dealing with the testing challenge for autonomous systems and the question regarding the applicability of available testing approaches, e.g., [23,24].

Let us have a look at an example. In Fig. 1, we depict a crash scenario of a Tesla car in Taiwan that happened in June 2020. In this scenario, a truck blocks the two left-most lanes on a highway. In automated mode, the Tesla approached this situation without stopping or reducing speed and finally crashed into the overturned truck. From such an example, we obtain several conclusions. First, the car was not reducing speed even in the case of a person in the lane on a highway, which clearly violates the corresponding UN regulations and laws. Second, it seems that the Tesla car did not recognize the overturned truck when approaching it. Third, the driver of the Tesla car was convinced that the car in automated mode worked as expected and did not take action to prevent the crash. The latter is a well-known tool overreliance problem that must be checked for and prevented. However, the most important conclusion is that such cases should be already identified during the verification and validation phase and not after deployment. Hence, there is a need to identify such critical scenarios in practice. For more information about finding and using critical scenarios in autonomous driving. We refer to [29].

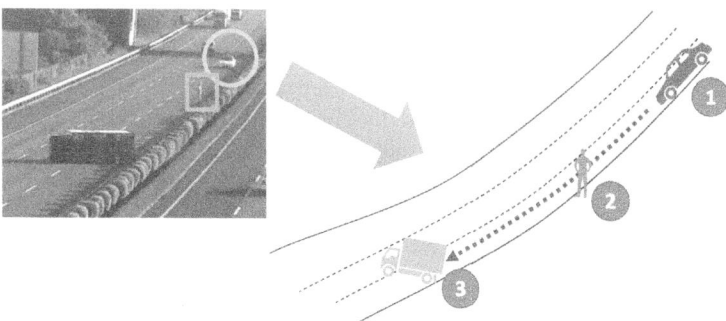

Taiwan, 2020-06-01: A Tesla car in automated driving mode is approaching an overthrown truck on the highway in Taiwan, which blocks two lanes. The truck driver stands on the left lane to warn other drivers. The Tesla 1 did not reduce speed or brake early enough (and latest when identifying a person on the highway 2) and finally crashed into the truck 3 violating the Convention on Road Traffic that emerged from the United Nations Conference on Road and Motor Transport in September 1949 in Geneva!

Fig. 1. A real crash scenario of a Tesla car in autonomous mode on a highway.

Scenario-based testing methods are not new in autonomous driving. [17] outlined such scenarios from which tests are generated, and [18] discussed the need for efficient and systematic test case generation combined with simulation for test execution. Using realistic simulation combining physical models together with 3D representation of the world brings in many advantages, for example, substantially reducing test execution, carrying out test cases that can hardly be executed in reality, and bringing testing to earlier phases of the development cycle. In order to implement such a testing approach, we have to ask ourselves how to formalize knowledge about scenarios. It is worth noting that focusing on identifying critical scenarios is different from ordinary test case generation from models, i.e., model-based testing, where the idea is to use a system model for obtaining a test suite. In the case of detection scenarios, we need to consider all interactions between the environment and the system. Hence, we need to formalize what the environment can do with the system, i.e., there is a need for environmental models to capture potential interactions. Note that interactions are not limited to direct calls of functions on the side of the system but also setting parameters that indirectly influence the system's behavior. For example, bad weather, e.g., heavy rain, influences object recognition, which may lead to the wrong decision on the side of an autonomous car.

In fact, there has been a lot of work on formalizing knowledge and coming up with general ontologies [5] that can be used for problem-solving. The idea behind using ontologies for testing exists already (e.g., [19]) and has gained attention for testing automated driving functions [6,27] including its use for generating scenarios [1]. However, in most of these publications, test cases or scenarios have to be manually extracted from ontologies. Moreover, these approaches are limited in providing test cases in a systematic manner for identifying critical interactions between the SUT, i.e., the autonomous system and its environment. To avoid these shortcomings, Wotawa and Li [26] introduced a test case generation approach that combines ontologies with combinatorial testing (CT) [12]. Although CT is a well-known testing technique, its application in combination with ontologies for test generation was new. Later, Li et al. [13] improved the work by loosening the coupling of ontologies to CT. It is worth noting that Wotawa et al. [25] named the approach ontology-based testing (OBT) for the first time. Various research papers showed the applicability of OBT in practice, e.g., [3,9,10].

In OBT, test cases are generated in a fully automated way. For this purpose, the ontologies are converted into an input model for combinatorial testing from which test cases are generated. These abstract test cases have to be mapped to concrete executable tests that can be executed automatically, for example, using simulation in the case of autonomous systems. In the following, we summarize our work by outlining the basic ideas and prerequisites and explaining how to use OBT in practice.

We organized the paper as follows: We first discuss the testing approach in Sect. 2. There, we introduce the basic definitions, discuss ontologies, and present an algorithm for converting ontologies into test input models, which is agnostic with respect to the underlying test case generation method. In particular, it generates an input model other testing approaches can use for automatic test suite generation. Afterward, in Sect. 3, we discuss the limitations and open research challenges behind OBT. Finally, we conclude the paper.

2 Testing Approach

In this section, we summarize and outline the basic foundations behind OBT. We assume a system under test (SUT) that interacts with its environment. Instead of generating test cases from the SUT's specification, which might be incomplete and not cover all potential interactions, we use the knowledge of the ontology for automated test suite generation. For the general architecture and principles behind it, let us have a look at Fig. 2. On the left side of this figure, we have an environmental ontology that captures the structure and also possible interactions of the SUT's surrounding environment. We assume to have a tool for generating such an ontology in the first step of the OBT approach. From the ontology, we generate an input model for test case generation in the second step. Unlike described in the original research paper [13, 26], we do not rely on a particular testing method anymore. Instead, we assume different test case generation methods to be used for generating scenarios that are executed using a simulator (Step 3). In Step 4, the outcome of the simulation is checked for correctness by a formalized test oracle. Note that a simple test oracle may only consider crashes or time to collision boundaries for distinguishing the passing from failing test cases. Hence, the proposed testing approach comprises automated test case generation and test execution. However, in the following, we focus on the test case generation part in more detail and outline the basic foundations.

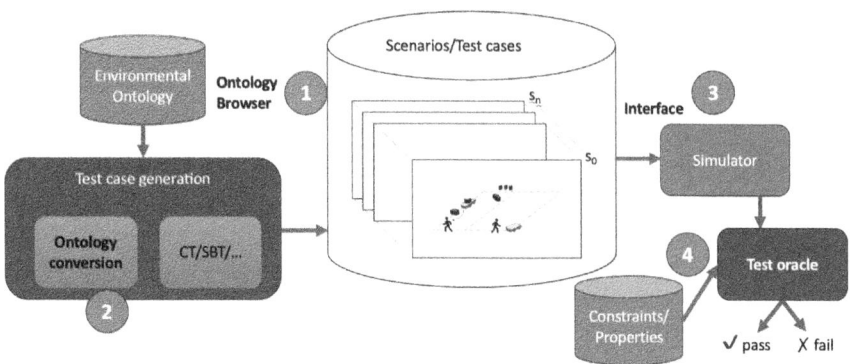

Fig. 2. The general principles behind ontology-based testing

We start by defining a test case for a given SUT, which is adapted to match our testing scenario. A *test case I* is a set of tuples (i, v) with i as an input, e.g., a variable or parameter from a finite set P, and v is a value from a finite domain D^1. A *test suite* is a set of test cases. To check for the correctness of the execution of the SUT, we further introduce an oracle function $o : P \times D \mapsto \{\sqrt{}, \times\}$ that maps the output of the SUT from P and its value from D to pass ($\sqrt{}$) or fail ($\times$). Note that these definitions do not explicitly consider using a sequence of actions or other inputs over time to stimulate the SUT. However, such a behavior can also be represented by introducing a separate input for a variable or parameter to be defined at different time steps. Hence, the definition is not limited to testing functions but also dynamic systems. In the dynamic case, we assume that the oracle function is called at specific time steps.

We can generate test cases or test suites either manually or automatically. In both cases, we need information, e.g., the requirements or the specification of the SUT, or at least the input parameters and their domains. The latter, we call an *input model* $(V, \{D_{v_1}, \ldots\}, C)$ comprising a set of input parameters $V \subseteq P$, a set of domains $D_x \subseteq D$ for each $x \in V$, and a set of constraints over V for restricting unwanted value combinations. Such an input model is used, for example, in CT [12], to come up with a test suite that fulfills certain criteria, like interaction strength (in CT). The interaction strength k in CT indicates the number of parameters where a test suite needs to specify all combinations for any subset of parameters k. Such a test suite is called k-way test suite or test suite of strength k. Note that such input models are not restricted to CT and can also be used as input for other test case generation methods like search-based testing or random testing. In the latter case, we utilize the information given in the input model to generate test cases by randomly choosing values for parameters from their domains.

As mentioned, the proposed OBT methodology is not directly based on input models but ontologies. Hence, we are going to discuss ontologies and their conversion to input models in the next subsections.

2.1 Ontologies

Ontologies [5] are formal, explicit specifications of shared conceptualizations that are characterized by high semantic expressiveness required for increased complexity, which captures concepts and their relationships. This informal definition states that ontologies describe concepts in a formal way as well as knowledge about these concepts, including relationships between concepts. In the following, we formalize the definition of ontologies restricting relations to composition and inheritance, which we take from [26].

An *ontology* is a tuple $(C, A, D, \omega, R, \tau, \psi)$ where C is a finite set of concepts, A is a finite set of attributes which characterize concepts, D is a finite

---

[1] We restrict parameters and domains to be finite in this paper. This assumption does not restrict the general applicability to any real-world SUT in the case of the parameter set. Considering only finite domains requires abstraction for continuous and infinite domains and refinement afterward for generating tests that the SUT can take as input. However, such an assumption is often introduced in test case generation methodologies like CT or model-based testing.

set of domain elements. Function ω maps concepts to a set of tuples specifying the attribute and its domain elements. R is a finite set of tuples stating that two concepts are related. Function τ assigns a type (i.e., composition (c) or inheritance (i)) to each relation. Function ψ maps composition relations to their minimum and maximum arity, specifying how many concepts another concept may comprise, ranging from 0 or 1 to any natural number. Hence, we assume that the maximum arity is finite and known. For example, we may want to state that a vehicle has 1 to 4 wheels, which can be captured using the minimum and maximum arity of a composition relation.

Let us illustrate this definition using the mobile robot example depicted in Fig. 3. From Fig. 3(a) we see that a mobile robot comprises wheels, allowing it to move from one location to another, where we can easily extract *MobileRobot*, *Wheel*, and *Location* as basic concepts. The relationship *comprise* semantically connects the concepts *MobileRobot* and *Wheel*. The relationship *isAt* connects *MobileRobot* with its location, i.e., the concept *Location*. Moreover, mobile robots are special cases of the concept *Robot* and therefore establish an inheritance relationship. However, for explaining the OBT approach, we use a different characterization of the mobile robot domain and, therefore, come up with an ontology that focuses on testing the robot considering its interactions with the environment.

In Fig. 3(a) on the left side, a robot comprises motors, wheels, and wheel encoders for obtaining the rotational speed of the wheels. In case the right wheel turns backward and the left wheel turns forward, the robot moves to the right. If the left wheel turns backward and the right wheel turns forward, the robot turns to the left. If both wheels turn forward using the same speed, the robot moves straight forward. A robot control software has to translate the commands regarding where to move to the speed settings of the engines. The wheel encoders can be used to see whether the wheels turn as expected. Now, let us consider that one wheel faces a more slippery surface than the other. In this case, one wheel, e.g., the left wheel, spins and does not provide any force to the ground. As illustrated in the right part of Fig. 3 (a), the robot is no longer taking the expected direction. A control software should detect such an unwanted situation and react accordingly, i.e., stop locomotion or decrease the speed to stop spinning.

For testing the control software, we have to provide test cases considering different surface conditions and steering commands. Therefore, we have to define an ontology that brings the corresponding concepts into relation. For this purpose, we introduce the concepts *RobotEnvironment*, *WheelSurface*, and *Robot-Command*. In Fig. 3(b), we depict the ontology, including relationships and further information. We know that a *RobotEnvironment* has to include a surface for each wheel and define at least one command for controlling the movement of the robot. Hence, the minimum and the maximum arity of the *has* relationship between `WheelSurface` and `RobotCommand` is set to 2. For the relationship between `RobotEnvironment` and `RobotCommand`, we allow 1 to 3 commands, which determines the given minimum and maximum arity. Of course, the max-

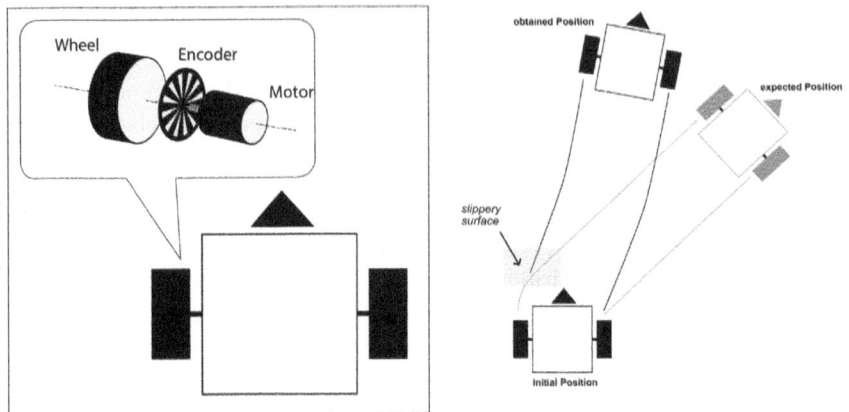

(a) Autonomous mobile robot with differential drive

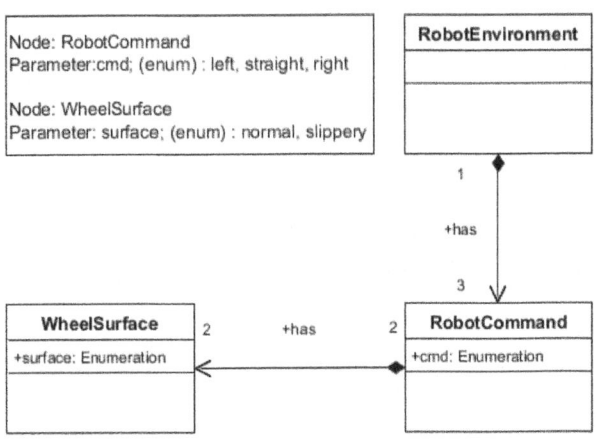

(b) Environmental ontology for the autonomous mobile robot

Fig. 3. The autonomous mobile robot example.

imum arity might be increased if we want to have more commands to be given to the robot. In addition, we define attributes like *surface* and *cmd* for the concepts *Wheel Surface* and *RobotCommand*, respectively, with their corresponding domains. This information is used to come up with test cases for *RobotEnvironment*.

Formally, we can express the environmental ontology $(C_M, A_M, D_M, \omega_M, R_M, \tau_M, \psi_M)$ for the mobile robot example as follows:

$C_M = \{\texttt{RobotEnvironment}, \texttt{RobotCommand}, \texttt{WheelSurface}\}$
$A_M = \{\texttt{cmd}, \texttt{surface}\}$
$D_M = \{\texttt{left}, \texttt{right}, \texttt{straight}, \texttt{normal}, \texttt{slippery}\}$
$\omega_M(\texttt{RobotEnvironment}) = \{\}$
$\omega_M(\texttt{RobotCommand}) = \{(\texttt{cmd}, \{\texttt{left}, \texttt{right}, \texttt{straight}\})\}$
$\omega_M(\texttt{WheelSurface}) = \{(\texttt{surface}, \{\texttt{normal}, \texttt{slippery}\})\}$
$R_M = \{(\texttt{RobotEnvironment}, \texttt{RobotCommand}), (\texttt{RobotCommand}, \texttt{WheelSurface})\}$
$\tau_M(\texttt{RobotEnvironment}, \texttt{RobotCommand}) = c$
$\tau_M(\texttt{RobotCommand}, \texttt{WheelSurface}) = c$
$\psi_M(\texttt{RobotEnvironment}, \texttt{RobotCommand}) = (1, 3)$
$\psi_M(\texttt{RobotCommand}, \texttt{WheelSurface}) = (1, 2)$

In summary, when using OBT, we have to define an environmental ontology that captures those parts of the environment of an SUT that may contribute to critical scenarios. In the case of mobile and autonomous systems, this includes surface conditions, the structure of roads, obstacles like pedestrians or other cars or robots, or the weather, i.e., everything that might influence the behavior of the SUT. In the next subsection, we discuss the conversion of ontologies into input models, which enables different test methodologies to utilize ontologies for automated test case generation.

2.2   Ontology Conversion

After introducing the basic ideas and definitions of ontologies, we now discuss the conversion of ontologies into an input model. The objective is that the input model covers the information stored in the ontology. The core components of an ontology are the concepts having attributes and their relationships. Hence, the resulting input model should store the information of attributes considering the structure of the ontology. Wotawa and Li [26] presented a first version of the ontology conversion, which was improved later by Li and colleagues [13]. In particular, the latter algorithm delivers an input model independently from the underlying test case generation method.

In contrast to previous work, we explain the algorithm onto2im considering previous work in relational databases [4,16]. There, the underlying entity-relationship (ER) model, which represents the data model, is converted into relations, i.e., database tables. In [20], this conversion is explained for different types of relationships. Although there are differences between ER models and ontologies, e.g., the former uniquely identifies instances of entities (which are a kind of concepts) using super keys where this is not the case for ontologies, there are also similarities, e.g., relationships between concepts or entities considering arity and a specific type like inheritance and composition.

Before discussing the handling of relationships, we first consider a single concept like RobotCommand from the ontology given in Fig. 3. This concept has a single attribute cmd, which can take left, right, or straight as values. Such a concept should be directly mapped to a parameter in the input model, i.e., the input model for RobotCommand is $(\{\texttt{RobotCommand.cmd}\}, \{(\texttt{RobotCommand.cmd}, \{\texttt{left}, \texttt{right}, \texttt{straight}\})\}, \{\})$ where $(\texttt{RobotCommand.cmd}, \{\texttt{left}, \texttt{right}, \texttt{straight}\})$ denotes the domain of parameter $\texttt{RobotCommand.cmd}$. In this example, we do not have any constraint, leaving the input model's constraint part empty. It is worth noting that we used the name of the concept together with the attribute name as the name of the parameter. With this naming schema, we can ensure that all variables representing parameters in the input model are unique. Hence, in the first step, we convert an attribute a for each concept C into a new variable $C.a$ having the same domain as given in the ontology.

The second case comprises the handling of relationships. We follow a similar idea for relationships used in relational databases. If $\texttt{RobotCommand}$ has to consider several $\texttt{WheelSurfaces}$, the relationship would be converted by adding one or more attributes representing the primary key of $\texttt{RobotCommand}$ to $\texttt{WheelSurface}$. In this way, the link from $\texttt{WheelSurface}$ to $\texttt{RobotCommand}$ is preserved. In the case of ontology conversion, we switch the view but also consider preserving the structure. For each $\texttt{WheelSurface}$, we add all its parameters to the parameters of $\texttt{RobotCommand}$. To ensure the uniqueness of parameter names, we further add an index value ranging from 1 to the maximum arity of the relationship. Hence, in the case of the robot example environment, we add the following parameters to $\texttt{RobotCommand}$: $\texttt{RobotCommand.1.WheelSurface.surface}$, $\texttt{RobotCommand.2.WheelSurface.surface}$. For the domains, we use the domains of the attribute $\texttt{surface}$ but need to modify them. In particular, we add the empty value represented by ϵ to all parameters with an index.

The rationale behind this modification is as follows: Let us assume two concepts C_1 and C_2 that are in a n-to-m relationship, where m is on the side of C_2, then we need at least n concepts C_2 to be stored in the input model that belongs to C_1. Because we are not storing the concept but its attributes, for each attribute a of C_2, we need at least n non-empty representations. In this example, we have $C_1.n.C_2.a$ to $C_1.m.C_2.a$ parameters (or representations) of attribute a. The maximum number of non-empty parameters is m, which is ensured considering m representations. However, we need to ensure that n attribute representations are non-empty. This can be done by adding a new constraint to the input model, i.e.:

$$\bigvee_{\{r_1,\dots,r_n\}\subseteq\{C_1.1.C_2.a,\dots,C_1.m.C_2.a\}} (r_1 \neq \epsilon \wedge \dots \wedge r_n \neq \epsilon)$$

This constraint states that at least for one subset of size n, there must be a value assignment not using ϵ.

The $\texttt{onto2im}$ algorithm now takes these two key ideas and brings them together in a recursive manner. The algorithm only works on tree-structured ontologies having a single root concept. It starts with the root concept of recursively adding new parameters, considering the other concepts in a relationship. For our mobile robot example, $\texttt{onto2im}$ computes the parameters for $\texttt{WheelSurface}$ and combines it with $\texttt{RobotCommand}$ as already discussed. Finally, we have another relationship between $\texttt{RobotEnvironment}$ and $\texttt{RobotCommand}$.

Hence, again, new parameters are generated considering different representations of all attributes in RobotCommand, finally leading to the following nine parameters:

```
RobotEnvironment.1.RobotCommand.1.WheelSurface.surface
RobotEnvironment.1.RobotCommand.2.WheelSurface.surface
RobotEnvironment.1.RobotCommand.cmd
RobotEnvironment.2.RobotCommand.1.WheelSurface.surface
RobotEnvironment.2.RobotCommand.2.WheelSurface.surface
RobotEnvironment.2.RobotCommand.cmd
RobotEnvironment.3.RobotCommand.1.WheelSurface.surface
RobotEnvironment.3.RobotCommand.2.WheelSurface.surface
RobotEnvironment.3.RobotCommand.cmd
```

The domains of all surface attributes are: $\{\epsilon, \mathtt{normal}, \mathtt{slippery}\}$. The one for cmd are: $\{\epsilon, \mathtt{left}, \mathtt{right}, \mathtt{straight}\}$. In addition, we have the following constraints:

$$
\mathtt{RobotEnvironment.1.RobotCommand.1.WheelSurface.surface} \neq \epsilon \wedge \\
\mathtt{RobotEnvironment.1.RobotCommand.2.WheelSurface.surface} \neq \epsilon \\
\mathtt{RobotEnvironment.2.RobotCommand.1.WheelSurface.surface} \neq \epsilon \wedge \\
\mathtt{RobotEnvironment.2.RobotCommand.2.WheelSurface.surface} \neq \epsilon \\
\mathtt{RobotEnvironment.3.RobotCommand.1.WheelSurface.surface} \neq \epsilon \wedge \\
\mathtt{RobotEnvironment.3.RobotCommand.2.WheelSurface.surface} \neq \epsilon
$$

$$
\mathtt{RobotEnvironment.1.RobotCommand.cmd} \neq \epsilon \vee \\
\mathtt{RobotEnvironment.2.RobotCommand.cmd} \neq \epsilon \vee \\
\mathtt{RobotEnvironment.3.RobotCommand.cmd} \neq \epsilon
$$

Such an input model can be used with test case generation tools like ACTS [28] for generating combinatorial test suites of arbitrary strength. However, what is still missing is a discussion on the handling of inheritance relationships in ontology conversion. We do not have any inheritance relationship in our robot example. But we are discussing inheritance handling by considering the following situation. Let A be a concept and B and C its sub-concepts. A has one attribute a, B another one b, and C an attribute c. Inheritance is based on generalization. Hence, A is the more general concept, whereas B and C are more specific ones. Hence, in a test case, we may not use A but one of its sub-concepts B or C. Therefore, we would like to replace any occurrence of A with either B or C. Because of the fact that this decision is taken during the generation of a test case, we add all parameters of sub-concepts B and C to the parameters of A and add a constraint stating that we either use the parameter of B and leave the parameter of C empty or vice versa. Formally, the parameters of A are $A.a$, $A.B.b$, and $A.C.c$, where the domains are the domains of the attributes with ϵ as an additional element. The constraint we add is:

$$(A.B.b = \epsilon \wedge A.C.c \neq \epsilon) \vee (A.B.b \neq \epsilon \wedge A.C.c = \epsilon)$$

Note that such a constraint must be added for each parameter of the sub-concepts. Note further that in the case of inheritance, we do not have any arity to consider. Furthermore, the conversion assumes that sub-concepts can be considered to be separated.

Algorithm 1. onto2im (O)

Require: *A well-formed ontology O.*
Ensure: *An input model for the root concept of the ontology O.*
1: Let r be the root concept of ontology O.
2: **return** the result of the call onto2im_h(r, O).

After discussing the onto2im algorithm, we formalize it in Algorithm 1. This algorithm calls Algorithm 2, i.e., onto2im_h, which recursively implements all the stated ideas of onto2im. In particular, it generates the parameters for the attributes of the concepts and also handles the inheritance and composition relationships. For the relationships, it also considers renaming parameters and adding necessary constraints.

3   Discussion

After introducing the basic ideas, foundations, and algorithms, this section discusses the approach's applicability in a practical setup. We report on existing research papers to answer the following questions: (i) Is the approach limited to testing autonomous and mobile systems? (ii) Is OBT capable of finding critical scenarios, and to what extent do we have an influence on the underlying test case generation method? And finally, (iii) what are the limitations and open research questions?

To answer the first question regarding the limitations of the application domain, we have a look at previous publications. In the original research paper of Wotawa and Li [26], the authors used a simple text ontology to explain the testing approach. In their example, the ontology describes sentences as a collection of words finalized by a delimiter, which is not related to autonomous systems. Moreover, Li and Wotawa [14] used OBT for test case generation in the domain of compiler testing. In particular, the ontology covered the grammar the compiler implements. Bozic et al. [3] used OBT for security testing of Web applications. There is plenty of work that utilizes ontologies and OTB for generating test cases. However, there might still be limitations. Developing a general ontology is not a simple task and requires a modeling methodology, which is currently not available. For some applications, developing ontologies might be easier than for others. Moreover, there are still some limitations of the approach, e.g., having only tree-structured ontology, which might restrict applicability. Finally, in some cases, the ontology might be too simple, and coming up with an input model directly might be less costly and, therefore, the better choice.

Algorithm 2. onto2im_h (c, O)

Require: *A concept c of a well-formed ontology O.*
Ensure: *An input model $(V, \{D_{v_1}, \ldots\}, C)$ for c.*

1: Let V and C be empty sets.
2: **for all** attributes $a \in \omega(c)$ **do**
3:      Add $c.a$ to V.
4:      Let $D_{c.a}$ be $dom(c, a)$.
5: **end for**
6: **if** c is not a leaf concept **then**
7:      Let $k = 1$ and let tmp and $subcon$ be both empty arrays ranging from 1 to the number n_i of inheritance relationships of c.
8:      **for all** relationships $(c, c') \in R$ with type $\tau(c, c') = $ i **do**
9:          Let $subcon[k]$ be c'.
10:         Store the result of onto2im(c', O) in $tmp[k]$.
11:         Let $(V', \{D'_{v_1}, \ldots\}, C')$ be $tmp[k]$.
12:         Add the content of V', $\{D'_{v_1}, \ldots\}$, and C' to V, $\{D_{v_1}, \ldots\}$, and C respectively, but rename all elements x of V' to $c.x$ before!
13:         $k = k + 1$
14:     **end for**
15:     **for all** Pairs (a, b) where $a, b \in \{1, \ldots, n_i\}$ **do**
16:         Let $tmp[a]$ be $(V^a, \{D^a_{v_1}, \ldots\}, C^a)$ and $tmp[b]$ be $(V^b, \{D^b_{v_1}, \ldots\}, C^b)$
17:         Let $subcon[a]$ be X, and $subcon[b]$ be Y.
18:         **for all** attributes x in V^a **do**
19:             **for all** attributes y in V^b **do**
20:                 Add $(c.X.x = \epsilon \wedge c.Y.y \neq \epsilon) \vee (c.X.x \neq \epsilon \wedge c.Y.y = \epsilon)$ to C.
21:             **end for**
22:         **end for**
23:     **end for**
24:     **for all** relationships $(c, c') \in R$ with type $\tau(c, c') = $ c **do**
25:         Call onto2im(c', O) and store the input model in $(V', \{D'_{v_1}, \ldots\}, C')$.
26:         Add ϵ to all the variable domains in $\{D'_{v_1}, \ldots\}$.
27:         Assume that $\psi(c, c') = (n, m)$
28:         **for** $i = 1$ to m **do**
29:             Add the content of V', $\{D'_{v_1}, \ldots\}$, and C' to V, $\{D_{v_1}, \ldots\}$, and C respectively, but rename all elements x of V' to $c.i.x$ before!
30:         **end for**
31:         **for all** attributes a in V' **do**
32:             **if** $n > 0$ **then**
33:                 Add $\bigvee_{\{r_1, \ldots, r_n\} \subseteq \{c.1.a, \ldots, c.m.a\}} (r_1 \neq \epsilon \wedge \ldots \wedge r_n \neq \epsilon)$ to C.
34:             **end if**
35:         **end for**
36:     **end for**
37: **end if**
38: **return** $(V, \{D_{v_1}, \ldots\}, C)$

We discuss the second question using a more recent work from Klück and colleagues [10]. In their paper, the authors compared the three different testing techniques: CT, Search-based Testing (SBT), and Random Testing (RT), utilizing the same input model, which originates from an ontology describing a set of driving scenarios for testing automated emergency braking (AEB) functions. In particular, the authors report the testing results for the three testing methodologies and two different AEB functions (AEB1 and AEB2). To be self-contained, we briefly summarize the obtained results.

Table 1 shows the summarized results for AEB1 and AEB2 separated by the testing methods SBT, RT, and CT. Note that CT2 and CT3 denote the test suite obtained using CT of strength 2 and 3, respectively. The comparison of the testing methods was carried out as follows. The test suites were obtained using different methods. Each test in the test suite was executed using a simulation environment. A test oracle was used to check whether a test case was passing or failing. The test oracle used the time-to-collision between the SUT and any other object in the simulation to determine a crash or non-crash behavior. For the analysis, the authors distinguished different categories of crashes, i.e., a front crash between the SUT and another vehicle (FCV), a front crash with the first pedestrian (FCP1) and the second pedestrian (FCP2), and a side crash with pedestrian 1 (SCP1) and 2 (SCP2). To be fair, the same number of test cases of CT2 were used when generating test cases for SBT and RT. Because of the non-determinism of SBT and RT, the test generation and execution phase was run 10 times. The results in Table 1 for SBT and RT are the minimum and maximum number of crashes that could be observed. For CT, the table states the number of crashes for the different crash categories that could be obtained from the test suites of strength 2 and 3, respectively.

Table 1. SBT, RT, and CT test generation and crash summary for AEB1 and AEB2 considering minimum and maximum values for 10 runs of SBT and RT.

AEB	Testing Method	FCV	FCP1	SCP1	FCP2	SCP2
AEB1	SBT	0..3	31..216	24..94	0..4	0..5
AEB1	RT	0..1	12..34	9..20	0..3	0..1
AEB1	CT2	0	12	18	4	2
AEB1	CT3	2	604	999	159	87
AEB2	SBT	2..84	10..65	10..65	0	0..17
AEB2	RT	10..27	1.13	2..28	0	0..1
AEB2	CT2	57	3	12	1	4
AEB2	CT3	1,356	122	96	286	38

Table 1 shows that all testing methods may reveal faults according to the different crash categories, with some exceptions. CT2 cannot detect an FCV for AEB1, whereas both SBT and RT cannot detect FCP2 in the case of AEB2. However, when increasing the combinatorial strength to 3, CT can always detect all different crashes. Hence, there is a dependency on the testing method when combined with an input model obtained from an ontology. In addition, we conclude that critical scenarios, i.e., scenarios that trigger a faulty behavior, can be identified.

Table 2. SBT, RT, and CT coverage results summary (2-way to 5-way). All values are average values over the 10 different testing runs for SBT and RT.

SUT	Testing Method	2-way coverage	3-way coverage	4-way coverage	5-way coverage
AEB1	SBT	75.7%	30.2%	8.9%	2.0%
AEB2	SBT	74.5%	29.2%	8.8%	2.0%
AEB1\ AEB2	RT	79.9%	34.1%	10.8%	3.0%
AEB1\ AEB2	CT2	100.0%	42.0%	13.0%	4.0%

What is now the reason behind the fault detection capabilities? From the environmental ontology, we can generate scenarios that capture interactions of parameter values and their corresponding concepts that have not been foreseen. From CT, we know that such interactions are required to reveal faults. The required interactions can be between 2 or more parameters. The degree of covering such interactions is called k-way coverage. Table 2 summarizes the average k-way coverage for the different testing methods, where k ranges from 2 to 5. CT comes with the highest k-way coverage, which explains its superiority of fault detection compared to the other methods.

The last question we want to discuss deals with limitations and open research questions of OBT. First and most important, *OBT needs an ontology* of the environment that captures concepts relevant to the SUT. In the case of autonomous mobile systems, the concepts include static parts, e.g., the street network or traffic lights, and dynamic ones, e.g., other vehicles or pedestrians. For other domains like compilers, the concepts are closely related to the grammar elements. However, there is no general ontology that captures all concepts for all SUTs. Hence, similar to other testing techniques like model-based testing, OBT comes with additional overhead and costs. In model-based testing, the overhead costs are due to the need for modeling the system.

Second, there is currently no *general modeling methodology* available, and modeling is always complicated. Hence, there is a need for such a methodology to bring OBT into practice. Besides the methodology, there need to be tools for constructing, sharing, and maintaining ontologies. Developing a methodology for constructing ontologies given an SUT is an open research question.

Third, the *current input model generation algorithm is limited* and only capable of handling well-structured ontologies. In particular, the algorithm assumes a tree-structured ontology where we have one root node and, at the maximum, one parent concept. These assumptions might be relaxed easily, but further investigations are needed. For example, if a parent concept has two relationships with its single child, we may duplicate the child concept and separate them, leading to a tree-structured ontology. Hence, there is a need to elaborate on ontology conversion techniques that allow the existing approach to be used for more elaborated ontologies. In addition, there are more types of relationships worth considering.

Fourth, when using OBT, we might run into a *combinatorial explosion problem*. From the ontology and the given conversion process, we might obtain far too many parameters, so using test case generation methods like CT is no longer feasible for more considerable strengths, which may be required. For example, in the automotive domain, OBT coupled with CT leads to the generation of millions of test cases. Executing the tests concurrently is possible and leads to acceptable total testing times. Nevertheless, OBT, in some situations, is computationally demanding, requiring high computing resources. It is worth noting that in such a case, we may couple OBT with other testing approaches like SBT or RT. From the first experimental results, we know that these approaches might not deliver the best outcome in terms of fault detection capabilities, but they can still find critical scenarios using the same input model.

Finally, the presented approach, similar to CT and model-based testing, needs *concretization*. In particular, OBT leads to abstract test cases, which can not always be directly used for stimulating the SUT. Hence, there is a need to replace abstract values of parameters with concrete ones, which is concretization, before executing the test cases. Coming up with abstract values is also an issue when developing the ontology, and if done right, it should not impact the fault detection capabilities. However, in our testing scenario, we do not know any details about the SUT or its requirements. Hence, abstractions are introduced only considering the environmental domain. For example, we might state that a low speed is less than 10 km per hour, whereas a high speed means more than 100. Everything in between is considered average speed. Such an abstraction might not well represent the internal decisions of the SUT. When only considering boundary values, we may not detect any faults. This is a known problem (see [2]), and concretization must handle this challenge.

Besides the mentioned current limitations, challenges, and related research questions, OBT has several benefits. A more general environmental ontology may be useful not only for one domain and SUT but also for many. Hence, we may re-use ontologies by selecting only relevant parts for a specific SUT. Such selection also leads to a further characterization of coverage, i.e., ontology coverage (similar to model coverage in model-based testing). Hence, we can evaluate the test suite with respect to how much of a general ontology is utilized and represented. Such ontology coverage can be beneficial in safety-critical systems domains like autonomous driving.

Furthermore, when using OBT, we may more easily be able to come up with input models that would be difficult to develop otherwise. For example, when having n actions over time, we need to introduce n parameters corresponding to an action in the input model. For a specific n, this is not a problem, but if we need to change n often, we need several different input models. In the case of OBT, this is not a problem. We introduce `Time` as a root concept that is in relationship with a concept `Action`. By varying the arity of the relationship, we automatically obtain a set of action-related parameters in the input model. Hence, modeling becomes more straightforward.

Finally, focusing on the environment and keeping the SUT a black box enables searching for interactions that might not be foreseen when considering the requirements and specifications of an SUT. Hence, there is a higher likelihood of finding critical scenarios, but it comes with additional computational and modeling costs.

4   Conclusions

Quality assurance is important, especially when dealing with safety-critical systems. In practice, many bugs are revealed during operation when users or the environment interact with a system in an unwanted and unforeseen way. This might not be problematic for ordinary software, but consider an autonomous car driving on our roads. In such a case, we want to ensure that no situation will lead to harming people or the environment. The presented ontology-based testing approach tackles the challenge by switching the focus of test case generation. In contrast to previous research like model-based testing, where models of the system are used to generate test suites, in ontology-based testing, we use models of the system's environment and consider the system under test as a black box. Note that ontology-based testing has been already successfully used in the automotive industry (see, e.g., [9, 10]).

In this paper, we present the foundations behind ontology-based testing, introduce an algorithm, and discuss the pros and cons of the approach in detail, considering the latest research literature. This discussion reveals that ontology-based testing is independent of the underlying test case generation method and domain agnostic. It allows for finding faults originating from critical interactions between the system under test and its environment. Furthermore, we discussed open research challenges of ontology-based testing to be tackled to further make the approach of interest for more general practical applications.

Acknowledgments. This paper is part of the A-IQ Ready project that has received funding within the CHIPS JU in collaboration with the European Union?s Horizon Framework Programme and National Authorities under grant agreement No. 101096658. The work was partially funded by the Austrian Federal Ministry of Climate Action, Environment, Energy, Mobility, Innovation and Technology under the FFG project FO999896574.

Disclosure of Interests. The author has no competing interests to declare relevant to this article's content.

References

1. Bagschik, G., Menzel, T., Maurer, M.: Ontology based scene creation for the development of automated vehicles. CoRR arXiv:1704.01006 (2017)
2. Baumann, C., Koroglu, Y., Wotawa, F.: On the impact of input models on the fault detection capabilities of combinatorial testing. SN Comput. Sci. **5**(821) (2024). https://doi.org/10.1007/s42979-024-03134-3
3. Bozic, J., Li, Y., Wotawa, F.: Ontology-driven security testing of web applications. In: IEEE International Conference On Artificial Intelligence Testing, AITest 2020, Oxford, UK, August 3–6, 2020, pp. 115–122. IEEE (2020). https://doi.org/10.1109/AITEST49225.2020.00024
4. Chen, P.: The entity-relationship model-toward a unified view of data. ACM Trans. Database Syst. **1**(1), 9–36 (1976). https://doi.org/10.1145/320434.320440
5. Feilmayr, C., Wöß, W.: An analysis of ontologies and their success factors for application to business. Data Knowl. Eng. 1–23 (2016). https://doi.org/10.1016/j.datak.2015.11.003
6. Geyer, S., et al.: Concept and development of a unified ontology for generating test and use-case catalogues for assisted and automated vehicle guidance. IET Intel. Transport Syst. **8**(3), 183–189 (2014). https://doi.org/10.1049/iet-its.2012.0188
7. Güdemann, M., Ortmeier, F., Reif, W.: Safety and dependability analysis of self-adaptive systems. In: Second International Symposium on Leveraging Applications of Formal Methods, Verification and Validation (isola 2006), pp. 177–184 (2006). https://doi.org/10.1109/ISoLA.2006.38
8. Kalra, N., Paddock, S.M.: Driving to safety: how many miles of driving would it take to demonstrate autonomous vehicle reliability? Transp. Res. A Policy Pract. **94**, 182–193 (2016). https://doi.org/10.1016/j.tra.2016.09.010, http://www.sciencedirect.com/science/article/pii/S0965856416302129
9. Klampfl, L., Klück, F., Nica, M., Tao, J., Wotawa, F.: Testing ADAS/ADS - from critical scenarios to automated testing oracles. Elektrotech. Informationstechnik **141**(6), 392–399 (2024). https://doi.org/10.1007/S00502-024-01242-9
10. Klück, F., Li, Y., Tao, J., Wotawa, F.: An empirical comparison of combinatorial testing and search-based testing in the context of automated and autonomous driving systems. Inf. Softw. Technol. **160**, 107225 (2023). https://doi.org/10.1016/J.INFSOF.2023.107225
11. Koopman, P., Wagner, M.: Challenges in autonomous vehicle testing and validation. SAE Int. J. Trans. Safety **4**, 15–24 (2016). https://doi.org/10.4271/2016-01-0128
12. Kuhn, D., Kacker, R., Lei, Y., Hunter, J.: Combinatorial software testing. Computer 94–96 (2009)
13. Li, Y., Tao, J., Wotawa, F.: Ontology-based test generation for automated and autonomous driving functions. Inf. Softw. Technol. **117** (2020)
14. Li, Y., Wotawa, F.: On using ontologies for testing compilers. In: 2020 IEEE International Conference on Software Testing, Verification and Validation Workshops (ICSTW), pp. 181–184 (2020). https://doi.org/10.1109/ICSTW50294.2020.00039
15. Liu, Q., Li, P., Zhao, W., Cai, W., Yu, S., Leung, V.: A survey on security threats and defensive techniques of machine learning: a data driven view. IEEE Access **6**, 12103–12117 (2018)
16. Maier, D.: Theory of Relational Databases. Computer Science Press (1983)
17. Menzel, T., Bagschik, G., Maurer, M.: Scenarios for development, test and validation of automated vehicles. arXiv:1801.08598 (2018). Accepted at the 2018 IEEE Intelligent Vehicles Symposium

18. Schuldt, F., Reschka, A., Maurer, M.: A method for an efficient, systematic test case generation for advanced driver assistance systems in virtual environments. In: Automotive Systems Engineering II, pp. 147–175. Springer (2018)

19. de Souza, E.F.: Knowledge management applied to software testing: an ontology based framework. Ph. D. thesis, National Institute for Space Research (2014)

20. Teorey, T.J., Yang, D., Fry, J.P.: A logical design methodology for relational databases using the extended entity-relationship model. ACM Comput. Surv. **18**(2), 197–222 (1986). https://doi.org/10.1145/7474.7475

21. Bernardo, M., Issarny, V. (eds.): SFM 2011. LNCS, vol. 6659. Springer, Heidelberg (2011). https://doi.org/10.1007/978-3-642-21455-4

22. Utting, M., Legeard, B.: Practical Model-Based Testing: A Tools Approach. Morgan Kaufmann (2006)

23. Wotawa, F.: Testing autonomous and highly configurable systems: challenges and feasible solutions. In: Watzenig, D., Horn, M. (eds.) Automated Driving. LNCS, pp. 519–532. Springer, Cham (2017). https://doi.org/10.1007/978-3-319-31895-0_22

24. Wotawa, F.: On the use of available testing methods for verification & validation of AI-based software and systems. In: Espinoza, H., et al. (eds.) Proceedings of the Workshop on Artificial Intelligence Safety 2021 (SafeAI 2021) co-located with the Thirty-Fifth AAAI Conference on Artificial Intelligence (AAAI 2021), Virtual, February 8, 2021. CEUR Workshop Proceedings, vol. 2808. CEUR-WS.org (2021). https://ceur-ws.org/Vol-2808/Paper_29.pdf

25. Wotawa, F., Bozic, J., Li, Y.: Ontology-based testing: an emerging paradigm for modeling and testing systems and software. In: 13th IEEE International Conference on Software Testing, Verification and Validation Workshops, ICSTW 2020, Porto, Portugal, October 24–28, 2020, pp. 14–17. IEEE (2020). https://doi.org/10.1109/ICSTW50294.2020.00020

26. Wotawa, F., Li, Y.: From ontologies to input models for combinatorial testing. In: Medina-Bulo, I., Merayo, M.G., Hierons, R.M. (eds.) Testing Software and Systems - 30th IFIP WG 6.1 International Conference, ICTSS 2018, Cádiz, Spain, October 1–3, 2018, Proceedings. Lecture Notes in Computer Science, vol. 11146, pp. 155–170. Springer (2018). https://doi.org/10.1007/978-3-319-99927-2_14

27. Xiong, Z., Dixit, V.V., Waller, S.T.: The development of an ontology for driving context modelling and reasoning. In: 2016 IEEE 19th International Conference on Intelligent Transportation Systems (ITSC), pp. 13–18 (2016). https://doi.org/10.1109/ITSC.2016.7795524

28. Yu, L., Lei, Y., Kacker, R., Kuhn, D.: Acts: a combinatorial test generation tool. In: IEEE Sixth International Conference on Software Testing, Verification and Validation (ICST), pp. 370–375 (2013)

29. Zhang, X., et al.: Finding critical scenarios for automated driving systems: a systematic mapping study. IEEE Trans. Software Eng. **49**(3), 991–1026 (2023). https://doi.org/10.1109/TSE.2022.3170122

Author Index

G. Ernst et al. (Eds.): Wolfgang Reif Festschrift, LNCS 15765, pp. 375–376, 2025.
https://doi.org/10.1007/978-3-031-92196-4

The manufacturer's authorised representative in the EU is Springer
Nature Customer Service Centre GmbH, Europaplatz 3, 69115 Heidelberg,
Germany. If you have any concerns regarding our products, please
contact ProductSafety@springernature.com

Printed and bound by CPI Group (UK) Ltd, Croydon, CR0 4YY
28/04/2026
02098521-0009